Random Growth Models

Proceedings of Symposia in
APPLIED MATHEMATICS

Volume 75

Random Growth Models

AMS Short Course
Random Growth Models
January 2–3, 2017
Atlanta, Georgia

Michael Damron
Firas Rassoul-Agha
Timo Seppäläinen
Editors

AMERICAN MATHEMATICAL SOCIETY
Providence, Rhode Island

LECTURE NOTES PREPARED FOR THE AMERICAN MATHEMATICAL SOCIETY SHORT COURSE ON RANDOM GROWTH MODELS HELD IN ATLANTA, GEORGIA, JANUARY 2–3, 2017

The AMS Short Course Series is sponsored by the Society's Program Committee for National Meetings. The series is under the direction of the Short Course Subcommittee of the Program Committee for National Meetings.

2010 *Mathematics Subject Classification.* Primary 60K35, 60K37, 82B43.

Library of Congress Cataloging-in-Publication Data

Names: Damron, Michael, 1981- editor. | Rassoul-Agha, Firas, editor. | Seppäläinen, Timo O., editor.

Title: Random growth models / Michael Damron, Firas Rassoul-Agha, Timo Seppäläinen, editors.

Description: Providence, Rhode Island : American Mathematical Society, [2018] | Series: Proceedings of symposia in applied mathematics ; volume 75 | Series: AMS short course | Lecture notes prepared for a course, January 2-3, 2017, Atlanta, Georgia. | Includes bibliographical references and index.

Identifiers: LCCN 2018023579 | ISBN 9781470435530 (alk. paper)

Subjects: LCSH: Random measures. | Stochastic processes–Mathematical models. | AMS: Probability theory and stochastic processes – Special processes – Interacting random processes; statistical mechanics type models; percolation theory. msc | Probability theory and stochastic processes – Special processes – Processes in random environments. msc | Statistical mechanics, structure of matter – Equilibrium statistical mechanics – Percolation. msc

Classification: LCC QA274.46 .R36 2018 | DDC 519.2/3–dc23

LC record available at https://lccn.loc.gov/2018023579

DOI: https://doi.org/10.1090/psapm/075

∞ The paper used in this book is acid-free and falls within the guidelines
established to ensure permanence and durability.
Visit the AMS home page at https://www.ams.org/

10 9 8 7 6 5 4 3 2 1 23 22 21 20 19 18

Contents

Preface

Irregular and stochastic growth is all around us: tumors, bacterial colonies, infections, fluid spreading in a porous medium, and propagating flame fronts. The study of simplified mathematical models of stochastic growth began in probability theory in the 1960s. Quite serendipitously, these models have turned out to be extremely hard to analyze. They have inspired innovative probability theory and have led to new connections between probability and other parts of mathematics.

Much progress has been made in a few exactly solvable special cases of random growth models. Out of this effort has arisen a new distinct subject, *integrable probability*, with a deep algebraic component. However, thus far the tools of integrable probability apply only in the exactly solvable setting where special structures are present. On the broader class of models, researchers have had success with approaches that combine probability with ideas of functional analytic and geometric nature. Examples include the use of concentration of measure and Busemann functions to study geodesics and prove fluctuation bounds, and curvature bounds in the derivation of the Kardar-Parisi-Zhang (KPZ) scaling relation.

This book consists of a collection of expanded notes from the AMS Short Course "Random Growth Models," which took place in Atlanta, GA, at the AMS Joint Mathematics Meetings in January 2017. The objective is to showcase recent breakthroughs in a class of stochastic growth models called first- and last-passage percolation and their relations to particle systems and the KPZ equation. The topics covered span aspects of both the general models and the exactly solvable ones.

This book is intended for an audience beyond specialists in the field. In particular, we hope to provide an opening into the area for graduate students and researchers who might consider studying these models. It is assumed that the reader is familiar with the basic notions of probability. A quick review of probability is given in [**2**].

Article [**2**] introduces various growth models and gives an overview of first- and last-passage percolation. After perusal of [**2**], the remaining chapters can be read independently of each other. Small changes in notation sometimes happen from chapter to chapter. In the remainder of this preface we highlight themes that run through more than one chapter and link together different parts of the book.

The first fundamental question of the subject is the limit shape of a growth model and the properties of the limit shape. Already here open questions are legion. Article [**2**] covers both known and conjectured general properties of the limit shape. These are then used and restated when needed in [**3**] and [**4**].

Rost's formula for the shape function of the corner growth model with exponential weights is stated in [**2**] and derived in both [**4**] and [**5**]. In [**4**], Rost's formula comes after a development of a theory for the corner growth model with general weights. Article [**5**] concentrates on the exactly solvable exponential case and derives the shape formula with a more direct argument. Continuity of the shape function up to the boundary is proved for directed last-passage percolation in [**4**]. Standard first-passage percolation is defined in the full space instead of a quadrant, so the boundary issue does not arise.

The second question is the question of fluctuations. There is a natural progression through the book. Article [**6**] treats the best known non-optimal bounds for general models with concentration inequalities. Article [**5**] proves sharp exponent bounds with coupling tools for the stationary version of an exactly solvable model. Finally [**1**] obtains the finest results with integrability techniques that go beyond probability and draw on ideas from quantum integrable systems. The fluctuation results proved in that chapter are widely believed to be universal and present in models for which the tools of exact solvability are no longer applicable. Though such a conjecture seems far from being proved, there is a weaker type of universality in which many systems under special scalings converge to the KPZ equation. Article [**1**] describes this weak KPZ universality conjecture and provides some of the stochastic analysis tools used in establishing it for certain models.

A third main theme is the structure of optimizing paths, or geodesics. These are studied in conjunction with Busemann functions. Article [**3**] takes up this topic in undirected first-passage percolation, and article [**4**] in directed last-passage percolation. The two chapters complement each other because most arguments can be transferred from one setting to the other. Both chapters show how Busemann functions are used to construct and study directional geodesics. The chapters take somewhat different approaches to the construction of Busemann functions. Article [**3**] uses subsequential weak limits in general dimensions while article [**4**] relies on a connection with queuing theory in two dimensions but shows how to go from the weak limit to an almost sure limit.

Both [**3**] and [**4**] utilize conjectured properties of the limit shape to derive results about Busemann functions. By contrast, in [**5**], the existence and precise distributional properties of Busemann functions can be proved without additional assumptions because the limit shape and invariant distributions of the exponential corner growth model are explicitly known.

We thank the American Mathematical Society for financial support of the 2017 AMS Short Course on Random Growth Models and for the opportunity to publish this proceedings volume.

Michael Damron
Firas Rassoul-Agha
Timo Seppäläinen

Bibliography

[1] Ivan Corwin, *Exactly solving the KPZ equation*, Proc. Sympos. Appl. Math **75** (2018).

[2] Michael Damron, *Random growth models: shape and convergence rate*, Proc. Sympos. Appl. Math. **75** (2018).

[3] Jack Hanson, *Infinite geodesics, asymptotic directions, and busemann functions in first-passage percolation*, Proc. Sympos. Appl. Math. **75** (2018).

[4] Firas Rassoul-Agha, *Busemann functions, geodesics, and the competition interface for directed last-passage percolation*, Proc. Sympos. Appl. Math. **75** (2018).

[5] Timo Seppäläinen, *The corner growth model with exponential weights*, Proc. Sympos. Appl. Math. **75** (2018).

[6] Philippe Sosoe, *Fluctuations in first-passage percolation*, Proc. Sympos. Appl. Math. **75** (2018).

Proceedings of Symposia in Applied Mathematics
Volume **75**, 2018
https://doi.org/10.1090/psapm/075/00656

Random growth models: Shape and convergence rate

Michael Damron

ABSTRACT. Random growth models are fundamental objects in modern probability theory, have given rise to new mathematics, and have numerous applications, including tumor growth and fluid flow in porous media. In this article, we introduce some of the typical models and the basic analytical questions and properties, like existence of asymptotic shapes, fluctuations of infection times, and relations to particle systems. We then specialize to models built on percolation (first-passage percolation and last-passage percolation) and give a self-contained treatment of the shape theorem, the subadditive ergodic theorem, and conjectured and proven properties of asymptotic shapes. We finish by discussing the rate of convergence to the limit shape, along with definitions of scaling exponents and a sketch of the proof of the KPZ scaling relation.

Contents

1. Introduction

1.1. Some typical growth models. The typical setting for a random growth model is as follows. We imagine an infection sitting at a vertex v of a connected graph $G = (V, E)$ with vertex set V and edge set E. The infection spreads along the edges of the graph according to some random rules, and each vertex of the graph is eventually infected. The infection takes time $T(v, w) < \infty$ to infect a vertex w, and at time t, a set of vertices

$$B(t) = \{x \in V : T(v, x) \leqslant t\}$$

is infected.

Eden model. One of the simplest examples is the Eden model, introduced by Eden [**20**] in '61, and gives a simplified version of cell reproduction. Eden considered

2010 *Mathematics Subject Classification.* 60K35, 60K37, 82B43.

This work was done with support from an NSF CAREER grant.

the two dimensional square lattice $\mathbb{Z}^2$ with its nearest-neighbor edges, but we could consider any graph, and we will generally take the d-dimensional cubic lattice

$$\mathbb{Z}^d = \{x = (x_1, \ldots, x_d) : x_i \in \mathbb{Z} \text{ for all } i\}$$

with the nearest-neighbor edges

$$\mathcal{E}^d = \{\{x, y\} : x, y \in \mathbb{Z}^d, \|x - y\|_1 = 1\},$$

and $\|x\|_1 = \sum_i |x_i|$ is the ℓ^1-norm. We begin with a cell occupying the site $0 = (0, \ldots, 0)$, so our occupied set at time 0 is $S_0 = \{0\}$. At time $n \geqslant 1$, we consider the edge boundary of S_{n-1}

$$\partial_e S_{n-1} = \{\{x, y\} \in \mathcal{E}^d : x \in S_{n-1}, y \notin S_{n-1}\}$$

and select some edge $e_n = \{x_n, y_n\}$ uniformly at random from it. Last, put

$$S_n = S_{n-1} \cup \{x_n, y_n\}.$$

Thus, in the Eden model, our cluster of cells S_{n-1} replicates on the boundary (uniformly at random) and produces a new cell directly outside of S_{n-1} to form our new cluster S_n. It turns out that $\cup_n S_n$ is all of $\mathbb{Z}^d$, and so the infection time from 0 to x (the first value of n such that S_n contains x) is finite. The Eden model can be rephrased in a larger framework, first-passage percolation (FPP), which we will meet soon. As a consequence, one can show a "shape theorem" for S_n: after proper scaling, the set S_n approaches a limiting shape. Although our selection rule appears not to bias any direction, the limiting shape is expected not to be rotationally invariant (that is, it is expected not to be a Euclidean ball). This statement is proved for high dimensions (see [**7, 28**] and the references in Section 3.1.3 below).

Diffusion-limited aggregation. In the Eden model, we select a boundary vertex y_n by picking a boundary edge uniformly at random. We could change this selection rule, and pick y_n according to some other distribution. In diffusion-limited-aggregation (DLA), introduced in [**45**], we select y_n according to the "harmonic measure from infinity." Roughly speaking, this corresponds to the hitting distribution of a random walk on the boundary of S_n started from a far away point. In this way we again obtain a sequence of growing sets (S_n). Since the DLA model is notoriously hard to analyze, a simplified version, internal DLA (IDLA) was introduced in [**38**], where the selection rule is more straightforward. We begin again with $S_0 = \{0\}$. At time n, we run a simple symmetric random walk started at 0, and set y_n (our selected vertex) to be the first vertex adjacent to S_{n-1} that the random walk touches. Precisely, at time n, we let $X_1^{(n)}, X_2^{(n)}, \ldots$ be an i.i.d. sequence of random vectors taking values

$$\mathbb{P}(X_i^{(n)} = \pm e_j) = 1/(2d) \text{ for } i \geqslant 1,\ j = 1, \ldots, d,$$

where (e_j) are the j standard basis vectors of $\mathbb{R}^d$, set

$$Y_0^{(n)} = 0, \text{ and } Y_i^{(n)} = \sum_{k=1}^{i} X_k^{(n)} \text{ for } i \geqslant 1,$$

and let y_n be the first element in the sequence $Y_0^{(n)}, Y_1^{(n)}, Y_2^{(n)}, \ldots$ that is not in S_{n-1}. There is no shape theorem proved for DLA, but there is one for IDLA [**31**] (convergence to a Euclidean ball), and even the rate of convergence to the shape is known [**24**].

First-passage percolation. FPP can be seen as a generalization of the Eden model, and was introduced [**23**] by Hammersley and Welsh in '65. In FPP, the infection spreads across edges according to explicit speeds. We let $(t_e)_{e\in\mathcal{E}^d}$ be a collection of i.i.d. nonnegative random variables. The variable t_e is thought of as the passage time of an edge; that is, the amount of time it takes for an infection to cross the edge. A path Γ is a sequence of edges $e_0, \ldots, e_n$ such that each pair e_i and e_{i+1} shares an endpoint, and the passage time of such a Γ is $T(\Gamma) = \sum_{e\in\Gamma} t_e$. The infection takes the path of minimal passage time, so we set the infection time, or passage time, from x to y, vertices in $\mathbb{Z}^d$, to be

$$T(x, y) = \inf_{\Gamma : x\to y} T(\Gamma),$$

where the infimum is over paths Γ starting at x and ending at y. (It is known that under general assumptions, for instance if $\mathbb{P}(t_e = 0) < p_c$, where p_c is the d-dimensional bond percolation threshold, then there is a unique minimizing path — a geodesic — from x to y. However, for some distributions with $\mathbb{P}(t_0 = 0) = p_c$ in dimensions $d \geqslant 3$, existence of geodesics is unknown.) In FPP, there is a shape theorem, but the limiting shape depends on the distribution of the (t_e)'s. Very little is known about the limit shapes for various distributions, apart from them being convex, compact, and having the symmetries of $\mathbb{Z}^d$. As we will see, it is expected that for most distributions, the limit shape is strictly convex, and certainly not a polygon, but strict convexity is not proved for any distribution, and there are only some two-dimensional examples of limit shapes that are not polygons. For a recent survey on FPP, see [**6**].

If the weights (t_e) have (rate 1) exponential distribution, the evolution of the ball $B(t)$ as t grows can be shown to be exactly the same as the evolution of the sets (S_n) in the Eden model. More precisely, using the "memoryless property" of the exponential distribution, one can show that the growth of $B(t)$ is the same as in the following algorithm: begin with $B(0) = \{0\}$ and assign i.i.d. exponential random variables to the edges in the set $\partial_e B(0)$. If τ_1 is the minimum of these variables, and it is assigned to edge $e = \{x, y\}$, then set $B(t) = B(0)$ for $t \in [0, \tau_1)$ and $B(\tau_1) = B(0) \cup \{x, y\}$. Next, generate new i.i.d. exponential random variables assigned to the edges in $\partial_e B(\tau_1)$, and set τ_2' to be the minimum of these variables, with $\tau_2 = \tau_1 + \tau_2'$. Once again, set $B(t) = B(\tau_1)$ for $t \in [\tau_1, \tau_2)$ and $B(\tau_2) = B(\tau_1) \cup \{x', y'\}$, where $\{x', y'\}$ is the edge in $\partial_e B(\tau_1)$ with minimal weight. We continue, and at each step, we sample i.i.d. exponentials for the boundary edges of our current set, choose the minimal weight edge (of weight τ), and add its endpoints into our set after we wait for time τ. Since the location of the minimum is uniformly distributed on the boundary, the sequence of sets $B(0), B(\tau_1), B(\tau_2), \ldots$ has the same distribution as the sequence of sets in the Eden model. There is no such representation of the DLA models in terms of FPP.

Last-passage percolation. LPP is a modification of FPP, introduced because of its relationship to the TASEP particle system. The typical setting is $\mathbb{Z}^d$, and one places i.i.d. nonnegative random variables (weights) $(t_v)_{v\in\mathbb{Z}^d}$ on the vertices. A path Γ is a sequence of vertices $v_0, \ldots, v_n$ such that $\|v_i - v_{i+1}\|_1 = 1$ for all i, and one assigns the passage time $T(\Gamma) = \sum_{v\in\Gamma} t_v$, as in FPP. The difference now is that in LPP we define the passage time between two vertices as the maximal passage time of any path between them. Of course this will generally be infinity unless we restrict ourselves to a finite set of paths, so we consider oriented paths;

that is, paths such that all the coordinates of the v_i's are nondecreasing (written $v_i \leqslant v_{i+1}$). So for any $v \leqslant w$, we set the infection time $T(v,w)$ to be the maximal passage time of any oriented path rom v to w. Once again there is a shape theorem; however, unlike in FPP, the limiting shape is generally compact but not convex. In two dimensions, it is believed that the boundary of the limit shape is the graph of a strictly concave function, but this is again not known. In LPP, however, it is known that the limit shape is not a polygon. For a survey of LPP, see [**35**].

1.2. Main questions. As we mentioned many times in the last section, a fundamental object of study in random growth models is the limit shape. In a certain sense to be described in the next section, we will have almost surely $B(t)/t \to \mathcal{B}$, where $\mathcal{B}$ is the limit shape and $B(t)$ is the set of infected sites at time t. Directly related to the limit shape are the following two questions.

(1) **Set of limit shapes.** What does the limit shape look like? Are there explicit descriptions? Is it a Euclidean ball? Is it the ℓ^1 or ℓ^∞ ball? In models (like FPP and LPP) where many limit shapes can arise, what is the collection of all possible limit shapes? What is the dependence in FPP and LPP of the limit shape on the weight distribution?
(2) **Convergence to the limit.** What is the convergence rate to the limit shape? Precisely, what is the set of functions $f(t)$ such that one has

$$(t - f(t))\mathcal{B} \subset B(t) \subset (t + f(t))\mathcal{B}$$

for t large?

We will focus here on these two questions, but we mention also some of the questions from the other articles in this proceedings volume.

(1) **Distributional limits of passage times.** Letting $T(x,y)$ be the infection time of y started at x, are there functions $a(x)$ and $b(x)$ such that

$$\frac{T(0,x) - a(x)}{b(x)} \Rightarrow X$$

as $\|x\|_1 \to \infty$ for some nondegenerate limiting distribution X? In two-dimensional FPP/LPP-type models, the answer is expected (and proved in a couple of cases) to be yes, and X should have the Tracy-Widom distribution from random matrix theory. $b(x)$ should be the order of fluctuations of $T(0,x)$, and embedded in this question is obviously the question: what is the order of fluctuations of $T(0,x)$? In the DLA model, fluctuations are of significantly lower order than in FPP/LPP-type models. Work on these questions has led into concentration of measure, particle systems, integrable probability and exactly solvable systems.
(2) **Structure of geodesics.** Optimal infection paths are called geodesics. What is the structure of the set of all geodesics? How different are geodesics from straight lines? Infinite geodesics are infinite paths all whose segments are geodesics. How many infinite geodesics are there? Are there doubly-infinite geodesics? These questions are all related to Busemann functions in metric geometry.

The existence (or nonexistence) of doubly-infinite geodesics mentioned above is directly related to the number of ground states of disordered ferromagnetic spin models. We explain the connection to the disordered Ising ferromagnet, which is a variant of the usual Ising model from statistical mechanics. We consider dimension two, and define the dual lattice

$$(\mathbb{Z}^2_*, \mathcal{E}^2_*) = (\mathbb{Z}^2, \mathcal{E}^2) + (1/2, 1/2),$$

which is the usual two-dimensional lattice shifted by $(1/2, 1/2)$. Each "dual edge" e^* in $\mathcal{E}^2_*$ bisects a unique edge e in $\mathcal{E}^2$, so we say that e^* is the edge dual to e. Define a "spin configuration" $\sigma = (\sigma_x)_{x \in \mathbb{Z}^2_*}$ to be an assignment of $+1$ and -1 to every dual vertex; that is, σ is an element of $\{-1, +1\}^{\mathbb{Z}^2_*}$. The interactions between spins are given by "couplings," which are variables assigned to the edges. Accordingly, let $(J_{x,y})_{\{x,y\} \in \mathcal{E}^2_*}$ be a family of independent random variables which are almost surely positive. For any σ and any finite $S \subset \mathbb{Z}^2_*$, we define the random energy of σ relative to the couplings and the set S as

$$\mathcal{H}_S(\sigma) = - \sum_{\{x,y\} \in \mathcal{E}^2_*, x \in S} J_{x,y} \sigma_x \sigma_y .$$

For the standard disordered Ising ferromagnet, the couplings $(J_{x,y})$ are typically chosen to be identically distributed. In this case, it is of great interest to determine the structure and number of ground states for the model. Precisely, a configuration σ is called a ground state for the couplings $(J_{x,y})$ if for each configuration $\tilde{\sigma}$ such that $\tilde{\sigma}_x \neq \sigma_x$ for only finitely many x, one has

$$\mathcal{H}_S(\sigma) \leqslant \mathcal{H}_S(\tilde{\sigma}) \text{ for all finite } S \subset \mathbb{Z}^2_*.$$

One can think of a ground state as a local minimizer of the energy functional. It is not known how many ground states there are for a given realization of couplings, but it is believed that in this two-dimensional model (and sufficiently low-dimensional analogues), there should be only two almost surely. These two are the all-plus and all-minus states.

If there exists a nonconstant ground state σ for couplings $(J_{x,y})$, then we can compare it to the all-plus ground state σ_+. We claim that σ and σ_+ cannot have any finite regions of disagreement: there is no finite S such that $\sigma_x = -1$ for all $x \in S$ and $\sigma_y = +1$ for all $y \in \partial S$ (here ∂S refers to the set of vertices in S^c with a neighbor in S). Indeed, if there were such an S, we could apply the energy minimization property to S to find that $\mathcal{H}_S(\sigma) = 0$. However, as long as the couplings are continuously distributed, one can argue that almost surely, there are no finite S with $\mathcal{H}_S(\sigma) = 0$ for some σ. This justifies the claim; from it we see that any nonconstant ground state must have a two-sided and circuitless original lattice path of edges whose dual edges $\{x, y\}$ satisfy $\sigma_x \neq \sigma_y$. That is, any nonconstant ground state can be associated to at least one such doubly-infinite path.

Conversely, one can construct nonconstant ground states if one assumes existence of doubly-infinite geodesics in a related FPP model. Namely, given couplings $(J_{x,y})$ as above, associated to the edges of the dual lattice, one defines a passage-time configuration (t_e) by setting $t_e = J_{x,y}$, where $\{x, y\}$ is the unique edge dual to e. Supposing that there is a doubly-infinite geodesic Γ (this is a doubly-infinite path each of whose subpaths is an optimizing path for T) in this configuration, one can set $\sigma_x = +1$ for all x on one side of Γ and -1 for all x on the other side of Γ, and such σ will be a nonconstant ground state for $(J_{x,y})$.

The relation between ground states and geodesics allows one to carry results between the two models. For example, if one can rule out existence of doubly-infinite geodesics in FPP models, one can deduce nonexistence of nonconstant ground states for the associated spin models.

2. A bit of basic probability

Before we begin, we review some of the notions from probability, and discuss typical limiting behavior, to compare to that which we see in random growth models. The main object of study is a random variable, say X, which is formally a function from a probability space to $\mathbb{R}$, but we think of it as a random number. A random variable is characterized by its distribution function F, which is a function $F : \mathbb{R} \to \mathbb{R}$ satisfying

(1) (nondecreasing) $F(x) \leqslant F(y)$ if $x \leqslant y$,
(2) (behavior at infinity)
$$\lim_{x \to -\infty} F(x) = 0 \text{ and } \lim_{x \to \infty} F(x) = 1,$$
(3) and (right continuity) F is a right continuous function.

We think of $F(x)$ as being the probability that the random variable X is no bigger than x:
$$F(x) = \mathbb{P}(X \leqslant x),$$
and so we can compute the probability that X lies in any given interval as
$$\mathbb{P}(X \in (a, b]) = F(b) - F(a).$$
It is a theorem that to every such function F there is a random variable X with these properties.

The multidimensional analogue of a random variable is usually called a random vector. A random vector with values in $\mathbb{R}^n$ is simply an n-tuple $X = (X_1, \ldots, X_n)$, each of whose coordinates are random variables (and all these variables are on the same probability space). Similarly to the one-dimensional case, a random vector is characterized by its distribution function, which is a function F on $\mathbb{R}^n$ (representing $F(x) = \mathbb{P}(X \leqslant x)$, where $X \leqslant x$ means $X_i \leqslant x_i$ for all i) which is nondecreasing ($F(x) \leqslant F(y)$ when $x \leqslant y$), satisfies
$$\lim_{x_i \to -\infty \text{ for all } i} F(x) = 0 \quad \text{and} \quad \lim_{x_i \to \infty \text{ for all } i} F(x) = 1,$$
and has the right continuity property $\lim_{x \downarrow y} F(x) = y$ for all y. Unlike in the one-dimensional case, an additional positivity property for F is required to guarantee that there is a random vector X with distribution function F. See [**18**, Section 3.9] for more details.

Back to the one-dimensional setting, in many cases of interest, F is actually differentiable, and we write its derivative $F'(x)$ as $f(x)$, calling f the density function of the random variable X. In this case, we can use the fundamental theorem of calculus to write
$$\mathbb{P}(X \in (a, b]) = F(b) - F(a) = \int_a^b f(x) \, \mathrm{d}x.$$
Note that by the above rules on F, one has $\int_{-\infty}^{\infty} f(x) \, \mathrm{d}x = \lim_{x \to \infty} F(x) = \mathbb{P}(X < \infty) = 1$, and f must be nonnegative. Furthermore we can compute the expectation of X (average value of X) as
$$\mathbb{E}X = \int_{-\infty}^{\infty} x \, \mathrm{d}F(x) = \int_{-\infty}^{\infty} x f(x) \, \mathrm{d}x.$$

Generally the expectation of a function g of X is calculated as

$$\mathbb{E}g(X) = \int_{-\infty}^{\infty} g(x)\ \mathrm{d}F(x) = \int_{-\infty}^{\infty} g(x)f(x)\ \mathrm{d}x.$$

Some typical random variables are the uniform on $[a, b]$, which has density

$$f(x) = \begin{cases} \frac{1}{b-a} & \text{if } x \in [a,b] \\ 0 & \text{otherwise} \end{cases}$$

(in this case we say X has uniform distribution on $[a, b]$) and exponential with parameter λ, which has density

$$f(x) = \begin{cases} \lambda e^{-\lambda x} & \text{if } x \geqslant 0 \\ 0 & \text{otherwise.} \end{cases}$$

From these density formulas, one can easily calculate, for instance, the "moments" of a random variable with the exponential distribution. That is, we can find the m-th moment of such an X as

$$\mathbb{E}X^m = \int_0^{\infty} x^m f(x)\ \mathrm{d}x = \lambda \int_0^{\infty} x^m e^{-\lambda x}\ \mathrm{d}x,$$

or the m-th central moment

$$\mathbb{E}|X - \mathbb{E}X|^m = \int_0^{\infty} |x - \mathbb{E}X|^m f(x)\ \mathrm{d}x.$$

We recall that the second central moment is typically called the variance of X:

$$\operatorname{Var} X = \mathbb{E}(X - \mathbb{E}X)^2 = \int_0^{\infty} (x - \mathbb{E}X)^2 f(x)\ \mathrm{d}x.$$

In a way, central moments measure how likely X is to vary from its mean, as we have Chebyshev's inequality: for $a > 0$,

$$\mathbb{P}(|X - \mathbb{E}X| \geqslant a) \leqslant \frac{1}{a^m}\mathbb{E}|X - \mathbb{E}X|^m.$$

For example, since the mean of the exponential distribution is λ^{-1}, we can compute

$$\mathbb{P}(|X - \lambda^{-1}| \geqslant a) = \mathbb{P}(|X - \lambda^{-1}|^2 \geqslant a^2) \leqslant \frac{1}{a^2}\operatorname{Var} X = \frac{1}{(a\lambda)^2}.$$

As $a \to \infty$, the right side converges to 0, and in a sense, this inequality states that X is reasonably approximated by its mean.

A fundamental notion in probability is that of independence. Informally, two variables X and Y are independent if given information about Y, the distribution of X is unchanged. Recalling that the conditional probability of an event B given A is

$$\mathbb{P}(B \mid A) = \frac{\mathbb{P}(A \cap B)}{\mathbb{P}(A)}$$

(when $\mathbb{P}(A) > 0$), we say that X, Y are independent if whenever $a_i \leqslant b_i$ for $i = 1, 2$, one has

$$\mathbb{P}(X \in [a_1, b_1] \mid Y \in [a_2, b_2]) = \mathbb{P}(X \in [a_1, b_1]),$$

or

$$\mathbb{P}(X \in [a_1, b_1], Y \in [a_2, b_2]) = \mathbb{P}(X \in [a_1, b_1])\mathbb{P}(Y \in [a_2, b_2]),$$

and the variables $X_1, \ldots, X_n$ are independent if whenever $a_i \leqslant b_i$ for $i = 1, \ldots, n$, one has

$$\mathbb{P}\left(\cap_{i=1}^n \{X_i \in [a_i, b_i]\}\right) = \prod_{i=1}^n \mathbb{P}\left(X_i \in [a_i, b_i]\right).$$

This condition carries over to expectation, and if $X_1, \ldots, X_n$ are independent, and $g_1, \ldots, g_n$ are integrable functions of the X_i's, then

$$\mathbb{E}g_1(X_1) \cdots g_n(X_n) = \mathbb{E}g_1(X_1) \cdots \mathbb{E}g_n(X_n).$$

One use of these factoring properties is to compute the variance of a sum of independent random variables (each of which is assumed to have finite second moment) as

$$\text{Var}(X_1 + \cdots + X_n) = \text{Var } X_1 + \cdots + \text{Var } X_n.$$

This in turn can be used to control the size of the sum of independent random variables using Chebyshev: if the X_i's all have the same distribution, for $a > 0$,

$$\begin{aligned}
\mathbb{P}\left(\left|\frac{X_1 + \cdots + X_n}{n} - \mathbb{E}X_1\right| \geqslant a\right) &= \mathbb{P}\left(\left|\sum_{i=1}^n X_i - \mathbb{E}\sum_{i=1}^n X_i\right| \geqslant an\right) \\
&\leqslant \frac{\text{Var } \sum_{i=1}^n X_i}{(an)^2} \\
&= \frac{\text{Var } X_1}{a^2 n},
\end{aligned}$$

which converges to 0 as $n \to \infty$. This is true for any $a > 0$ (however small), so we have proved informally that if $X_1, \ldots, X_n$ are i.i.d. (independent and identically distributed), all with finite second moments, then the Weak Law of Large Numbers (WLLN) holds: one has $\frac{X_1 + \cdots + X_n}{n} \to \mathbb{E}X_1$ in probability, or, for any $a > 0$,

$$\mathbb{P}\left(\left|\frac{X_1 + \cdots + X_n}{n} - \mathbb{E}X_1\right| \geqslant a\right) \to 0.$$

In many cases, there is even a rate of convergence on the right side. For instance, we would say that $X_1 + \cdots + X_n - n\mathbb{E}X_1$ is "exponentially concentrated on scale $\psi(n)$" if for some $c > 0$ and all $t \geqslant 0$,

$$\mathbb{P}\left(|X_1 + \cdots + X_n - n\mathbb{E}X_1| \geqslant t\psi(n)\right) \leqslant e^{-ct}.$$

If $\psi(n)$ here can be taken to be of smaller order than n, then this is an improved WLLN, as we can take $t = n/\psi(n)$, so that the right side converges to 0 quickly. Exponential concentration is often written in the following form, from which the above follows through a Chebyshev-like inequality: for some $\alpha, C > 0$ and all n,

$$\mathbb{E}\exp\left(\alpha \frac{|X_1 + \cdots + X_n - n\mathbb{E}X_1|}{\psi(n)}\right) \leqslant C < \infty.$$

There are stronger versions of the above convergence, and the most famous is the Strong Law of Large Numbers (SLLN), which gives an almost sure convergence result: if (X_i) are i.i.d. with finite first moment ($\mathbb{E}|X_1| < \infty$), then one has $\frac{X_1 + \cdots + X_n}{n} \to \mathbb{E}X_1$ almost surely; that is,

$$\mathbb{P}\left(\frac{X_1 + \cdots + X_n}{n} \to \mathbb{E}X_1\right) = 1.$$

Strictly speaking, for this last result to hold, we need to know that we can always find a probability space on which there is an infinite sequence of independent random variables, with specified distributions. This fact is given to us by a technical theorem of Kolmogorov:

THEOREM 2.1. *Given distribution functions $F_1, F_2, \ldots$, there exists a probability space on which there are independent random variables $X_1, X_2, \ldots$ such that*

(1) *the X_i's are independent: each finite collection $(X_j)_{j\in J}$ (where the set $J \subset \{1, 2, \ldots\}$ is finite) of the X_i's is independent, and*
(2) *for each i, X_i has distribution function F_i.*

This result is used all throughout probability, and we can use it to build our random growth models, like FPP, by generating independent edge-weights (t_e) with specified distributions, one for each edge on an infinite graph, so long as there are countably many edges.

The SLLN tells us that the asymptotic behavior of a sum $S_n = X_1 + \cdots + X_n$ of i.i.d. random variables with finite first moment satisfies almost surely

$$S_n = n\mathbb{E}X_1 + o(n),$$

but how large is this error term? Since it is random, the simplest way to quantify it is to find sequences of numbers (b_n) such that

$$\mathbb{P}\left(S_n - n\mathbb{E}X_1 \geqslant b_n\right) \to 0 \text{ or } \to 1.$$

In the case $\to 1$, we would say that $S_n - n\mathbb{E}X_1$ is typically larger than b_n, and in the case $\to 0$, smaller than b_n. To find the exact scale of $S_n - n\mathbb{E}X_1$, another way is to search for sequences (b_n) such that the convergence above is to a number strictly between 0 and 1. Formally, we can look for and a fixed random variable Z (the "limiting random variable") and a sequence (b_n) such that for any $t \in \mathbb{R}$,

$$\mathbb{P}\left(\frac{S_n - n\mathbb{E}X_1}{b_n} \geqslant t\right) \to \mathbb{P}(Z \geqslant t).$$

In this case, we say that $\frac{S_n - n\mathbb{E}X_1}{b_n}$ converges in distribution to Z, write

$$\frac{S_n - n\mathbb{E}X_1}{b_n} \Rightarrow Z,$$

and we content ourselves by saying that $S_n - n\mathbb{E}X_1$ "fluctuates on the scale b_n." (A technical point here is that we do not need this convergence at all t, but just those at which the distribution function of Z is continuous.) This is precisely what the Central Limit Theorem (CLT) says:

THEOREM 2.2 (CLT). *Let (X_i) be i.i.d. such that the X_i's have finite second moment. Then, writing $\mu = \mathbb{E}X_1$ and $\sigma^2 = \mathrm{Var}\, X_1$, one has*

$$\frac{S_n - n\mu}{\sqrt{n}\sigma} \Rightarrow Z,$$

where Z has the standard Gaussian distribution:

$$\mathbb{P}(Z \geqslant t) = \frac{1}{\sqrt{2\pi}} \int_t^\infty e^{-\frac{x^2}{2}}\, dx.$$

In the CLT, we see that sums of i.i.d. random variables "fluctuate on scale $\sqrt{n}$," and the limiting distribution is Gaussian. These two statements are quite typical of classical limit theorems in probability. We will see however, in the rest of the proceedings volume, that in random growth models, scales of fluctuation and limiting distributions are not typically of order $\sqrt{n}$ "diffusive" and Gaussian, but rather they are of order $n^{1/2-\epsilon}$ for some $\epsilon > 0$ "subdiffusive" and related to random matrix theory (the Tracy-Widom distribution).

3. Limit shapes

For the rest of the article, we focus on the first two main growth model questions (limit shapes and convergence rate), and only in the context of FPP and LPP. First we will describe results from FPP including shape theorems and properties of limit shapes. Afterward, we switch to LPP to show which results are similar (shape theorems) and which are different (non-polygonal shapes and exactly solvable cases).

3.1. FPP. We begin with the shape theorem in FPP. First let's recall the model. We are given the d-dimensional cubic lattice $\mathbb{Z}^d$ with the set of its nearest-neighbor edges $\mathcal{E}^d$. We place i.i.d. nonnegative random variables (edge-weights) $(t_e)_{e\in\mathcal{E}^d}$ with common distribution function F on the edges. A path is a sequence $(v_0, e_0, v_1, e_1, \ldots, e_{n-1}, v_n)$ of vertices and edges such that e_i has endpoints v_i and v_{i+1}. The passage time of a path Γ is

$$T(\Gamma) = \sum_{e\in\Gamma} t_e$$

and the passage time between vertices $x, y \in \mathbb{Z}^d$ is

$$T(x,y) = \inf_{\Gamma : x\to y} T(\Gamma),$$

where the infimum is over all paths from x to y.

3.1.1. *Shape theorem in FPP.* If $\mathbb{P}(t_e = 0) = 0$ then T is almost surely a metric on $\mathbb{Z}^d$, as it is nonnegative, one has $T(x,y) = 0$ only when $x = y$, and T satisfies the triangle inequality $T(x,y) \leqslant T(x,z) + T(z,y)$. (When edge-weights can be zero, T is a pseudometric.) For convenience, we also extend T to real points; that is, for $x \in \mathbb{R}^d$, we set $[x]$ to be the unique point in $\mathbb{Z}^d$ with $x \in [x] + [0,1)^d$. Then the shape theorem is a type law of large numbers for the set

$$B(t) = \{x \in \mathbb{R}^d : T(0,x) \leqslant t\},$$

saying that $B(t)/t$ approaches a limiting set $\mathcal{B}$.

In the statement of the shape theorem below, $p_c = p_c(d)$ is the threshold for Bernoulli bond percolation on $\mathbb{Z}^d$ and can be defined as follows. Let $\mathbb{P}_p$ be the probability measure under which edge-weights are Bernoulli with parameter $1-p$:

$$\mathbb{P}_p(t_e = 0) = p \quad \text{and} \quad \mathbb{P}_p(t_e = 1) = 1 - p.$$

Putting

$$\theta(p) = \mathbb{P}_p(0 \text{ is in an infinite self-avoiding path of edges } e \text{ with } t_e = 0),$$

one can then use a coupling argument to show that θ is nondecreasing in p. Therefore the number

$$p_c = \sup\{p \in [0,1] : \theta(p) = 0\}$$

has the properties $\theta(p) > 0$ for $p > p_c$ and $\theta(p) = 0$ for $p < p_c$. It is an important result that $p_c \in (0,1)$ for all dimensions $d \geqslant 2$, and we can quickly argue at least that $p_c > 0$. (The argument for $p_c < 1$ can be found in [**21**], for example.). Write A_n for the event that there is a self-avoiding path starting from 0 with n edges e, all satisfying $t_e = 0$. Then for any n,

$$\theta(p) \leqslant \mathbb{P}_p(A_n) \leqslant \sum_{\gamma:\#\gamma=n} \mathbb{P}_p(t_e = 0 \text{ for all } e \in \gamma),$$

where the sum is over all self-avoiding paths γ starting from 0 with n edges. Each such γ satisfies

$$\mathbb{P}_p(t_e = 0 \text{ for all } e \in \gamma) = p^{\#\gamma},$$

and since there are at most $2d(2d-1)^n$ many such paths with n edges (we first choose an edge touching 0, and then at every subsequent step choose an adjacent edge we have not chosen before), we obtain

$$\theta(p) \leqslant 2d(2d-1)^n p^n.$$

This converges to 0 if $p < 1/(2d-1)$, and so $p_c \geqslant 1/(2d-1)$.

Returning to the shape theorem, the condition $\mathbb{P}(t_e = 0) < p_c$ below is there to ensure that there are not so many zero-weight edges that $T(0,x)$ will grow sublinearly in x. (This occurs for $\mathbb{P}(t_e = 0) \geqslant p_c$.)

THEOREM 3.1 (Richardson [**40**], Cox-Durrett [**12**], Kesten [**28**]). *Assume that* $\mathbb{E}\min\{t_1, \ldots, t_{2d}\}^d < \infty$, *where* t_i *are i.i.d. copies of* t_e *and* $\mathbb{P}(t_e = 0) < p_c$. *There exists a deterministic, convex, compact set in* $\mathbb{R}^d$, *symmetric about the axes and with nonempty interior, such that for any* $\epsilon > 0$,

$$\mathbb{P}\left((1-\epsilon)\mathcal{B} \subset B(t)/t \subset (1+\epsilon)\mathcal{B} \text{ for all large } t\right) = 1.$$

There is also a version for ergodic distributions by Boivin [**8**], first done in less generality by Derriennic, as reported in [**28**, (9.25)]. One can show that the above is equivalent to: there is a norm g on $\mathbb{R}^d$ such that $\mathcal{B}$ is the unit ball of g, and

$$\limsup_{x \to \infty} \frac{|T(0,x) - g(x)|}{\|x\|_1} = 0 \text{ almost surely}.$$

This can be thought of as

$$T(0,x) = g(x) + o(\|x\|_1) \text{ as } x \in \mathbb{R}^d \to \infty.$$

We will take this approach in the proof: build the norm g and then show the limsup statement above.

PROOF OF SHAPE THEOREM. The idea of the proof is to first show "radial" convergence; that is, for a fixed $x \in \mathbb{Z}^d$, to show that

$$g(x) := \lim_n \frac{T(0,nx)}{n} \text{ exists}.$$

To do this, we appeal to the subadditive ergodic theorem. Then we "patch" together convergence in many different directions x to get a uniform convergence. We can see quickly that this limit should at least exist when we take expectations, since by invariance under lattice translations,

$$\mathbb{E}T(x,y) = \mathbb{E}T(x+z,y+z) \text{ for all } x,y,z \in \mathbb{Z}^d,$$

and so by the triangle inequality, for $0 \leqslant m \leqslant n$,

$$\mathbb{E}T(0,nx) \leqslant \mathbb{E}T(0,mx) + \mathbb{E}T(mx,nx) = \mathbb{E}T(0,mx) + \mathbb{E}T(0,(n-m)x).$$

Thus the sequence (a_n) given by $a_n = \mathbb{E}T(0,nx)$ is subadditive and a_n/n must have a limit from the following standard argument (referred to usually as Fekete's lemma): for fixed k and all $n \geqslant k$,

$$a_n \leqslant a_k + a_{n-k} \leqslant \cdots \leqslant \lfloor n/k \rfloor a_k + a_{n-\lfloor n/k \rfloor k}.$$

Since $n - \lfloor n/k \rfloor k$ is bounded as $n \to \infty$, we divide by n to obtain

$$\limsup_n a_n/n \leqslant \limsup_n \frac{\lfloor n/k \rfloor}{n} a_k = a_k/k.$$

This is true for all k, so

$$\limsup_n a_n/n \leqslant \inf_k a_k/k \leqslant \liminf_k a_k/k,$$

meaning $\lim_n a_n/n$ exists and equals $\inf_k a_k/k$.

To show the limit without expectation is much harder, and we will have to appeal to some machinery that was invented specifically for this problem. Liggett's version [**33**, Theorem 1.10] of Kingman's subadditive ergodic theorem states:

THEOREM 3.2. *Let $\{X_{m,n} : 0 \leqslant m < n\}$ is an array of random variables satisfying the following assumptions:*

(1) *for each n, $\mathbb{E}|X_{0,n}| < \infty$ and $\mathbb{E}X_{0,n} \geqslant -cn$ for some constant $c > 0$,*
(2) *$X_{0,n} \leqslant X_{0,m} + X_{m,n}$ for $0 < m < n$,*
(3) *for each $m \geqslant 0$, the sequence $\{X_{m+1,m+k+1} : k \geqslant 1\}$ is equal in distribution to the sequence $\{X_{m,m+k} : k \geqslant 1\}$, and*
(4) *for each $k \geqslant 1$, $\{X_{nk,(n+1)k} : n \geqslant 1\}$ is a stationary ergodic process.*

Then

$$g := \lim_n \frac{1}{n} \mathbb{E}X_{0,n} = \inf_n \frac{1}{n}\mathbb{E}X_{0,n} \text{ exists,}$$

and

$$\lim_n \frac{1}{n} X_{0,n} = g \text{ a.s. and in } L^1.$$

PROOF. We will give a proof of a version of the subadditive ergodic theorem in the next section. □

We apply this theorem for a fixed $x \in \mathbb{Z}^d$ to the sequence

$$X_{m,n} = T(mx,nx).$$

Item 2 holds by the triangle inequality, whereas 3 and 4 hold by invariance of the environment under integer translations. In item 1 we can take any $c > 0$, since $T \geqslant 0$ a.s. The only thing to check is that $\mathbb{E}T(0,nx) < \infty$ for each n. By subadditivity, and symmetry, it suffices to check that $\mathbb{E}T(0,e_1) < \infty$. To do this, we construct $2d$ edge-disjoint deterministic paths $\gamma_1, \ldots, \gamma_{2d}$ from 0 to e_1 and note

$$\mathbb{E}T(0,e_1) \leqslant \mathbb{E}\min\{T(\gamma_1), \ldots, T(\gamma_{2d})\}.$$

Now we can check the following result from [**12**]:

LEMMA 3.3. *If* $\mathbb{E}\min\{t_1, \ldots, t_{2d}\} < \infty$ *for i.i.d. edge-weights* t_i*, then the right side above is finite.*

PROOF. Let k be the maximal number of edges of any γ_i and note that for each i,

$$\mathbb{P}\left(T(\gamma_i) \geqslant \lambda\right) \leqslant k\mathbb{P}(t_e \geqslant \lambda/k),$$

so that if we put $M = \min\{T(\gamma_1), \ldots, T(\gamma_{2d})\}$,

$$\begin{aligned}\mathbb{P}(M \geqslant \lambda) = \prod_{i=1}^{2d} \mathbb{P}(T(\gamma_i) \geqslant \lambda) &\leqslant k^{2d} \left(\mathbb{P}(t_e \geqslant \lambda/k)\right)^{2d} \\ &= k^{2d}\mathbb{P}\left(\min\{t_1, \ldots, t_{2d}\} \geqslant \lambda/k\right).\end{aligned}$$

Therefore

$$\mathbb{E}M = \int_0^\infty \mathbb{P}(M \geqslant \lambda)\,\mathrm{d}\lambda \leqslant k^{2d} \int_0^\infty \mathbb{P}\left(\min\{t_1, \ldots, t_{2d}\} \geqslant \lambda/k\right)\,\mathrm{d}\lambda < \infty. \qquad \square$$

By the subadditive ergodic theorem, then, we define

$$g(x) = \lim_n \frac{T(0, nx)}{n},$$

which is a.s. constant. We next extend g to $\mathbb{Q}^d$ by taking such a rational x and letting $m \in \mathbb{N}$ be such that mx is an integer point. Then set

$$g(x) = \lim_n \frac{T(0, mnx)}{mn} = \frac{1}{m} \lim_n \frac{T(0, n(mx))}{n} = \frac{1}{m} g(mx).$$

g thus defined on $\mathbb{Q}^d$ satisfies the following properties: for $x, y \in \mathbb{Q}^d$,

(1) $g(x + y) \leqslant g(x) + g(y)$,
(2) g is uniformly continuous on bounded sets,
(3) for $q \in \mathbb{Q}$, $g(qx) = |q|g(x)$,
(4) g is symmetric about the axes.

Item 1 follows from the triangle inequality for T, and 4 follows from symmetries of the edge weights. Item 3 is an easy exercise, and 2 follows from 1: for $h = (h_1, \ldots, h_d)$,

$$\begin{aligned}|g(z) - g(z + h)| &\leqslant \max\{g(h), g(-h)\} = g(h) = g(h_1e_1 + \cdots + h_de_d) \\ &\leqslant g(e_1)(|h_1| + \cdots + |h_d|) \leqslant \|h\|_1\mathbb{E}T(0, e_1).\end{aligned}$$

Then g has a continuous extension to $\mathbb{R}^d$. The above properties extend to real arguments, so g is a seminorm. (A norm, except it could have $g(x) = 0$ for some $x \neq 0$.) It is a result of Kesten [**28**, Theorem 6.1] that g is a norm when $\mathbb{P}(t_e = 0) < p_c$.

Now that we have "radial" convergence to a norm g, we need to patch together convergence in every direction to a type of uniform convergence. Here we do this under the unnecessary but simplifying assumption:

$$\mathbb{P}(t_e \in [a, b]) = 1, \text{ where } 0 < a < b < \infty.$$

This implies

$$T(0, x), g(x) \in [a\|x\|_1, b\|x\|_1] \text{ for } x \in \mathbb{Z}^d.$$

So define the event

$$\Omega' = \{t_e \in [a,b] \text{ for all } e\} \cap \left\{ \lim_n \frac{T(0,nx)}{n} = g(x) \text{ for all } x \in \mathbb{Z}^d \right\}.$$

Note that on the above event we actually have

$$\lim_{\alpha\to\infty} \frac{T(0,\alpha x)}{\alpha} = g(x) \text{ for all } x \in \mathbb{Z}^d. \tag{3.1}$$

(Here α is real instead of just being an integer.) The reason is that we can bound

$$\begin{aligned} &\left| \frac{T(0,\alpha x)}{\alpha} - g(x) \right| \\ &\quad \leqslant \left| \frac{T(0,\lfloor\alpha\rfloor x)}{\lfloor\alpha\rfloor} - \frac{T(0,\alpha x)}{\lfloor\alpha\rfloor} \right| + \left| \frac{T(0,\alpha x)}{\lfloor\alpha\rfloor} - \frac{T(0,\alpha x)}{\alpha} \right| + \left| \frac{T(0,\lfloor\alpha\rfloor x)}{\lfloor\alpha\rfloor} - g(x) \right| \\ &\quad \leqslant \frac{T(\lfloor\alpha\rfloor x, \alpha x)}{\lfloor\alpha\rfloor} + T(0,\alpha x)\left| \frac{1}{\lfloor\alpha\rfloor} - \frac{1}{\alpha} \right| + \left| \frac{T(0,\lfloor\alpha\rfloor x)}{\lfloor\alpha\rfloor} - g(x) \right|. \end{aligned}$$

By the assumed inequality $t_e \leqslant b$, the first two terms go to zero as $\alpha \to \infty$. The last term approaches zero because of condition of the second intersected event in the definition of Ω'.

Fix $\omega \in \Omega'$, a set which has probability one, for the rest of the argument. We will show the equivalent statement

$$\limsup_{x\to\infty} \frac{|T(0,x) - g(x)|}{\|x\|_1} = 0,$$

so suppose it fails: for some $\epsilon > 0$, there is a sequence (x_n) going to infinity such that

$$|T(0,x_n) - g(x_n)| \geqslant \epsilon \|x_n\|_1 \text{ for all } n.$$

We may assume by compactness that $x_n/\|x_n\|_1 \to z$ for some z with $\|z\|_1 = 1$. We will show that for some x in a nearby direction to z, one cannot have $T(0,nx)/n \to g(x)$. (See Figure 3.1 for an illustration.)
Fix

$$\delta \in \left(0, \frac{\epsilon}{4b+1}\right)$$

and choose $x \in \mathbb{Z}^d$ such that

$$\left\| \frac{x}{\|x\|_1} - z \right\|_1 < \delta,$$

so that for n large,

$$\left\| \frac{x_n}{\|x_n\|_1} - \frac{x}{\|x\|_1} \right\|_1 < 2\delta.$$

We will compare the passage time from 0 to x_n to the passage time to the "nearby" point $\|x_n\|_1 x/\|x\|_1$. Then

$$\begin{aligned} |T(0,x_n) - g(x_n)| \leqslant &\left| T(0,x_n) - T\left(0, \|x_n\|_1 \frac{x}{\|x\|_1}\right) \right| \\ &+ \left| T\left(0, \|x_n\|_1 \frac{x}{\|x\|_1}\right) - g\left(\|x_n\|_1 \frac{x}{\|x\|_1}\right) \right| \\ &+ \left| g\left(\|x_n\|_1 \frac{x}{\|x\|_1}\right) - g(x_n) \right|. \end{aligned}$$

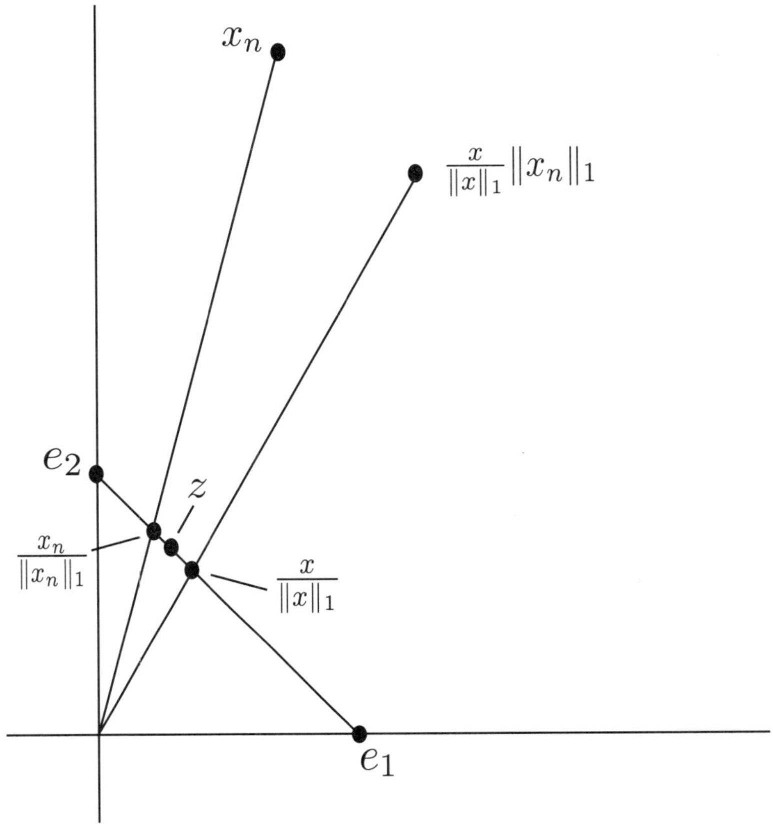

FIGURE 3.1. Illustration of the argument given in the proof of the shape theorem. The points $\frac{x_n}{\|x_n\|_1}$, z, and $\frac{x}{\|x\|_1}$ are all of ℓ^1-norm 1 and within ℓ^1-distance 2δ of each other. The point x is chosen in $\mathbb{Z}^d$ so that we have radial convergence of $T(0,nx)/n$ to $g(x)$ in this direction. Because x_n is close in direction to x, and our weights are bounded between a and b, the passage time $T(0,x_n)$ cannot be too different (on a linear scale) than passage time from 0 to the "nearby point" $\frac{x}{\|x\|_1}\|x_n\|_1$. The latter passage time is controlled, since it is from 0 to a multiple of x, so this prohibits the sequence $T(0,x_n)$ from fluctuating too much.

The first and last terms are bounded by

$$b\left\|\|x_n\|_1\frac{x}{\|x\|_1}-x_n\right\|_1 = b\|x_n\|_1\left\|\frac{x}{\|x\|_1}-\frac{x_n}{\|x_n\|_1}\right\|_1 < 2b\delta\|x_n\|_1.$$

However since we have radial convergence in direction x, (3.1) gives

$$T\left(0,\|x_n\|_1\frac{x}{\|x\|_1}\right) = g\left(\|x_n\|_1\frac{x}{\|x\|_1}\right) + o(\|x_n\|_1),$$

so for n large, the second term is bounded by $\delta\|x_n\|_1$. In total,

$$\epsilon\|x_n\|_1 \leqslant |T(0,x_n) - g(x_n)| \leqslant (4b+1)\delta\|x_n\|_1 < \epsilon\|x_n\|_1,$$

a contradiction. □

3.1.2. *Proof of subadditive ergodic theorem.* In this section, we prove a version of the subadditive ergodic theorem using ideas from [**13**] (which turn out to be similar to those of Liggett's proof). The current version developed through my discussions with F. Rassoul-Agha, and it is his notes we give here.

We prove the following version of the theorem, which is weaker than that stated in the last section, but sufficient for our purposes in FPP.

THEOREM 3.4. *Let $(X_{m,n} : m,n \in \mathbb{Z})$ be an array of random variables satisfying the following assumptions.*

1. *For all $m,n \in \mathbb{Z}$, $X_{n,n} = 0$ and $X_{n,m} = X_{m,n}$,*
2. *for each $n \in \mathbb{Z}$, $\mathbb{E}|X_{0,n}| < \infty$,*
3. *for all $m,n,l \in \mathbb{Z}$, $X_{l,n} \leqslant X_{l,m} + X_{m,n}$,*
4. *for each $m \in \mathbb{Z}$, the sequence $(X_{m+1,m+k+1} : k \in \mathbb{Z})$ is equal in distribution to the sequence $(X_{m,m+k} : k \in \mathbb{Z})$, and*
5. *for each $k \geqslant 1$, the process $(X_{nk,(n+1)k} : n \in \mathbb{Z})$ is stationary and ergodic.*

Then

$$g := \lim_{n\to\infty} \frac{\mathbb{E}X_{0,n}}{n} = \inf_{n\geqslant 1} \frac{\mathbb{E}X_{0,n}}{n} \text{ exists} \tag{3.2}$$

and

$$\lim_{n\to\infty} \frac{X_{0,n}}{n} = g \text{ a.s. and in } L^1.$$

PROOF. Note first that (3.2) holds by subadditivity of the sequence $\mathbb{E}X_{0,n}$ (which itself follows from stationarity in item (4)), and Fekete's lemma (mentioned above Theorem 3.2). Also observe that (1) and (3) imply that $0 = X_{n,n} \leqslant X_{n,m} + X_{m,n} = 2X_{m,n}$; that is, the random variables are nonnegative.

For $k \geqslant 1$, let μ_k be the distribution of

$$(X_{m,n}, X_{r,k} - X_{l,k})_{m,n,r,l\in\mathbb{Z}}.$$

This is a probability measure on the space of sequences

$$\widetilde{\Omega} = \{(x_{m,n}, b_{r,l})_{m,n,r,l\in\mathbb{Z}}\},$$

equipped with the product topology and its Borel sigma-field. Define the measure ν_L as

$$\nu_L = \frac{1}{L}\sum_{k=1}^{L} \mu_k.$$

By subadditivity, we have

$$X_{m,k} \leqslant X_{m,n} + X_{n,k} \quad \text{and} \quad X_{n,k} \leqslant X_{n,m} + X_{m,k} = X_{m,n} + X_{m,k},$$

so

$$|X_{m,k} - X_{n,k}| \leqslant X_{m,n} \text{ for all } k. \tag{3.3}$$

Exercise. Prove that $(\nu_L : L \geqslant 1)$ is a tight family of probability measures.

By the above exercise and Prohorov's theorem (see, for instance [**18**, Theorem 3.2.7]), there is a subsequence (L_j) such that ν_{L_j} converges weakly to a limiting probability measure ν on $\widetilde{\Omega}$. This measure has the property that its marginal on the sequences $(x_{m,n} : m, n \in \mathbb{Z})$ is precisely the distribution of $(X_{m,n} : m, n \in \mathbb{Z})$. Indeed, for bounded continuous $f = f(x_{m,n} : m, n \in \mathbb{Z})$, we have

$$\begin{aligned}\mathbb{E}_\nu f(X_{m,n} : m, n \in \mathbb{Z}) &= \lim_{j\to\infty} \frac{1}{L_j} \sum_{k=1}^{L_j} \mathbb{E}_{\mu_k} f(X_{m,n} : m, n \in \mathbb{Z}) \\ &= \lim_{j\to\infty} \frac{1}{L_j} \sum_{k=1}^{L_j} \mathbb{E} f(X_{m,n} : m, n \in \mathbb{Z}) \\ &= \mathbb{E} f(X_{m,n} : m, n \in \mathbb{Z}).\end{aligned}$$

The probability measure ν has the following properties (writing as above $X_{m,n}$, $B_{m,n}$ as the coordinate random variables on $\widetilde{\Omega}$)

(a) $|B_{m,n}| \leqslant X_{m,n}$, ν-a.s.,
(b) $B_{l,n} = B_{l,m} + B_{m,n}$, ν-a.s., and for all $l, m, n \in \mathbb{Z}$,
(c) for any $N \geqslant 1$, $(B_{nN,(n+1)N} : n \in \mathbb{Z})$ is stationary, and
(d) $\mathbb{E}_\nu B_{0,1} = g$.

We will now (mostly) justify the above four points. We will use the following fact:

Exercise. Suppose that (A_n, B_n) converge in distribution to (A, B). Assume also that $A_n \leqslant B_n$ a.s. Show that $A \leqslant B$ a.s. Prove the same result for $\leqslant$ replaced by $=$.

Property (a) comes from (3.3) and applying the above exercise. Property (b) comes by another application of the exercise and telescoping the sum

$$X_{l,k} - X_{n,k} = X_{l,k} - X_{m,k} + X_{m,k} - X_{n,k}.$$

Property (c) is a consequence of stationarity (item 4) in the statement of the theorem, and it is why averaging was used to define ν_L. More precisely, if $f = f(b_{nk,(n+1)k} : n \in \mathbb{Z})$ is bounded and continuous, then

$$\begin{aligned}\mathbb{E}_\nu f(B_{nN,(n+1)N} : n \in \mathbb{Z}) &= \lim_{j\to\infty} \frac{1}{L_j} \sum_{k=1}^{L_j} \mathbb{E}_{\mu_k} f(B_{nN,(n+1)N} : n \in \mathbb{Z}) \\ &= \lim_{j\to\infty} \frac{1}{L_j} \sum_{k=1}^{L_j} \mathbb{E} f(X_{nN,k} - X_{(n+1)N,k} : n \in \mathbb{Z}) \\ &= \lim_{j\to\infty} \frac{1}{L_j} \sum_{k=1-N}^{L_j-N} \mathbb{E} f(X_{nN,k} - X_{(n+1)N,k} : n \in \mathbb{Z}) \\ &= \lim_{j\to\infty} \frac{1}{L_j} \sum_{k=1-N}^{L_j-N} \mathbb{E} f(X_{(n+1)N,k+N} - X_{(n+2)N,k+N} : n \in \mathbb{Z}) \\ &= \lim_{j\to\infty} \frac{1}{L_j} \sum_{k=1}^{L_j} \mathbb{E} f(X_{(n+1)N,k} - X_{(n+2)N,k} : n \in \mathbb{Z}) \\ &= \lim_{j\to\infty} \frac{1}{L_j} \sum_{k=1}^{L_j} \mathbb{E}_{\mu_k} f(B_{(n+1)N,(n+2)N} : n \in \mathbb{Z}) \\ &= \mathbb{E}_\nu f(B_{(n+1)N,(n+2)N} : n \in \mathbb{Z}).\end{aligned}$$

The third equality follows from the fact that f is bounded, and so changing finitely many terms in the sum does not affect the limit. The fourth equality is from item 4 in the statement of the theorem.

To show item (d), we will give the idea, and the rest is an exercise. If we knew that the coordinates $B_{m,n}$ were bounded, then weak convergence would give us

$$\begin{aligned}
\mathbb{E}_\nu B_{0,1} = \lim_{j\to\infty} \frac{1}{L_j}\sum_{k=1}^{L_j} \mathbb{E}_{\mu_k} B_{0,1} &= \lim_{j\to\infty} \frac{1}{L_j}\sum_{k=1}^{L_j} \mathbb{E}(X_{0,k} - X_{1,k}) \\
&= \lim_{j\to\infty} \frac{1}{L_j}\left(\sum_{k=1}^{L_j} \mathbb{E}X_{0,k} - \sum_{k=1}^{L_j} \mathbb{E}X_{1,k}\right) \\
&= \lim_{j\to\infty} \frac{1}{L_j}\left(\sum_{k=1}^{L_j} \mathbb{E}X_{0,k} - \sum_{k=1}^{L_j} \mathbb{E}X_{0,k-1}\right) \\
&= \lim_{j\to\infty} \frac{1}{L_j}\mathbb{E}X_{0,L_j} \\
&= g.
\end{aligned}$$

(The last limit holds by (3.2).) However, f is not bounded.

Exercise. Fix the above argument by truncating and using (3.3).

Now, abbreviate $m_n = \lfloor n/N \rfloor$. Subadditivity implies that

$$X_{0,n} \leqslant X_{0,N} + X_{N,2N} + \cdots + X_{(m_n-1)N,m_nN} + X_{m_nN,nN}.$$

Exercise. Prove that if f is an L^1 function defined on a probability space whose measure is invariant under a measurable map θ, then $f \circ \theta^n / n$ converges to 0 a.s.

Since the process $(X_{nN,(n+1)N} : n \in \mathbb{Z})$ is assumed to be ergodic, by the above exercise and the ergodic theorem,

$$\begin{aligned}
\limsup_{n\to\infty} \frac{1}{n} X_{0,n} &\leqslant \lim_{n\to\infty} \frac{m_n}{n} \frac{X_{0,N} + \cdots + X_{(m_n-1)N,m_nN}}{m_n} \\
&\quad + \lim_{n\to\infty} \frac{m_n}{n} \frac{\max_{1\leqslant l\leqslant N} X_{0,l} \circ \theta^{m_nN}}{m_n} \\
&= \frac{1}{N}\mathbb{E}X_{0,N}.
\end{aligned}$$

Here, the map θ acts on a sequence $a = (a_n)_{n\in\mathbb{Z}}$ by shifting to the left: $(\theta a)_n = a_{n+1}$. The maximum comes because $n - N \leqslant m_nN \leqslant n$, and so $n - m_nN \in \{1, \ldots, N\}$. Taking infimum over all N and applying (3.2), we have

$$\limsup_{n\to\infty} \frac{1}{n} X_{0,n} \leqslant g. \tag{3.4}$$

Then by property (a) of the process $(B_{m,n})$, we have ν-a.s.

$$\lim_{n\to\infty} \frac{B_{0,n}}{n} \leqslant \limsup_{n\to\infty} \frac{X_{0,n}}{n} \leqslant g.$$

Since $(B_{n,n+1})$ is stationary, additivity (item (b)) and the ergodic theorem imply that ν-a.s.,

$$\lim_{n\to\infty} \frac{B_{0,n}}{n} = \lim_{n\to\infty} \frac{1}{n}\sum_{k=0}^{n-1} B_{k,k+1} = \mathbb{E}_\nu[B_{0,1} \mid \mathcal{I}],$$

where $\mathcal{I}$ is the invariant sigma-field. Putting these two displays together, we obtain

$$\mathbb{E}_\nu[B_{0,1} \mid \mathcal{I}] \leqslant g.$$

However, property (d) gives

$$\mathbb{E}_\nu\left[\mathbb{E}_\nu[B_{0,1} \mid \mathcal{I}]\right] = \mathbb{E}_\nu B_{0,1} = g,$$

and so

$$\lim_{n\to\infty} \frac{1}{n} B_{0,n} = g, \ \nu\text{-a.s.}.$$

Apply item (a) again, combined with (3.4), to obtain

$$g = \lim_{n\to\infty} \frac{1}{n} B_{0,n} \leqslant \liminf_{n\to\infty} \frac{1}{n} X_{0,n} \leqslant \limsup_{n\to\infty} \frac{1}{n} X_{0,n} \leqslant g.$$

This means that $\lim_{n\to\infty} \frac{1}{n} X_{0,n} = g$ ν-a.s. Since this is the same as $\mathbb{P}$-a.s., we are done. □

REMARK 3.5. If we knew the existence of a.s. limits $B_{m,n} = \lim_{k\to\infty}(X_{m,k} - X_{n,k})$, then the proof would be more direct. In other words, there would be no need for the measures μ_k and ν_L, and one could prove properties (a)-(d) and deduce the result similarly to the above.

REMARK 3.6. In the version of the theorem in the previous section, one only takes an array $(X_{m,n})$ with $n > m$. In this case, one needs an extra assumption which in a sense replaces nonnegativity, and it appears as the second part of condition (1): $\mathbb{E}X_{0,n} \geqslant -cn$ for a constant $c > 0$. Then one needs an additional argument for tightness in the proof.

3.1.3. *High dimensions.* In FPP, the shape theorem gives that $\mathcal{B}$ is convex, compact, with nonempty interior, and with all the symmetries of $\mathbb{Z}^d$. Many different sets have this property, in particular all the ℓ^p balls for $p \in [1, \infty]$. So the shape theorem allows in principle a cube (ℓ^∞ ball) and a diamond (ℓ^1 ball), and therefore says nothing about strict convexity, flat edges, corners, or whether the shape is a polygon.

For (t_e) that are i.i.d. and, say, continuous, the following properties are expected for $\mathcal{B}$.

(1) $\mathcal{B}$ is strictly convex. That is, for any distinct $x, y \in \mathcal{B}$ and $\lambda \in (0, 1)$ the point $\lambda x + (1 - \lambda)y$ is in the interior of $\mathcal{B}$. Thus $\mathcal{B}$ should have no flat facets (like a polygon has).

(2) $\mathcal{B}$ has no "corners." That is, the boundary of $\mathcal{B}$ should be differentiable. One way to say this is in terms of supporting hyperplanes. A hyperplane in $\mathbb{R}^d$ is a set of the form $\{x = (x_1, \ldots, x_d) : x \cdot y = a\}$ for some $y \in \mathbb{R}^d$ and $a \in \mathbb{R}$ (where '$\cdot$' is the standard dot product $x \cdot y = \sum_{i=1}^d x_i y_i$). A hyperplane H is supporting for $\mathcal{B}$ at $z \in \partial\mathcal{B}$ if H contains z but $\mathcal{B}$ intersects at most one component of H^c. By the Hahn-Banach Theorem, since $\mathcal{B}$ is convex and bounded, each $z \in \partial\mathcal{B}$ has a supporting hyperplane. We say that $\partial\mathcal{B}$ is differentiable if each $z \in \partial\mathcal{B}$ has exactly one supporting hyperplane. (This rules out "corners" for $\mathcal{B}$, as in the case of a polygon.)

(3) Combining the above two cases, but weaker, the set $\mathcal{B}$ should not be a polygon. To state this precisely, we say that $\mathcal{B}$ should have infinitely many extreme points. An extreme point $x \in \mathcal{B}$ is a point which is not on the interior of a line segment with endpoints in $\mathcal{B}$. Precisely, whenever we

write $x = \lambda z + (1-\lambda)y$ with $z, y \in \mathcal{B}$ and $\lambda \in (0,1)$, we have $z = y$. So $\mathcal{B}$ has infinitely many extreme points.

(4) The boundary $\partial\mathcal{B}$ should have uniformly positive curvature. In other words, near every boundary point, the boundary should locally look like the boundary of a Euclidean ball with bounded radius.

Here we state the main high-dimensional asymptotic result. It says that for distributions with no atom at 0 but with 0 in the support (like exponential), for high dimensions, the limit shape is not a Euclidean ball, an ℓ^1-ball, or an ℓ^∞-ball. Since we saw that the Eden model is equivalent to FPP with exponential weights, this shows the same result for the Eden model in high dimensions. For the statement, we define the balls $\mathbf{D}, \mathbf{B}, \mathbf{C}$ to be the ℓ^1, ℓ^2, and ℓ^∞-balls of radius $g(e_1)^{-1}$. We will also make the following assumptions from [**7**]:

$$\mathbb{E}t_e < \infty, \tag{3.5}$$

and for some $a \geqslant 0$,

$$F(0) = 0 \text{ and } \left|\frac{\mathbb{P}(t_e \leqslant x)}{x} - a\right| = O(|\log x|^{-1}) \text{ as } x \downarrow 0. \tag{3.6}$$

The following theorem is from Auffinger-Tang [**7**], which weakens various assumptions (widens the class of distributions in particular) of the work of previous authors. Some earlier work was done by Kesten [**28**], Dhar [**16**], Couronné-Enriquez-Gerin [**11**], and Martinsson [**37**].

Theorem 3.7. *Assume* (3.5) *and* (3.6). *For all large d, in FPP on $\mathbb{Z}^d$ with fixed weights (t_e), one has*

$$\mathbf{D} \subset \mathcal{B} \subset \mathbf{C} \textit{ and } \mathcal{B} \neq \mathbf{B}, \mathbf{C}, \textit{ or } \mathbf{D}.$$

- The proof proceeds by showing the asymptotic
$$\lim_{d\to\infty} \frac{g(e_1)d}{\log d} = \frac{1}{2a}.$$
For example, if $\mathcal{B}$ were the ℓ^1-ball, one would have for $e = (1, \ldots, 1)$,
$$g(e) = dg(1/d, \ldots, 1/d) = dg(e_1) \sim \frac{\log d}{2a}.$$
However one can show that $g(e) \leqslant C$ for some constant C. In fact, combining the work of Martinsson [**37**] and Couronné-Enriquez-Gerin [**11**], one has can show that $\lim_{d\to\infty} g(e)$ exists for the exponential distribution and is related to the nonzero solution of $\coth \alpha = \alpha$.
- It is not hard to give the bound for exponential weights $g(e) \leqslant 1$. Construct a path γ with vertices $x_0, x_1, \ldots$ as follows: set $x_0 = 0$ and for $n \geqslant 0$, starting from x_n, take the minimal weight edge of the d different edges leading in directions $e_1, \ldots, e_d$ (the positive coordinate directions) and call its endpoint x_{n+1}. We see that if $X_1, X_2, \ldots$ are the weights of the first edge, second edge, and so on, then X_i is the minimum of d i.i.d. exponential random variables, so is an exponential with mean $1/d$. Thus putting $H_n = \{x = (x_1, \ldots, x_d) : \sum_i x_i = n\}$ one obtains
$$\mathbb{E}T(0, H_n) \leqslant \sum_{i=1}^n \mathbb{E}X_i = n/d.$$

But one can show using the shape theorem that

$$g(e) = \lim_n \frac{\mathbb{E}T(0, ne)}{n} = \lim_n \frac{\mathbb{E}T(0, H_{dn})}{n} \leqslant \lim_n \frac{\sum_{i=1}^{nd} \mathbb{E}X_i}{n} = 1$$

3.1.4. *Flat edges.* A main question in FPP is: which compact convex sets are realizable as limit shapes? The question above is completely open in the i.i.d. setting. Interestingly, though, this is solved by Häggström and Meester '95 [**22**] in the case of stationary (not necessarily i.i.d.) passage times.

THEOREM 3.8 (Häggström-Meester). *Any non-empty compact, convex set C that has the symmetries of $\mathbb{Z}^d$ is a limit shape for some FPP model with weights distributed according to a stationary (under translations of $\mathbb{Z}^d$) and ergodic measure.*

From this result, we see for example that there are stationary-ergodic models of FPP that have limit shapes equal to polygons or Euclidean balls. So certainly conjectured properties from the i.i.d. setting like strict convexity are not true in such generality.

The previous section focused on high dimensions. What can we say about low dimensions? For one special class of distributions, Durrett and Liggett [**19**] were able to say much more: that the limit shape is not strictly convex. Of course we believe strict convexity in the continuous weight case, so these distributions have atoms. Their "flat edge" result holds for higher dimensions as well, but we can give a precise description of it in two dimensions, and further work has been done by Marchand [**34**], Zhang [**47**, **48**], and Auffinger-Damron [**2**].

Recalling that F is the distribution of our weights, let $\mathcal{M}_p$ be the set of distributions F that satisfy the following:

(1) $F(x) = 0$ for all $x < 1$ and $F(1) = p \geqslant \vec{p}_c$, where $\vec{p}_c$ is the two-dimensional oriented bond percolation threshold (approximately .70548), and
(2) $\int x \, \mathrm{d}F(x) = \mathbb{E}t_e < \infty$.

In [**19**], it was shown that if $F \in \mathcal{M}_p$ then the limit shape $\mathcal{B}$ has some flat edges. The precise location of these edges was found in [**34**]. To describe this, write $\mathbf{B}_1$ for the closed ℓ^1 unit ball:

$$\mathbf{B}_1 = \{(x, y) \in \mathbb{R}^2 \; : \; |x| + |y| \leqslant 1\}$$

and write *int* $\mathbf{B}_1$ for its interior. For $p > \vec{p}_c$ let α_p be the asymptotic speed of oriented percolation (see [**17**]), define the points

$$M_p = \left(\frac{1}{2} - \frac{\alpha_p}{\sqrt{2}}, \frac{1}{2} + \frac{\alpha_p}{\sqrt{2}}\right) \text{ and } N_p = \left(\frac{1}{2} + \frac{\alpha_p}{\sqrt{2}}, \frac{1}{2} - \frac{\alpha_p}{\sqrt{2}}\right) \tag{3.7}$$

and let $[M_p, N_p]$ be the line segment in $\mathbb{R}^2$ with endpoints M_p and N_p. For symmetry reasons, the following theorem is stated only for the first quadrant.

THEOREM 3.9 (Durrett-Liggett [**19**], Marchand [**34**]). *Let $F \in \mathcal{M}_p$ in two dimensions.*

(1) $\mathcal{B} \subset \mathbf{B}_1$.
(2) *If* $p < \vec{p}_c$ *then* $\mathcal{B} \subset$ *int* $\mathbf{B}_1$.
(3) *If* $p > \vec{p}_c$ *then* $\mathcal{B} \cap [0, \infty)^2 \cap \partial \mathbf{B}_1 = [M_p, N_p]$.
(4) *If* $p = \vec{p}_c$ *then* $\mathcal{B} \cap [0, \infty)^2 \cap \partial \mathbf{B}_1 = (1/2, 1/2)$.

The angles corresponding to points in the line segment $[M_p, N_p]$ are said to be in the *percolation cone*. The existence of a flat edge for the limit shape can be explained heuristically. The definition of the oriented percolation threshold $\vec{p}_c$ is as follows, taking $\mathbb{P}_p$ to be the distribution of i.i.d. edge-weights (η_e) on $\mathcal{E}^2$ with probability p to be 1 and $1-p$ to be 2:

$$\vec{p}_c = \sup\{p : \mathbb{P}_p(\exists \text{ infinite oriented path } \Gamma \text{ with } \eta_e = 1 \text{ for } e \in \Gamma) = 0\}.$$

(Here, oriented means as usual that the vertices of the path, in order, have non-decreasing coordinates.) By monotonicity of the probability in p, one has that for $p > \vec{p}_c$, if $F \in \mathcal{M}_p$, then there is positive probability of existence of an infinite oriented path of all 1-weights starting from 0. In fact one can even show the stronger statement that for any angle θ in the percolation cone, there is positive probability for an infinite oriented path of 1-weights starting from 0 and going in direction θ (the angles of the vertices on the path converge to θ). Since no edge-weights have value below 1, any finite segment of such an infinite path must be a minimizing path (geodesic). But distance along these geodesics correspond to the ℓ^1-distance on $\mathbb{Z}^2$, so in such directions, the limit shape must correspond to the ℓ^1-ball. It is important to point out that it is unknown (but expected to be false) whether there are other distributions whose limit shapes have flat edges. Even in the case of $\mathcal{M}_p$, the flat part of the percolation cone ends at M_p and N_p; however, that does not exclude the limit shape from having further flat spots. This is not expected though.

Let $\beta_p := 1/2 + \alpha_p/\sqrt{2}$, that is, define β_p as the x coordinate of N_p. Convexity and symmetry of the limit shape imply that $1/g(e_1) \geqslant \beta_p$. A non-trivial statement about the edge of the percolation cone came in 2002 when Marchand [**34**, Theorem 1.4] proved that this inequality is in fact strict:

$$1/g(e_1) > 1/2 + \alpha_p/\sqrt{2}.$$

In other words, Marchand's result says that the line that goes through N_p and is orthogonal to the e_1-axis is not a tangent line of $\partial\mathcal{B}$. The following theorem builds on Marchand's result and technique and says that at the edge of the percolation cone, one cannot have a corner.

THEOREM 3.10 (Auffinger-Damron [**2**]). *Let $F \in \mathcal{M}_p$ for $p \in [\vec{p}_c, 1)$. The boundary $\partial\mathcal{B}$ is differentiable at M_p and N_p.*

The theorem above shows that any distribution in $\mathcal{M}_p$ has a non-polygonal limit shape. The first example of a non-polygonal limit shape was discovered by Damron-Hochman [**14**].

3.2. LPP. LPP is defined similarly to FPP, but taking the maximum over oriented paths, instead of the infimum over all paths. On $\mathbb{Z}^d$, we assign i.i.d. site-weights $(t_v)_{v\in\mathbb{Z}^d}$ with common distribution F. These weights no longer need to be nonnegative, but for simplicity we will take them to be so. An oriented path Γ with vertices $x_0, x_1, \ldots, x_n$ has the property that all coordinates of x_i are less or equal to the corresponding coordinates of x_{i+1} (this is written $x_i \leqslant x_{i+1}$); that is, for all standard basis vectors e_k, and all i, one has $x_i \cdot e_k \leqslant x_{i+1} \cdot e_k$. By convention, we identify the path Γ with its vertices, but we exclude the initial point. The passage time of such a path is $T(\Gamma) = \sum_{v\in\Gamma} t_v$, and the last-passage time $T(x,y)$ between vertices $x \leqslant y$ is

$$T(x,y) = \max_{\Gamma: x\to y} T(\Gamma),$$

where the maximum is over oriented paths from x to y. Note that there are only finitely many paths under consideration, so we can take a maximum. It is important to note that when x, y do not satisfy $x \leqslant y$ or $y \leqslant x$, then $T(x, y)$ is not defined. Also, by convention, $T(x, x) = 0$ for all x.

Due to directedness of the model and the fact that we are taking a maximum, T has somewhat different properties than those in FPP. One still has for $x \leqslant y$, $T(x, y) \geqslant 0$ and if $t_v > 0$ for all v, then $T(x, y) > 0$ when $x \neq y$. Due to excluding the initial point from all our paths, we have a super-additivity property of T that corresponds to the triangle inequality in FPP:

$$\text{for } x \leqslant y \leqslant z, \ T(x, z) \geqslant T(x, y) + T(y, z).$$

Due to this super-additivity, the limiting shape in LPP is not convex, since the corresponding "shape function" g will be super-additive. For its definition, we again appeal to the subadditive ergodic theorem, noting that $-T$ is subadditive. The only difficulty is to come up with conditions under which the limit is finite. The following version of "radial convergence" for T to a limit comes from Martin [**35**].

THEOREM 3.11. *Suppose that*

$$\int_0^\infty (1 - F(x))^{1/d} \, dx < \infty. \tag{3.8}$$

Then for each $x \geqslant 0$ in $\mathbb{R}^d$, the following (deterministic) limit exists a.s. and in L^1:

$$g(x) = \lim_n \frac{T(0, [nx])}{n} < \infty,$$

where $[nx]$ is the point of $\mathbb{Z}^d$ with $nx \in [nx] + [0, 1)^d$. The function g is continuous on $\{x : x \geqslant 0\}$ (including at the boundaries), satisfies $g(x + y) \geqslant g(x) + g(y)$, is invariant under permutations of the coordinates, and $g(ax) = ag(x)$ for $a \geqslant 0$.

Although $\mathbb{E}t_v < \infty$ is sufficient to guarantee that $\mathbb{E}T(0, x) < \infty$ for all $x \geqslant 0$ (just bound $T(0, x)$ above by the sum of all weights t_y with $0 \leqslant y \leqslant x$), the subadditive ergodic theorem this time gives

$$g(x) = \sup_{n \geqslant 1} \frac{\mathbb{E}T(0, nx)}{n},$$

and to deduce that $g(x) < \infty$, we need to know that the means $\mathbb{E}T(0, nx)$ do not grow faster than linearly. The corresponding issue in FPP is that the means $\mathbb{E}T(0, nx)$ do not grow sublinearly, which is guaranteed by $F(0) < p_c$. Here the corresponding condition is not on $F(0)$ but, as we are in LPP, it is on the tail of F, and is the integrability condition (3.8).

As in FPP, one can upgrade radial convergence to a sort of shape theorem [**35**]. As before, put

$$\mathcal{B} = \{x \geqslant 0 : g(x) \leqslant 1\}.$$

THEOREM 3.12 (LPP shape theorem). *Assume* (3.8) *and put $B(t) = \{x \geqslant 0 : T(0, x) \leqslant t\}$. For any $\epsilon > 0$, one has*

$$\mathbb{P}\left((1 - \epsilon)\mathcal{B} \subset B(t)/t \subset (1 + \epsilon)\mathcal{B} \text{ for all large } t\right) = 1.$$

3.2.1. *Near the boundary.* Just as in FPP, not much is known about the limiting shape $\mathcal{B}$. It is expected as before to have differentiable boundary (at least when the weights are continuous) with positive curvature, and certainly not to be a polygon. The flat edge result from FPP carries over to LPP: the analogous condition on our distribution F is

$$F(1) = 1, \ F(1^-) = 1 - p, \text{ where } p > \vec{p}_c.$$

Such distributions F of weights t_v have $\mathbb{P}(t_v > 1) = 0$ but $\mathbb{P}(t_v = 1) = p > \vec{p}_c$. The value c is the highest weight of an edge, and any oriented path of all c-weights will be an optimal infection path. Thus again in the percolation cone, the boundary of the limit shape will agree with that of the ℓ^1-ball.

Somewhat surprisingly, in two dimensions, the limit shape is shown not to be a polygon for most distributions. This is in contrast to the situation in FPP, where this is only known for distributions in $\mathcal{M}_p$. The LPP result is a corollary of a "universality" of the shape function g near the boundary. The following result, proved by Martin [**36**], shows that the asymptotics of g near the boundary of the quarter plane $\{x : x \geqslant 0\}$ are explicit, and only depend on the mean and variance of F. From these asymptotics we can extract non-polygonality of the limit shape.

THEOREM 3.13. *Consider $d = 2$. Write μ for the mean of F and σ^2 for the variance of F. If F satisfies* (3.8), *then*

$$g(1, a) = \mu + 2\sigma\sqrt{a} + o(\sqrt{a}) \ \text{as } a \downarrow 0.$$

Here, $g(1, a)$ is the function g evaluated at the point $(1, a)$, and as $a \downarrow 0$, this point approaches the boundary of the quarter plane. Note that $g(1, 0) = \mu$, since the passage time from 0 to ne_1 must be achieved along the one oriented path connecting these points, and its passage time is the sum of $n + 1$ i.i.d. random variables with distribution F. Thus the law of large numbers gives the value of g at $(1, 0)$. Therefore the above result says:

$$g(e_1 + ae_2) - g(e_1) = 2\sigma\sqrt{a} + o(\sqrt{a}) \text{ as } a \downarrow 0.$$

To see that the above result implies that the limit shape is not a polygon, suppose for a contradiction that the limit shape is a polygon. Then it must have finitely many extreme points, and the boundary of the shape between the extreme points consists of straight line segments. If the limit shape is a triangle, then by symmetry, g must be a multiple of the ℓ^1-norm, but then $g(e_1 + ae_2) - g(e_1) = a$ for $a > 0$ and the above asymptotics are violated. Otherwise, there is a closest extreme point w to the point $(1, a)$, and the limit shape boundary must be a line segment between w and $(1, a)$. In this case, $g(e_1 + ae_2) - g(e_1) = ca$ for some real c and a small enough. But this again violates the above asymptotics.

3.2.2. *Exactly solvable cases in two dimensions.* The most famous case of LPP is when the distribution F of the site-weights is exponential in two dimensions. Here, there is a direct mapping from the growth of $B(t)$ to a particle system called the Totally Asymmetric Simple Exclusion Process (TASEP). TASEP is defined loosely as follows. We imagine that at each site z of $\mathbb{Z}$ with $z \leqslant 0$, there sits a particle at time 0. Associated to each particle is a Poisson process, and when the process increments, the particle attempts to move to the site directly to the right. If there is already a particle there, the move is suppressed, and the particle stays in its current location. The particle that is initially at site 0 is allowed to move unrestricted (since there are never any particles to the right of it), but the other particles may sometimes be blocked by particles to their rights. Our convention is that the particle at 0 immediately moves to the right at time zero. That is, at time zero, there is a particle at site 1, and particles at sites $-k$ for $k \geqslant 1$.

What is the relation between TASEP and LPP with exponential weights? To begin, the procession of the first particle in TASEP is the same as the infection in LPP along the positive e_1-axis from 0. Indeed, the infection appears at site 0 at time 0, just as the first particle in TASEP moves to the right. It then infects e_1 after an independent exponential time, just as the same particle in TASEP moves again to the right. Generally, the infection time from 0 to ne_1 is achieved through the path that proceeds directly down the positive e_1-axis, and occurs when the first particle in TASEP reaches site $n+1$.

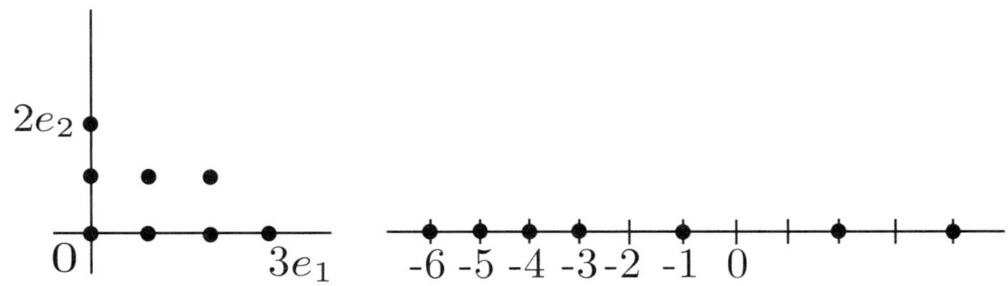

FIGURE 3.2. Illustration of the correspondence between LPP and TASEP. In the left (LPP), the infection has moved 4 steps to the right on the e_1-axis, and so on the right (TASEP) first particle has moved four steps to the right, from 0 to 4. Similarly, the infection has taken three steps at the second level, and the second particle has moved three steps to the right from -1 to 2. Last, the third TASEP particle has taken one step from -2 to -1.

At the second level, the infection of site e_2 occurs an independent exponential time after the infection appears at 0. This corresponds to the second particle in TASEP moving into the space left open after the first particle moves. Generally, the n-th step of the k-th particle in TASEP corresponds to the site $ke_2 + (n-1)e_1$ being infected from 0. To see this, we can derive the following relation in LPP: for $x_1, x_2 > 0$, one has

$$T(0,(x_1,x_2)) = t_{(x_1,x_2)} + \max\{T(0,(x_1-1,x_2)), T(0,x_1,x_2-1)\}.$$

This is because the infection from 0 reaches (x_1, x_2) through either (x_1-1, x_2) or (x_1, x_2-1) (whichever is infected last), and after the one of these sites with maximal passage time from 0 is infected, (x_1, x_2) must wait $t_{(x_1,x_2)}$ additional time. Similarly, in TASEP, for the k-th particle to make its n-th step, it must wait an independent exponential time after both of the following events occur: (a) the $k-1$-st particle makes its n-th step and (b) the k-th particle makes its $n-1$-st step. See Figure 3.2 for an example.

Because of this coupling, we can represent the passage times in LPP in terms of "currents" in TASEP. For example,

$$T(0,(n,n)) = \text{ time for } n\text{-th particle to make } n+1\text{-st step.}$$

But this is exactly the time needed for n particles to pass through the origin. Thus if we define c_t as the "current through zero at time n"; that is, the number of particles having passed through 0 by time t, one has

$$T(0,(n,n)) \leqslant t \Leftrightarrow c_t \geqslant n.$$

In a pioneering work, Rost [**41**] showed in '81 exact asymptotics for variables like this current, and this translates directly to an exact formula for the limit shape in exponential LPP.

THEOREM 3.14. *Let (t_v) be exponentially distributed with mean 1. Then*

$$g(x) = g(x_1, x_2) = (\sqrt{x_1} + \sqrt{x_2})^2.$$

Therefore the limit shape boundary in the above case is

$$\left\{(x,y) \in \mathbb{R}^2 : x, y \geqslant 0 \text{ and } \sqrt{x} + \sqrt{y} = 1\right\}.$$

So we can see directly that the limit shape is not a polygon, contains no flat segments, and has no "corners."

In the case of geometrically distributed weights with parameter p [**10, 25, 42**], there is an exact formula as well, showing

$$g(x_1, x_2) = \frac{1}{p}(x_1 + x_2 + 2\sqrt{x_1 x_2 (1-p)}).$$

In both of these cases, finer asymptotics are available [**26**] by relating the law of the passage time to the largest eigenvalue of a random matrix ensemble. Specifically for $x = (x_1, x_2)$ with $x_1 > 0$ and $x_2 > 0$,

$$\frac{T(0, nx) - ng(x)}{n^{1/3}} \Rightarrow Z \text{ as } n \to \infty$$

for a nondegenerate variable Z. This shows that the fluctuations of T about g are of order $n^{1/3}$, and this is expected for general distributions in both LPP and FPP in two dimensions. Unfortunately, despite decades of work, the best available bounds for $|T(0, nx) - ng(x)|$ are of order $\sqrt{n}$.

4. Rate of convergence and scaling exponents

In this section, we restrict to the case of FPP for simplicity, although similar theorems are provable in LPP.

4.1. Decomposition of error and strengthened shape theorems in FPP. The question we address here is: is it possible to improve the shape theorem to one with ϵ that depends on t? In other words, can we find ϵ_t such that $\epsilon_t \downarrow 0$ quickly and

$$\mathbb{P}\left((1-\epsilon_t)\mathcal{B} \subset B(t)/t \subset (1+\epsilon_t)\mathcal{B} \text{ for all large } t\right) = 1?$$

This is a question about the rate of convergence in the shape theorem. Restating it in terms of the norm g, we can ask the question in the following form: writing $T(0,x) = g(x) + o(\|x\|)$, how large is the term $o(\|x\|)$? The standard way to study this is to decompose

$$o(\|x\|) = [T(0,x) - \mathbb{E}T(0,x)] \ + \ [\mathbb{E}T(0,x) - g(x)]$$

into a random fluctuation term and a nonrandom fluctuation term.

The random fluctuation term is typically analyzed using techniques from concentration of measure – there many techniques (transportation inequalities, entropy methods, influence inequalities, isoperimetry, etc.) have been developed to study deviations of a function $f(X_1, X_2, \ldots)$ of independent variables (X_n) away from its mean or its median. Despite all these methods, however, current bounds on the first term are far from the predictions (and we will see this in the articles on fluctuations).

The nonrandom term is purely deterministic, and can be written as

$$\|x\|\left[\frac{\mathbb{E}T(0,x)}{\|x\|}-g\left(\frac{x}{\|x\|}\right)\right]\geqslant 0.$$

The term in the parenthesis is of the form $\mathbb{E}T(0,nx)/n-g(x)$ for x on the unit circle. We have seen that for $x\in\mathbb{Z}^d$, the sequence $(\mathbb{E}T(0,nx))$ is subadditive and so, when divided by n, converges. So quantifying this nonrandom error is really a problem of estimating the rate of convergence of a_n/n to its limit for a subadditive sequence (a_n). Unfortunately there are no general methods for this, but in the context of lattice models (like FPP), techniques have been developed to bound these nonrandom fluctuations in terms of the random ones. Specifically, if one has a concentration inequality of the form

$$\mathbb{P}\left(T(0,x)-\mathbb{E}T(0,x)<-\lambda\|x\|^{\alpha}\right)\leqslant\exp\left(-C\lambda^{\beta}\right),\ \lambda\geqslant 0$$

for constants $C,\alpha,\beta>0$, then one can show an upper bound of the type

$$\mathbb{E}T(0,x)-g(x)\leqslant C\|x\|^{\alpha}(\log\|x\|)^{\delta},$$

implying that nonrandom fluctuations should be no larger than random fluctuations. Indeed, using Gaussian concentration inequalities, one has the following version of Alexander's [**1**] result from Damron-Kubota [**15**].

THEOREM 4.1. *Assume that* $\mathbb{E}\min\{t_1,\ldots,t_d\}^2<\infty$ *and* $\mathbb{P}(t_e=0)<p_c$. *Then for some* $C>0$ *one has*

$$g(x)\leqslant\mathbb{E}T(0,x)\leqslant g(x)+C\sqrt{\|x\|\log\|x\|}\ \textit{for all}\ \|x\|>1.$$

On the other hand, evidence from [**5**] implies that random fluctuations should be no larger than nonrandom fluctuations. Therefore if we posit the existence of exponents such that

$$T(0,x)-\mathbb{E}T(0,x)\sim\|x\|^{\chi}\text{ and }\mathbb{E}T(0,x)-g(x)\sim\|x\|^{\gamma}$$

then one should have $\chi=\gamma$. (Note that Theorem 4.1 is a version of $\gamma\leqslant 1/2$. Also, since the first term is random, it may be measured as $\mathrm{Var}\ T(0,x)\sim\|x\|^{2\chi}$ or in terms of a concentration inequality.) Unfortunately this is far from the state of art, as under various general assumptions (exponential moments for the passage times, for instance), the best existing bounds are

$$0\leqslant\chi\leqslant 1/2\text{ and }-1/2\leqslant\gamma\leqslant 1/2.$$

(Under strong assumptions on existence of a fluctuation exponent $\chi<1/2$, one can show $\chi=\gamma$ [**5**].) As we saw in the section on exactly solvable models of LPP, we believe that in two dimensions, $\chi=\gamma=1/3$. It is reasonable to expect that these (equal) numbers decrease strictly with dimension, and approach 0 as dimension tends to infinity. Much more on the random fluctuation term and concentration estimates will be given in article [**43**]

In summary, we have a strengthened shape theorem of the following type. Some improvements to the assumptions have been made by Tessera [**44**] and Damron-Kubota [**15**] more recently, in particular generally replacing the log with $\sqrt{\log}$. The term $\sqrt{t}$ comes from the bounds $\chi,\gamma\leqslant 1/2$.

THEOREM 4.2 (Rate of convergence bound in the shape theorem). *Assume that* $\mathbb{E}e^{\alpha t_e} < \infty$ *for some* $\alpha > 0$. *There is* $C > 0$ *such that*

$$\mathbb{P}\left((t - C\sqrt{t\log t})\mathcal{B} \subset B(t) \subset (t + C\sqrt{t\log t}) \text{ for all large } t\right) = 1.$$

It is reasonable to believe, as we saw above, that a stronger shape theorem may hold, with $\sqrt{t}$ replaced by $t^\chi = t^\gamma$ for the fluctuation exponents explained above.

4.2. Scaling exponents and the KPZ relation. There are no accepted definitions of the exponents χ and γ from the last section. One can define, as in [**5**], directional p-fluctuation exponents ($p \geqslant 1$) in direction $x \in \mathbb{Z}^d$ as

$$\underline{\chi}_p(x) = \liminf_{n\to\infty} \frac{\log \|T(0,nx) - \mathbb{E}T(0,nx)\|_p}{\log n}$$

and

$$\overline{\chi}_p(x) = \limsup_{n\to\infty} \frac{\log \|T(0,nx) - \mathbb{E}T(0,nx)\|_p}{\log n},$$

where $\|X\|_p = (\mathbb{E}|X|^p)^{1/p}$, and nonrandom fluctuation exponents

$$\underline{\gamma}(x) = \liminf_{n\to\infty} \frac{\log(\mathbb{E}T(0,nx) - g(nx))}{\log n}$$

and

$$\overline{\gamma}(x) = \limsup_{n\to\infty} \frac{\log(\mathbb{E}T(0,nx) - g(nx))}{\log n}.$$

At the moment, it is not known if $\underline{\chi}_p(x) = \overline{\chi}_p(x)$ for any p or x generally, or if $\underline{\gamma}(x) = \overline{\gamma}(x)$. We can precisely state the bounds from the last section in terms of these exponents: it is known, combining work of Alexander [**1**], Kesten [**29**], and Auffinger-Damron-Hanson [**5**] that under an exponential moment assumption (this can be weakened in various cases) that

$$0 \leqslant \underline{\chi}_p(x) \leqslant \overline{\chi}_p(x) \leqslant 1/2 \text{ for all } p \geqslant 1 \text{ and } x \neq 0$$

and

$$-1 \leqslant \underline{\gamma}(x) \text{ and } -1/2 \leqslant \overline{\gamma}(x) \leqslant 1/2 \text{ for all } x \neq 0.$$

Our lack of information on the model leads to other (stronger) direction-independent definitions. We will present these below while discussing the so-called KPZ scaling relation.

The last exponent we need is the "wandering exponent," which measures the maximal displacement of the (random) geodesics from Euclidean straight lines. Following Chatterjee [**9**], for any nonzero x, let $D(0,x)$ be the maximal Euclidean distance between any point on a geodesic (minimizing path for T) from 0 to x and the line segment connecting 0 and x. It is reasonable to believe that

$$\mathbb{E}D(0,x) \sim \|x\|^\xi$$

for some (dimension-dependent) $\xi = \xi(d)$. The predictions are that in two dimensions, $\xi = 2/3$, that ξ decreases with d to $1/2$, but might always be $> 1/2$. In fact, these statements can be read directly off of the similar predictions for χ, along with the conjectured KPZ scaling relation

$$\chi = 2\xi - 1.$$

This relation is expected to be "universal." That is, it does not depend on the distribution of the (t_e)'s, and does not even depend on the dimension d, as long as the distribution of the (t_e)'s is reasonable, say with no atoms, and enough moments. There are heuristic arguments from physics for this relation in [**30**], but to date, there is no proof that is valid in full generality; that is, there is no proof under only edge-weight assumptions. One major difficulty is that there is no accepted definition of exponents which is usable. For instance, the exponents $\overline{\chi}_p$ and $\underline{\chi}_p$ defined above always exist, but unless one can show they are equal, they are not so helpful. Furthermore, the current bounds on the exponent ξ depend even on which definition is taken! In Newman-Piza [**39**] and Licea-Newman-Piza [**32**], the only work giving general bounds on ξ, one has versions of

$$\xi \geqslant 1/(d+1),\ \xi \geqslant 1/2,\ \xi \geqslant 3/5, \quad \text{and } \xi \leqslant 3/4,$$

depending on the definition of ξ. For example, the first bound is valid for a quite general definition of ξ, the second for a point-to-line geodesic wandering exponent, the third for a more restricted point-to-line exponent, and the fourth only in directions of "positive curvature" of the limiting shape. So, for instance, it is not even known at this point if

$$\mathbb{E}D(0, ne_1) = o(n).$$

Regardless of the precise definition, Newman-Piza provided the first rigorous argument for the inequality $\chi \geqslant 2\xi - 1$, and it is essentially from this inequality and the known bound $\chi \leqslant 1/2$ that they derive $\xi \leqslant 3/4$. It took another 16 years for a rigorous argument, due to Chatterjee, for the other inequality, $\chi \leqslant 2\xi - 1$, even under strong assumptions of existence of exponents. A couple of months later, a simplified proof due to Auffinger-Damron [**4**] appeared, which removed a technical assumption on the valid class of distributions. We stress though that these inequalities are still conditional, in that they assume existence of exponents χ and ξ.

Below we will give a proof sketch for the KPZ relation. We begin with Chatterjee's exponents.

DEFINITION 4.3. χ_u is the smallest number such that for any $\chi' > \chi_u$, there exists $\alpha > 0$ such that

$$\sup_{x \neq 0} \mathbb{E} \exp\left(\alpha \frac{|T(0,x) - \mathbb{E}T(0,x)|}{\|x\|^{\chi'}}\right) < \infty,$$

and χ_ℓ is the largest number such that for any $\chi'' < \chi_\ell$, one has

$$\inf_{x \neq 0} \frac{\text{Var } T(0,x)}{\|x\|^{\chi''}} > 0.$$

ξ_u is the smallest number such that for any $\xi' > \xi_u$, there exists $\beta > 0$ such that

$$\sup_{x \neq 0} \mathbb{E} \exp\left(\beta \frac{D(0,x)}{\|x\|^{\xi'}}\right) < \infty$$

and ξ_ℓ is the largest number such that for any $\xi'' < \xi_\ell$,

$$\inf_{x \neq 0} \frac{\mathbb{E}D(0,x)}{\|x\|^{\xi''}} > 0.$$

There are two important things to notice about these definitions. First, they are not directional, as they all take supremum or infimum over all directions x. So, for instance, if we assume that $\chi_u = \chi_\ell$ (as we will in the theorem below), then we are assuming that random fluctuations are the same in all directions, which rules out the case of the class $\mathcal{M}_p$ from Section 3.1.4 (see [**48**] for more details). For continuous distributions, though, this should be true. Next, the upper exponents are somewhat stronger than we might want, as they incorporate information about exponential concentration of the variables $T(0,x)$ and $D(0,x)$. In other words, these definitions imply (by an application of Markov's inequality for the exponential): for $\chi' > \chi_u$, there exists $C_1, C_2 > 0$ such that for all x

$$\mathbb{P}\left(|T(0,x) - \mathbb{E}T(0,x)| > \|x\|^{\chi'}\right) \leqslant C_1 \exp\left(-\|x\|^{C_2}\right),$$

and for $\xi' > \xi_u$, there are C_3, C_4 such that for all x,

$$\mathbb{P}\left(D(0,x) \geqslant \|x\|^{\xi'}\right) \leqslant C_3 \exp\left(-\|x\|^{C_4}\right). \tag{4.1}$$

In fact, by using Alexander's technique (as shown in [**9**]), the first inequality can be upgraded to

$$\mathbb{P}\left(|T(0,x) - g(x)| > \|x\|^{\chi'}\right) \leqslant C_1 \exp\left(-\|x\|^{C_2}\right). \tag{4.2}$$

Now we state the KPZ relation that has been rigorously proved to date, combining the results of Chatterjee [**9**] and Auffinger-Damron [**4**].

THEOREM 4.4 (Scaling relation). *Assume that $\mathbb{P}(t_e = 0) < p_c$. Then if $\chi := \chi_\ell = \chi_u$ and $\xi := \xi_\ell = \xi_u$, one has $\chi = 2\xi - 1$.*

Before the sketch, we mention some comments. The above relation has been extended to a positive temperature model, directed polymers in a random environment [**3**]. Next, curvature exponents κ for the boundary of the limit shape have been defined in [**4**], and it is believed that $\kappa = 2$. This corresponds to a boundary which is locally like Euclidean ball. For other curvature exponents, arguments from [**4**] suggest that a different relation holds: $\chi = \kappa\xi - (\kappa - 1)$. Last, versions of this relation have been proved in two dimensions in the exactly solvable version of LPP and other continuum models, where it is known that $\chi = 1/3$ and $\xi = 2/3$. In fact, the argument of [**4**] for the upper bound on χ is quite similar to the one by Wüthrich [**46**] and later by Johansson [**27**].

SKETCH OF PROOF OF THEOREM 4.4. We will give the proofs in the e_1 direction for simplicity. For the lower bound, suppose for a contradiction that $\xi > \frac{1+\chi}{2}$ and choose $\xi' \in \left(\frac{1+\chi}{2}, \xi\right)$. We will show that

$$\mathbb{P}(D(0, ne_1) > n^{\xi'}) \leqslant e^{-n^c} \tag{4.3}$$

for some $c > 0$. This itself suggests a contradiction, since $\xi' < \xi$. A more careful argument would actually give $\mathbb{E}D(0, ne_1) = O(n^{\xi'})$, which contradicts $\xi' < \xi = \xi_\ell$.

To show (4.3), put L to be the line segment connecting 0 and ne_1, and consider the set

$$S = \{x \in \mathbb{Z}^d : \|x - u\| < n^{\xi'} \text{ for some } u \in L\},$$

with

$$\partial S = \{y \in \mathbb{Z}^d \backslash S : \|y - x\|_1 = 1 \text{ for some } x \in S\}.$$

If $D(0, ne_1) > n^{\xi'}$, then there is $z \in \partial S$ such that z is on a geodesic from 0 to ne_1, and so $A_z = \{T(0, ne_1) = T(0,z) + T(z, ne_1)\}$ occurs. Therefore

$$\mathbb{P}(D(0, ne_1) > n^{\xi'}) \leqslant \sum_{z \in \partial S} \mathbb{P}(A_z).$$

One can now show that there is $c > 0$ depending on χ' such that

$$\mathbb{P}(A_z) \leqslant e^{-n^c} \text{ for all } z \in \partial S. \tag{4.4}$$

Summing over all z, we then obtain (4.3).

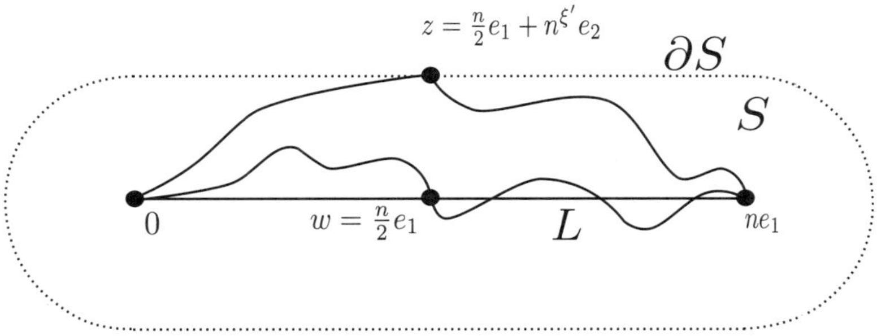

FIGURE 4.1. Illustration of the argument for the inequality $\chi \geqslant 2\xi - 1$. S is the set of vertices within distance $n^{\xi'}$ of L, the line segment connecting 0 and ne_1, and its boundary ∂S contains z. The idea is to show that with high probability, the passage time from 0 to ne_1 through z is larger than the passage time through $(n/2)e_1$. When this occurs, z cannot be on a geodesic from 0 to ne_1.

We will not show (4.4) for all z, but only for $z = (n/2)e_1 + n^{\xi'}e_2$, which may not be an integer point, but we will pretend it is, to avoid notation. In this case, if A_z occurs, we write

$$\begin{aligned} 0 &\geqslant T(0, ne_1) - T(0, (n/2)e_1) - T((n/2)e_1, ne_1) \\ &= [T(0,z) - T(0,(n/2)e_1)] + [T(z, ne_1) - T((n/2)e_1, ne_1)]. \end{aligned} \tag{4.5}$$

(See Figure 4.1.) These bracketed terms have the same distribution, so we will only bound one of them. Write $w = (n/2)e_1$ and estimate

$$T(0,z) - T(0,w) = g(z) - g(w) + [T(0,z) - g(z)] + [T(0,w) - g(w)]. \tag{4.6}$$

To estimate $g(z) - g(w)$, we write it as

$$g\left((n/2)e_1 + n^{\xi'}e_2\right) - g((n/2)e_1) = \frac{n}{2}\left[g\left(e_1 + \frac{2n^{\xi'}}{n}e_2\right) - g(e_1)\right].$$

Here we must use information about the limit shape, namely that in certain directions it is known to have "positive curvature." It is known (see [**9**]) that there exists a point on the limit shape boundary near which the boundary is locally positively curved, and instead of doing the argument in that direction, we assume this condition in direction e_1. It amounts to the statement that there exists $\delta, C_1 > 0$ such that if v is a vector with $\|v\| < \delta$ and $v \perp e_1$, then

$$g(e_1 + v) - g(e_1) \geqslant C_1\|v\|^2.$$

We apply this in the above equation with $v = \frac{2n^{\xi'}}{n}e_2$. This is less than δ in norm for n large because we may assume that $\xi' < 1$. If not, then $\xi \geqslant 1$, and in this case, the KPZ inequality reads $\chi \geqslant 1$, which is false, as known exponential concentration bounds imply that $\chi \leqslant 1/2$. So we obtain

$$g(z) - g(w) \geqslant C_1 \frac{n}{2} \left(\frac{2n^{\xi'}}{n} \right)^2 = C_2 n^{2\xi' - 1}.$$

Returning to (4.6), we obtain

$$T(0,z) - T(0,w) \geqslant C_2 n^{2\xi'-1} + [T(0,z) - g(z)] + [T(0,w) - g(w)].$$

The same development works for the other term of (4.5), and we find that if A_z occurs, then

$$\begin{aligned} 0 \geqslant [T(0,z) - g(z)] + [T(0,w) - g(w)] + [T(z, ne_1) - g(ne_1 - z)] \\ + [T(w, ne_1) - g(ne_1 - w)] + 2C_2 n^{2\xi'-1}. \end{aligned}$$

This means that at least one of the four bracketed terms is at least $(C_2/2)n^{2\xi'-1}$ in absolute value. Thus using symmetry,

$$\mathbb{P}(A_z) \leqslant 2\mathbb{P}(|T(0,z) - g(z)| \geqslant (C_2/2)n^{2\xi'-1}) + 2\mathbb{P}(|T(0,w) - g(w)| \geqslant (C_2/2)n^{2\xi'-1}).$$

By our choice of ξ', we have $2\xi' - 1 > \chi$. We have assumed exponential concentration above scale χ' (see (4.2)), so these probabilities are (stretched) exponentially small in n. In other words, each one is smaller than e^{-n^c} for some $c > 0$. This shows (4.4) and completes the sketch of the bound $\chi \geqslant 2\xi - 1$.

We turn to the other inequality, $\chi \leqslant 2\xi - 1$. For technical reasons, we assume that $\chi > 0$; the other case can be proved using a different argument [**9**]. Suppose it is false and chose χ', χ'', ξ' such that

$$2\xi - 1 < 2\xi' - 1 < \chi' < \chi < \chi''. \tag{4.7}$$

We define the variable δT, which first appeared in Licea-Newman-Piza [**32**]: $\delta T = T - T'$, where

$$T = T(0, ne_1) \text{ and } T' = T(n^{\xi'} e_2, ne_1 + n^{\xi'} e_2).$$

Because $\xi' > \xi$, these passage times are nearly independent, as they are with high probability equal to two passage times restricted to disjoint sets of edges ("tubes" of width $n^{\xi'}/2$ centered on the straight lines connecting their endpoints). Using the exponential concentration assumption on $D(0, ne_1)$ (from (4.1)), we then obtain if T'' is an independent copy of T,

$$\text{Var } T = \frac{1}{2}\mathbb{E}(T - T'')^2 \leqslant C_3 \mathbb{E}(T - T')^2.$$

Since $\chi' < \chi$, we find

$$n^{2\chi'} \leqslant C_4 \mathbb{E}(T - T')^2. \tag{4.8}$$

The next step is to upper bound $(T - T')^2$. Let w be the first intersection of a geodesic from $n^{\xi'} e_2$ to $n^{\xi'} e_2 + ne_1$ with the hyperplane $H_1 = \{y : y \cdot e_1 = n^\beta\}$ (for some $\beta < 1$ but very close to 1), and let z be the first intersection of this geodesic

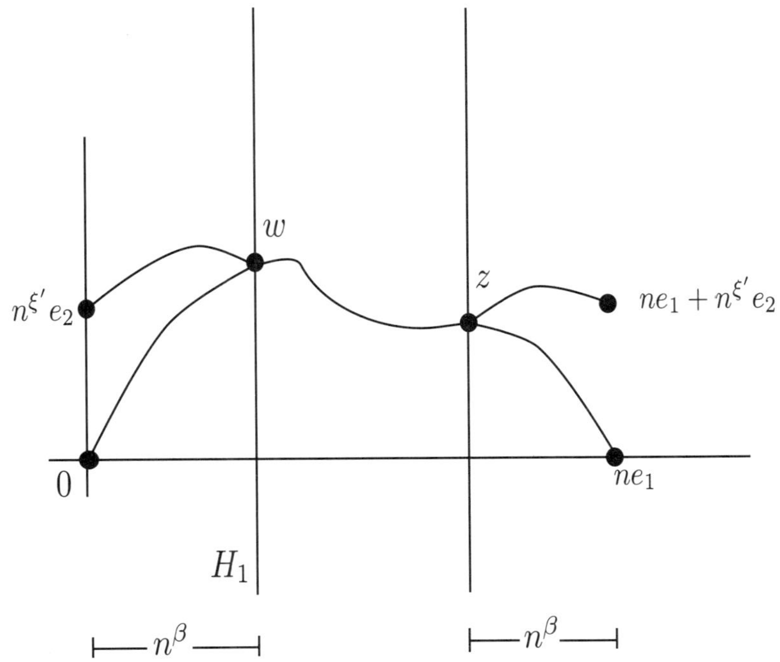

FIGURE 4.2. Illustration of the argument for the inequality $\chi \leqslant 2\xi - 1$. The path that connects $n^{\xi'}e_2$ to $ne_1 + n^{\xi'}e_2$ through w (its first intersection with H_1) and z is a geodesic. The path that connects 0 to ne_1 through w and z is a possibly suboptimal path, and this produces inequality (4.9).

with the hyperplane $\{y : y \cdot e_1 = n - n^\beta\}$. Note that

$$\begin{aligned} T = T(0, ne_1) &\leqslant T(0,w) + T(w,z) + T(z, ne_1) \\ &= T(0,w) - T(n^{\xi'}e_2, w) + T(n^{\xi'}e_2, ne_1 + n^{\xi'}e_2) \\ &\quad + T(z, ne_1) - T(z, ne_1 + n^{\xi'}e_2), \end{aligned}$$

so that

$$T - T' \leqslant \left[T(0,w) - T(n^{\xi'}e_2, w)\right] + \left[T(z, ne_1) - T(z, ne_1 + n^{\xi'}e_2)\right]. \tag{4.9}$$

With (exponentially) high probability, w and z are not further than distance $C_5 n^{\xi'}$ from the e_1-axis, since $\xi' > \xi$ (see (4.1)). So using symmetry, we obtain with high probability

$$T - T' \leqslant A + B,$$

where A and B have the same distribution as

$$D = \max\left\{|T(0,w) - T(n^{\xi'}e_2, w)| : w \in H_1, \|w\| \leqslant C_5 n^{\xi'}\right\}.$$

Reversing the roles of T and T', we obtain the same inequality for $T' - T$, and so

$$\mathbb{E}(T - T')^2 \leqslant C_6 \mathbb{E} D^2.$$

We last have to bound $\mathbb{E}D^2$. This is a maximum over many different passage times (to all $w \in H_1$ with $\|w\| \leqslant C_5 n^{\xi'}$). However, since we have assumed that the fluctuation exponent χ exists in a strong sense (there is exponential concentration — see (4.2)), it is possible to replace this maximum with simply one passage time, and $\mathbb{E}D^2$ will increase by only a logarithmic factor. Thus we can write

$$\mathbb{E}(T-T')^2 \leqslant C_7(\log n)\mathbb{E}\left[T(0,w)-T(n^{\xi'}e_2,w)\right]^2, \tag{4.10}$$

where $w = n^\beta e_1 + n^{\xi'} e_2$. As in the proof of the other inequality, we decompose this difference as

$$\begin{aligned} T(0,w) - T(n^{\xi'}e_2,w) &= [T(0,w)-\mathbb{E}T(0,w)] - [T(n^{\xi'}e_2,w) - \mathbb{E}T(n^{\xi'}e_2,w)] \\ &\quad + [\mathbb{E}T(0,w) - g(w)] - [\mathbb{E}T(n^{\xi'}e_2,w) - g(w-n^{\xi'}e_2)] \\ &\quad + g(w) - g(w-n^{\xi'}e_2). \end{aligned}$$

And once again, our exponential concentration assumption (4.2) allows us to upper bound all the terms in the first two lines (with high probability) by $\|w\|^{\chi''}$. If $\xi \geqslant 1$, then our main inequality $\chi \leqslant 2\xi - 1$ is simply $\chi \leqslant 2\xi - 1$, where $2\xi - 1$ is $\geqslant 1$, and we already know this to be true (as $\chi \leqslant 1/2$), so we can assume that $\xi < 1$. In this case, we can also enforce

$$\xi < \xi' < \beta < 1, \tag{4.11}$$

and we obtain that $\|w\| = \|n^\beta e_1 + n^{\xi'}e_2\| \leqslant C_8 n^\beta$. Therefore from (4.8) and (4.10),

$$n^{2\chi'} \leqslant C_9(\log n)\left(C_{10}n^{\beta\chi''} + g(w) - g\left(w - n^{\xi'}e_2\right)\right)^2. \tag{4.12}$$

Again, we analyze the difference in g by mandating a curvature assumption. We calculate

$$\begin{aligned} g(w) + g(w-n^{\xi'}e_2) &= g\left(n^\beta e_1 + n^{\xi'}e_2\right) - g(n^\beta e_1) \\ &= n^\beta\left(g\left(e_1 + n^{\xi'-\beta}e_2\right) - g(e_1)\right). \end{aligned}$$

Our curvature condition here is the opposite as in the previous inequality $\chi \leqslant 2\xi - 1$. That is, we assume that there are $C_{11}, \delta > 0$ such that if u satisfies $\|u\| < \delta$ and $u \perp e_1$, then

$$g(e_1+u) - g(e_1) \leqslant C_{11}\|u\|^2.$$

Fortunately since the limit shape is convex, one can show that this inequality holds in almost every direction, so we will assume it in the e_1 direction, as it is written. Since $\xi' < \beta$, the term $n^{\xi'-\beta} < \delta$ for large n, and we obtain

$$g(w) - g(w - n^{\xi'}e_2) \leqslant C_{11}n^\beta n^{2\xi'-2\beta'} = C_{11}n^{2\xi'-\beta}.$$

Last we plug this back into (4.12) for

$$n^{2\chi'} \leqslant C_9(\log n)\left(C_{10}n^{2\beta\chi''} + C_{11}n^{2(2\xi'-\beta)}\right).$$

This is true for all n large, so

$$2\chi' \leqslant \max\left\{2\beta\chi'', 2(2\xi'-\beta)\right\}.$$

This holds for all $\beta, \chi', \chi'', \xi'$ satisfying (4.7) and (4.11).

So take $\chi'' \downarrow \chi$ and $\chi' \uparrow \chi$ for fixed β, ξ' for

$$2\chi \leqslant \max\{2\beta\chi, 2(2\xi'-\beta)\}.$$

As $\beta < 1$, we find

$$\chi \leqslant 2\xi' - \beta.$$

Now take $\beta \uparrow 1$ and $\xi' \downarrow \xi$ to obtain $\chi \leqslant 2\xi - 1$. $\square$

Acknowledgments. I thank F. Rassoul-Agha for discussions about the subadditive ergodic theorem, and for the notes on the proof from his course that appeared in this article.

References

[1] Kenneth S. Alexander, *Approximation of subadditive functions and convergence rates in limiting-shape results*, Ann. Probab. **25** (1997), no. 1, 30–55, DOI 10.1214/aop/1024404277. MR1428498

[2] Antonio Auffinger and Michael Damron, *Differentiability at the edge of the percolation cone and related results in first-passage percolation*, Probab. Theory Related Fields **156** (2013), no. 1-2, 193–227, DOI 10.1007/s00440-012-0425-4. MR3055257

[3] Antonio Auffinger and Michael Damron, *The scaling relation $\chi = 2\xi - 1$ for directed polymers in a random environment*, ALEA Lat. Am. J. Probab. Math. Stat. **10** (2013), no. 2, 857–880. MR3141825

[4] Antonio Auffinger and Michael Damron, *A simplified proof of the relation between scaling exponents in first-passage percolation*, Ann. Probab. **42** (2014), no. 3, 1197–1211, DOI 10.1214/13-AOP854. MR3189069

[5] Antonio Auffinger, Michael Damron, and Jack Hanson, *Rate of convergence of the mean for sub-additive ergodic sequences*, Adv. Math. **285** (2015), 138–181, DOI 10.1016/j.aim.2015.07.028. MR3406498

[6] Antonio Auffinger, Michael Damron, and Jack Hanson, *50 years of first-passage percolation*, University Lecture Series, vol. 68, American Mathematical Society, Providence, RI, 2017. MR3729447

[7] Antonio Auffinger and Si Tang, *On the time constant of high dimensional first passage percolation*, Electron. J. Probab. **21** (2016), Paper No. 24, 23, DOI 10.1214/16-EJP1. MR3485366

[8] Daniel Boivin, *First passage percolation: the stationary case*, Probab. Theory Related Fields **86** (1990), no. 4, 491–499, DOI 10.1007/BF01198171. MR1074741

[9] Sourav Chatterjee, *The universal relation between scaling exponents in first-passage percolation*, Ann. of Math. (2) **177** (2013), no. 2, 663–697, DOI 10.4007/annals.2013.177.2.7. MR3010809

[10] Henry Cohn, Noam Elkies, and James Propp, *Local statistics for random domino tilings of the Aztec diamond*, Duke Math. J. **85** (1996), no. 1, 117–166, DOI 10.1215/S0012-7094-96-08506-3. MR1412441

[11] Olivier Couronné, Nathanaël Enriquez, and Lucas Gerin, *Construction of a short path in high-dimensional first passage percolation*, Electron. Commun. Probab. **16** (2011), 22–28, DOI 10.1214/ECP.v16-1595. MR2753301

[12] J. Theodore Cox and Richard Durrett, *Some limit theorems for percolation processes with necessary and sufficient conditions*, Ann. Probab. **9** (1981), no. 4, 583–603. MR624685

[13] Michael Damron and Jack Hanson, *Busemann functions and infinite geodesics in two-dimensional first-passage percolation*, Comm. Math. Phys. **325** (2014), no. 3, 917–963, DOI 10.1007/s00220-013-1875-y. MR3152744

[14] Michael Damron and Michael Hochman, *Examples of nonpolygonal limit shapes in i.i.d. first-passage percolation and infinite coexistence in spatial growth models*, Ann. Appl. Probab. **23** (2013), no. 3, 1074–1085, DOI 10.1214/12-AAP864. MR3076678

[15] Michael Damron and Naoki Kubota, *Rate of convergence in first-passage percolation under low moments*, Stochastic Process. Appl. **126** (2016), no. 10, 3065–3076, DOI 10.1016/j.spa.2016.04.001. MR3542626

[16] H. Eugene Stanley and Nicole Ostrowsky (eds.), *On growth and form*, NATO Advanced Science Institutes Series E: Applied Sciences, vol. 100, Martinus Nijhoff Publishers, Dordrecht, 1986. Fractal and nonfractal patterns in physics; Papers from a course given as part of the NATO-ASI series at the Cargèse summer school, Cargèse, June 26–July 6, 1985. MR837661

[17] Richard Durrett, *Oriented percolation in two dimensions*, Ann. Probab. **12** (1984), no. 4, 999–1040. MR757768

[18] Richard Durrett, *Probability: theory and examples*, 4th ed., Cambridge Series in Statistical and Probabilistic Mathematics, vol. 31, Cambridge University Press, Cambridge, 2010. MR2722836

[19] Richard Durrett and Thomas M. Liggett, *The shape of the limit set in Richardson's growth model*, Ann. Probab. **9** (1981), no. 2, 186–193. MR606981

[20] Murray Eden, *A two-dimensional growth process*, Proc. 4th Berkeley Sympos. Math. Statist. and Prob., Vol. IV, Univ. California Press, Berkeley, Calif., 1961, pp. 223–239. MR0136460

[21] Geoffrey Grimmett, *Percolation*, 2nd ed., Grundlehren der Mathematischen Wissenschaften [Fundamental Principles of Mathematical Sciences], vol. 321, Springer-Verlag, Berlin, 1999. MR1707339

[22] Olle Häggström and Ronald Meester, *Asymptotic shapes for stationary first passage percolation*, Ann. Probab. **23** (1995), no. 4, 1511–1522. MR1379157

[23] J. M. Hammersley and D. J. A. Welsh, *First-passage percolation, subadditive processes, stochastic networks, and generalized renewal theory*, Proc. Internat. Res. Semin., Statist. Lab., Univ. California, Berkeley, Calif., Springer-Verlag, New York, 1965, pp. 61–110. MR0198576

[24] David Jerison, Lionel Levine, and Scott Sheffield, *Logarithmic fluctuations for internal DLA*, J. Amer. Math. Soc. **25** (2012), no. 1, 271–301, DOI 10.1090/S0894-0347-2011-00716-9. MR2833484

[25] W. Jockusch, J. Propp, and P. Shor. (1998). Random domino tilings and the arctic circle theorem. preprint available at *arXiv:9801068*.

[26] Kurt Johansson, *Shape fluctuations and random matrices*, Comm. Math. Phys. **209** (2000), no. 2, 437–476, DOI 10.1007/s002200050027. MR1737991

[27] Kurt Johansson, *Transversal fluctuations for increasing subsequences on the plane*, Probab. Theory Related Fields **116** (2000), no. 4, 445–456, DOI 10.1007/s004400050258. MR1757595

[28] Harry Kesten, *Aspects of first passage percolation*, École d'été de probabilités de Saint-Flour, XIV—1984, Lecture Notes in Math., vol. 1180, Springer, Berlin, 1986, pp. 125–264, DOI 10.1007/BFb0074919. MR876084

[29] Harry Kesten, *On the speed of convergence in first-passage percolation*, Ann. Appl. Probab. **3** (1993), no. 2, 296–338. MR1221154

[30] J. Krug, *Scaling relation for a growing interface,* Phys. Rev. A, **36** (1987), 5465-5466.

[31] Gregory F. Lawler, Maury Bramson, and David Griffeath, *Internal diffusion limited aggregation*, Ann. Probab. **20** (1992), no. 4, 2117–2140. MR1188055

[32] C. Licea, C. M. Newman, and M. S. T. Piza, *Superdiffusivity in first-passage percolation*, Probab. Theory Related Fields **106** (1996), no. 4, 559–591, DOI 10.1007/s004400050075. MR1421992

[33] Thomas M. Liggett, *An improved subadditive ergodic theorem*, Ann. Probab. **13** (1985), no. 4, 1279–1285. MR806224

[34] R. Marchand, *Strict inequalities for the time constant in first passage percolation*, Ann. Appl. Probab. **12** (2002), no. 3, 1001–1038, DOI 10.1214/aoap/1031863179. MR1925450

[35] J. B. Martin, *Last-passage percolation with general weight distribution*, Markov Process. Related Fields **12** (2006), no. 2, 273–299. MR2249632

[36] James B. Martin, *Limiting shape for directed percolation models*, Ann. Probab. **32** (2004), no. 4, 2908–2937, DOI 10.1214/009117904000000838. MR2094434

[37] Anders Martinsson, *First-passage percolation on Cartesian power graphs*, Ann. Probab. **46** (2018), no. 2, 1004–1041, DOI 10.1214/17-AOP1199. MR3773379

[38] P. Meakin and J. M. Deutch. (1986). The formation of surfaces by diffusion-limited annihilation. *J. Chem. Phys.*, 85, 2320.

[39] Charles M. Newman and Marcelo S. T. Piza, *Divergence of shape fluctuations in two dimensions*, Ann. Probab. **23** (1995), no. 3, 977–1005. MR1349159

[40] Daniel Richardson, *Random growth in a tessellation*, Proc. Cambridge Philos. Soc. **74** (1973), 515–528. MR0329079

[41] H. Rost, *Nonequilibrium behaviour of a many particle process: density profile and local equilibria*, Z. Wahrsch. Verw. Gebiete **58** (1981), no. 1, 41–53, DOI 10.1007/BF00536194. MR635270

[42] T. Seppäläinen, *Hydrodynamic scaling, convex duality and asymptotic shapes of growth models*, Markov Process. Related Fields **4** (1998), no. 1, 1–26. MR1625007

[43] Philippe Sosoe, *Fluctuations in first-passage percolation*, Proc. Sympos. Appl. Math. **75** (2018).

[44] Romain Tessera, *Speed of convergence in first passage percolation and geodesicity of the average distance*, Ann. Inst. Henri Poincaré Probab. Stat. **54** (2018), no. 2, 569–586, DOI 10.1214/16-AIHP810. MR3795059

[45] T. A. Witten and L. M. Sander, *Diffusion-limited aggregation*, Phys. Rev. B (3) **27** (1983), no. 9, 5686–5697. MR704464

[46] Mario V. Wüthrich, *Scaling identity for crossing Brownian motion in a Poissonian potential*, Probab. Theory Related Fields **112** (1998), no. 3, 299–319, DOI 10.1007/s004400050192. MR1660910

[47] Yu Zhang, *Supercritical behaviors in first-passage percolation*, Stochastic Process. Appl. **59** (1995), no. 2, 251–266, DOI 10.1016/0304-4149(95)00051-8. MR1357654

[48] Yu Zhang, *Shape fluctuations are different in different directions*, Ann. Probab. **36** (2008), no. 1, 331–362, DOI 10.1214/009117907000000213. MR2370607

School of Mathematics, Georgia Institute of Technology, 686 Cherry St., Atlanta, Georgia 30332

Email address: mdamron6@gatech.edu

Proceedings of Symposia in Applied Mathematics
Volume **75**, 2018
https://doi.org/10.1090/psapm/075/00657

Infinite geodesics, asymptotic directions, and Busemann functions in first-passage percolation

Jack Hanson

ABSTRACT. We show existence, uniqueness, and directedness properties for infinite geodesics in the FPP model. After giving the fundamental definitions, we describe results by Newman and collaborators giving existence and uniqueness of directed geodesics under an unproven curvature assumption. We then give two proofs of the existence of at least two infinite geodesics under no unproven assumptions. In the final two sections, we give proofs of directedness statements for infinite geodesics using more recent methods which give information even under no unproven assumptions and prove a generalized uniqueness statement for infinite geodesics.

CONTENTS

1. Motivation

We will concern ourselves with the usual FPP model. In this chapter, we consider the following broad question: how do long geodesics behave? Recall that (see [**5**, Section 4]) it is expected that the geodesic from 0 to nx should deviate from the straight line parallel to x by a distance of order n^ξ. One should reasonably expect that this exponent gives uniform control of the geodesic close to its endpoints, and that n^ξ is an over-estimate of wandering far from the midpoint of the geodesic. For instance, if $m \ll n$, the first m vertices on the geodesic should lie within distance m^ξ of the line parallel to x. This would represent a form of tightness for the family of geodesics from 0 to nx as $n \to \infty$, leading naturally to the question of whether these geodesics converge. That is, whether for fixed K, the first K steps of the geodesics to each nx are identical for all large n. The resulting limiting path is referred to

2010 *Mathematics Subject Classification.* 60K35, 60K37, 82B43.

as an infinite geodesic (some precise definitions are given below in Section 2.1, and others are recalled from [**5**]).

More generally, one can ask about what classes of limits are possible for different sequences of finite geodesics. A path from 0 to ne_1 which "backtracks" too far in the $-e_1$ direction seems unlikely to be a geodesic, since it should be likely to accumulate excessive passage time the further it backtracks. In particular, the family of geodesics from 0 to ne_1 and the family from 0 to $-ne_1$ should admit subsequential limits which are distinct in a strong sense: sharing only finitely many initial edges or vertices. Analogous conjectures for roughly parallel sequences of points seem trickier to make. Should the geodesic from 0 to $ne_1 + \sqrt{n}e_2$ produce yet another distinct limit with positive probability? We shall see that, under strong assumptions on the model's limit shape, one expects a unique limiting geodesic corresponding to each direction in the plane, so the preceding question should have a negative answer in two dimensions.

A further motivation for these questions comes from their relationship to other important questions in the model. As we will discuss, properties of infinite geodesics are closely related to those of the FPP limiting shape. Because of the relationship between geodesic wandering and fluctuations of the passage time (recall the scaling relation $\chi = 2\xi - 1$), one also could hope that techniques for controlling geodesic wandering will be useful for estimating fluctuations. Another issue already discussed (see [**5**, Section 1.2]) is the low-temperature behavior of the so-called "disordered ferromagnet" model from statistical physics. According to scaling arguments, the two-dimensional disordered ferromagnet should have a unique pair of ground states, which would imply that 2d FPP admits no doubly infinite geodesic. While doubly infinite geodesics will not be a focus of this chapter, the tools we develop have direct application to this question, since a doubly infinite geodesic should resemble a singly infinite geodesic near its midpoint.

Busemann functions (and certain "Busemann-like" objects) will be a central tool in our study. These functions were originally introduced in the context of non-random metric geometry, and were brought to the study of FPP by Hoffman [**10**]. A Busemann function has the form $B(x, y) = \lim_{n\to\infty}[T(x, x_n) - T(y, x_n)]$, where (x_n) is the sequence of vertices in some infinite geodesic, ordered as they appear on the geodesic. $B(x, y)$ can be interpreted as the relative distance to infinity of x and y along this geodesic. By bounding the growth of B, one can control where limiting geodesics are allowed to wander. Limits of the form $\lim_{n\to\infty}[T(x, ne_1) - T(y, ne_1)]$ (note the sequence (ne_1) is typically not an infinite geodesic) were studied earlier by Newman [**12**], since such a limit gives information about the "microstructure" of the surface of the growing region in the FPP growth model.

2. Geodesics and infinite geodesics

2.1. Setting and assumptions. We consider the setting of FPP on $(\mathbb{Z}^d, \mathcal{E}^d)$ with $d \geqslant 2$ — we recall some of the notation here for readability. For each $e \in \mathcal{E}^d$, there is an associated edge weight $t_e \geqslant 0$. We assume for definiteness that the collection (t_e) is independent and identically distributed with common distribution F. However, it is important to note that in many of the results below, the i.i.d. restriction can be considerably relaxed to more general translation-invariant cases. We generally write $\mathbf{P}$ for the joint distribution of the edge weights.

In all cases that follow, we consider F having at least the following properties.

ASSUMPTION 2.1 (Standing assumptions). Unless otherwise specified, from now on we assume:

(1) $\mathbf{E}\min\{t_1, \ldots, t_{2d}\}^d < \infty$.
(2) F is continuous — i.e., $\mathbf{P}(t_e = a) = 0$ for each real a.

Note that (1) from Assumption 2.1 is the moment assumption made in the statement of the shape theorem (see [**5**, Section 3]). Item (2) is stronger than the assumption $\mathbf{P}(t_e = 0) < p_c$ from that theorem, so these assumptions give us the existence of a limit shape $\mathcal{B}$ and a norm g such that $\mathcal{B} = \{x : g(x) \leqslant 1\}$.

For some results (the results of Newman below — see Theorem 3.1), we will need the following strong assumptions in addition to the above.

ASSUMPTION 2.2 (Strong assumptions).

(ExpM) There is some $\alpha > 0$ such that $\mathbf{E}\exp(\alpha t_e) < \infty$.

(Curve) The limit shape $\mathcal{B}$ is uniformly curved, in the following sense. There exists a uniform constant $c = c_{\mathcal{B}}$ such that, for any z_1 and z_2 having $g(z_1) = g(z_2) = 1$ and any $a \in [0, 1]$,

$$g(z) \leqslant 1 - c\left[\min\{g(z - z_1), g(z - z_2)\}\right]^2 ,$$

where $z = az_1 + (1 - a)z_2$.

Some comments on these assumptions are in order. Assumption (ExpM) is straightforward but strong, and it is (as we will see) unnecessary for many of the results on infinite geodesics. In fact, it is mainly applied to guarantee concentration results for passage times hold, and any result which relies too heavily on these will not adapt well to the non-i.i.d. settings mentioned above.

Assumption (Curve) is widely believed to hold for most "reasonable" edge weight distributions, and this claim is bolstered by comparison to some exactly solvable models closely related to FPP. But as it stands, (Curve) is far from proof. Indeed, as previously discussed, it is not even currently known that there exists some F such that $\mathcal{B}$ is strictly convex for i.i.d. edge weights, though this is also highly plausible.

2.2. Types of geodesics. We recall the definition of a geodesic between two terminal points x and y. In what follows, we identify a path in $\mathbb{Z}^d$ with either the sequence of vertices or edges it traverses, depending on which is more convenient in context.

DEFINITION 2.3. A nearest-neighbor path $\gamma = (x = x_0, x_1, \ldots, x_n = y)$ is called a **(finite) geodesic** between x and y if $T(\gamma) = T(x, y)$.

LEMMA 2.4. *Under Assumption* 2.1, *there a.s. exists a unique finite geodesic between each x and y in $\mathbb{Z}^d$. Moreover, for any two finite paths $\gamma_1 \neq \gamma_2$, we a.s. have $T(\gamma_1) \neq T(\gamma_2)$* .

We call the event on that all finite paths have distinct passage times and that there exists a unique geodesic between every pair of points of $\mathbb{Z}^d$ by the name Ω_u; in this language, Lemma 2.4 tells us that $\mathbf{P}(\Omega_u) = 1$. In many of the arguments that follow, we work conditionally on the event Ω_u without further comment.

PROOF. The existence claim follows easily from the shape theorem; we sketch the argument. Namely, we have under our assumptions that (a.s.) $T(x, z) \to \infty$

as $|z| \to \infty$. On the other hand, fixing a nonrandom shortest lattice path (in the $\mathbb{Z}^d$ metric, not the FPP metric) γ from x to y, we have that $T(x,y) \leqslant T(\gamma)$. In particular, for a.e. realization of the edge weights, there is some Euclidean ball such that $T(x,z) > T(x,y)$ for any z outside this ball. This implies that the infimum over paths in the definition of $T(x,y)$ is a.s. effectively over a finite set and ensures the existence of the geodesic.

We show uniqueness. Suppose that $\gamma_1 \neq \gamma_2$ are two finite nearest-neighbor paths from x to y — in particular, that some edge $e \in \gamma_1 \backslash \gamma_2$. For both γ_1 and γ_2 to be geodesics from x to y, we must have $T(\gamma_1) = T(\gamma_2)$; we show that this has probability zero of occurring.

Condition on the values of the weights of all edges other than e, and view $h = T(\gamma_1) - T(\gamma_2)$ conditionally as a function of t_e. We have $h'(t_e) = 1$ for all values of t_e, so there exists at most one value a such that $h(a) = 0$. But $\mathbf{P}(t_e = a) = 0$, so there is conditional probability 0 that $T(\gamma_1) = T(\gamma_2)$. □

DEFINITION 2.5. The unique geodesic from x to y whose existence is guaranteed by Lemma 2.4 is denoted $G(x,y)$.

Of course, a subpath of a geodesic is also a geodesic. In particular, if x_i and x_j are vertices of $G(x,y)$, then the subpath of $G(x,y)$ between them is identical to $G(x_i, x_j)$.

One can ask many interesting questions about the behavior of $G(x,y)$ — for instance, how far does it typically deviate or "wander" from the straight line segment between x and y? One such question of particular importance, which is closely related to the preceding wandering question, is:

- For what sequences $(y_n)_n \subseteq \mathbb{Z}^d$ with $|y_n| \to \infty$ does the sequence of geodesics $(G(x, y_n))_n$ converge?

Here we say a sequence of (nearest-neighbor) paths $(\gamma_n)_n$ converges to a path γ if for each fixed k and all large n, the first k edges of γ_n are identical with the first k edges of γ. It is not hard to see (and we make precise in the proof of Proposition 2.7 below) that the following holds for almost every edge weight configuration: if there are a vertex x, a sequence $(y_n)_n$, and some infinite path γ such that $\lim_n G(x, y_n) = \gamma$, then γ must be an infinite geodesic, in the sense of the next definition.

DEFINITION 2.6. An infinite path $\gamma = (x_0, x_1, \ldots)$ is an **infinite geodesic** or **unigeodesic** if its finite subsegments are all finite geodesics — that is, for every $i < j$,

$$(x_i, x_{i+1}, \ldots, x_j) = G(x_i, x_j) .$$

We will sometimes omit the words "finite" or "infinite," simply saying "geodesic" to mean a finite or infinite geodesic, when the meaning is clear by context.

Of course, simpler than convergence statements would just be existence statements, and this is mainly what we will focus on. Do there exist unigeodesics? How many? And what can we say about their wandering?

PROPOSITION 2.7. *With probability one, there exists an infinite geodesic* $\gamma = (0 = x_0, x_1, \ldots)$.

PROOF. This is a standard "diagonal argument". Start with the sequence $(G(0, ne_1))_n$. The second vertex of each $G(0, ne_1)$ is an element of $\{\pm e_i\}_{i=1}^d$, a finite set; there thus exists an $e \in \{\pm e_i\}$ and a sequence $n_k \to \infty$ such that the first two vertices of $G(0, n_k)$ are identical for all k — i.e., $G(0, n_k e_1) = (0, e, \ldots)$ for all k. Repeating this argument on the subsequence $(G(0, n_k e_1))$ and inducting produces a subsequential limit γ.

Since any finite subsegment of γ is also a finite subsegment of $G(0, ne_1)$ for some n, and since finite subsegments of geodesics are also geodesics, we see that γ is a unigeodesic. □

The proof of Proposition 2.7 uses practically nothing from the model, but for this reason it is difficult to improve. One obvious consequence of translation-invariance is that we can replace 0 by any x and get an infinite geodesic from x as well. But perhaps every unigeodesic is essentially the same.

To clarify, let us call unigeodesics γ_1 and γ_2 **distinct** if $|\gamma_1 \triangle \gamma_2| = \infty$. It is easy to see that uniqueness implies that unigeodesics cannot touch, subsequently separate, then touch again, as this would imply that the separated segments are two different geodesics between vertices where they touch. This can be formalized as:

PROPOSITION 2.8. *On the probability one event Ω_u of Lemma 2.4, geodesics γ_1, γ_2 are distinct if and only if they touch at most finitely often: $|\gamma_1 \cap \gamma_2| < \infty$. More specifically, if $\gamma_1 = (z_1, z_2, \ldots)$ and $\gamma_2 = (z_1', z_2', \ldots)$ are distinct, the intersection is a single subpath: for instance, $\gamma_1 \cap \gamma_2 = (z_K, \ldots, z_{K+\ell})$.*

In particular, geodesics γ_1 and γ_2 (almost surely) cannot touch infinitely often without **merging** or **coalescing**: becoming asymptotically identical. We leave the formal proof of Proposition 2.8 as an exercise, since it gives some intuition about the "treelike" behavior of geodesics when they are known to be unique.

It is not clear at all a priori how to show that there exist even two distinct unigeodesics. It is intuitively plausible, however, that subsequential limits of (for instance) $(G(0, ne_1))_{n>0}$ and $(G(0, -ne_1))_{n>0}$ should look very different. If some γ arose as a subsequential limit of both sequences, it would have to in some sense represent a "fast path" in both the e_1 and $-e_1$ directions, which is difficult to imagine.

Let $\mathcal{N}$ denote the random number of distinct unigeodesics (the supremal cardinality among collections of distinct unigeodesics). The above reasoning suggests that $\mathbf{P}(\mathcal{N} \geqslant 2) > 0$. It also suggests that limits of $G(0, ne_1)$ and $G(0, -ne_1)$ could be differently directed, in the following sense:

DEFINITION 2.9. Let $\gamma = (x_0, x_1, \ldots)$ be a unigeodesic. We say that θ is a limiting direction of γ if θ is a limit point of the sequence $(x_i/|x_i|)_i$. The set of all limiting directions for γ shall be denoted $\Theta(\gamma)$.

We will now focus on the following questions:

QUESTION 1. Can we make the above reasoning precise to show $\mathbf{P}(\mathcal{N} \geqslant 2) > 0$? What is the largest value $\mathcal{N}$ can take — for instance, is $\mathbf{P}(\mathcal{N} = \infty) > 0$?

QUESTION 2. What $\Theta(\gamma)$ are realizable for unigeodesics γ? Can we find γ such that $\Theta(\gamma)$ is strictly smaller than $\mathbb{S}^{d-1}$? Or indeed, is there positive probability that there exists γ with $\Theta(\gamma) = \mathbb{S}^{d-1}$?

3. Geodesic trees and the curvature condition

The first results on the above existence questions are due to Newman [**12**] (and related results with collaborators C. Licea [**11**] and M. Piza [**13**]). We say that a path γ has **direction** $\theta \in \mathbb{S}^{d-1}$ if $\Theta(\gamma) = \{\theta\}$. That is, if $\gamma = (x_0, x_1, \ldots)$, then γ has direction θ if and only if $x_i/|x_i| \to \theta$.

THEOREM 3.1 ([**12**]). *Assume (ExpM) and (Curve) in addition to the usual assumptions. Then:*

(1) *With probability one, for each $\theta \in \mathbb{S}^{d-1}$ there is a unigeodesic beginning at 0 having direction θ.*
(2) *With probability one, each unigeodesic has direction; i.e., for each geodesic γ, there exists a $\theta \in \mathbb{S}^{d-1}$ such that $\Theta(\gamma) = \{\theta\}$.*

In addition to the above existence questions, one is often interested in the relationship of various geodesics from 0 to one another: how large is the intersection between two infinite or long finite geodesics. In fact, we can simultaneously consider the collection of all geodesics $\{G(0,x)\}$ as a graph in its own right. This is the perspective taken by Newman [**12**], which will be useful in the study of properties of infinite geodesics.

For any $x \in \mathbb{Z}^d$, define the **geodesic tree** $\mathcal{T}(x)$ as the directed graph given by the union $\bigcup_{y\in\mathbb{Z}^2} G(x,y)$, where each $G(x,y)$ is considered as a path directed away from x. In other words, $\mathcal{T}(x)$ is the graph whose vertex set is $\mathbb{Z}^2$ and whose (directed) edge set is the set of all ordered pairs (z_i, z_{i+1}) which appear in $G(x,y) = (z_0, z_1, \ldots, z_n)$ for some y. As the name anticipates, this graph is a tree:

LEMMA 3.2. *Almost surely, each finite directed path in $\mathcal{T}(x)$ is a geodesic. Moreover, considering $\mathcal{T}(x)$ as an undirected graph, it is a tree.*

PROOF. We consider an outcome in the existence and uniqueness event Ω_u from Lemma 2.4. We first prove that each directed path in $\mathcal{T}(x)$ is a geodesic. We inductively prove the following stronger statement: each directed path $(y_0, y_1, \ldots, y_n)$ in $\mathcal{T}(x)$ is the terminal segment of $G(x, y_n)$. For the base case $n = 1$, we consider a single directed edge (y_0, y_1) in $\mathcal{T}(x)$. By definition, (y_0, y_1) appears as a directed edge in some geodesic $G(x,y)$. Since subpaths of geodesics are geodesics, the segment of $G(x,y)$ up to the appearance of y_1 is the geodesic $G(x, y_1)$, proving the statement for $n = 1$.

Assume the statement holds for a positive integer n, and consider a directed path $(y_0, \ldots, y_{n+1})$ in $\mathcal{T}(x)$. By the inductive hypothesis, $(y_1, \ldots, y_{n+1})$ is itself the terminal segment of $G(x, y_{n+1})$. The segment of $G(x, y_{n+1})$ up to y_1 must be $G(x, y_1)$ by uniqueness of geodesics. But since (y_0, y_1) is a directed edge of $\mathcal{T}(x)$, it must be the final directed edge of $G(x, y_1)$ by the induction hypothesis. Thus, $G(x, y_{n+1})$ is the concatenation of $G(x, y_0)$ with the directed path $(y_0, \ldots, y_{n+1})$, completing the inductive step.

We now prove the claim that (as an undirected graph) $\mathcal{T}(x)$ is a tree. Clearly it is connected, since there is a directed path from 0 to each y. Suppose for the sake of contradiction that there were some simple cycle $(z_1, z_2, \ldots, z_n = z_1)$ in $\mathcal{T}(x)$. If this were also a directed cycle in either orientation, it would be $G(z_1, z_1)$ — a contradiction to the uniqueness of geodesics. So we may assume that the orientation of the edges reverses as we traverse this cycle.

Without loss of generality, we assume (z_1, z_2) is a directed edge, and let $1 < K$ be the smallest integer such that (z_K, z_{K+1}) is not a directed edge. Then (z_{K+1}, z_K) is a directed edge, and so is (z_{K-1}, z_K). In particular, both of these must be the last edge of $G(x, z_K)$. This clearly leads to a contradiction, since $(z_1, \ldots, z_n)$ is a simple cycle. □

Since geodesics in distinct directions must be distinct, Theorem 3.1 suggests that "reasonable" edge weight distributions have $\mathbf{P}(\mathcal{N} = \infty) = 1$. In other words, the tree $\mathcal{T}(0)$ a.s. has infinitely many ends. The proof of the theorem amounts to showing that $\mathcal{T}(0)$ is "straight", in the sense that finite paths in $\mathcal{T}(0)$ stay close to to the Euclidean line segment between their endpoints. The result then follows using the limiting procedure described after Proposition 2.7, by taking limits of finite geodesics $G(0, v_n)$ with $v_n/|v_n| \to \theta$. These (or a subsequence thereof) converge to an infinite directed path in $\mathcal{T}(0)$, which is an infinite geodesic by Lemma 3.2.

Implementing the strategy amounts to rigorously establishing a strong version of our heuristic that geodesics to points far off in direction θ should not be geodesics for points in direction $-\theta$ ("no backtracking"). However, one must show a version of this statement not just for $-\theta$, but also for directions even very close to θ.

We will not attempt to prove the theorem here, but will give a very basic idea of the connections between regularity properties of $\mathcal{B}$ and statements of a "no backtracking" flavor. Suppose we wish to show a much weaker statement; namely, that the geodesic from 0 to $ne_1 + ne_2$ is not likely to be identical to the concatenation of the geodesic $G(0, ne_1)$ with $G(ne_1, ne_1 + ne_2)$. One way to do this would be to show that

$$T(0, ne_1 + ne_2) \ll T(0, ne_1) + T(ne_1, ne_1 + ne_2). \tag{3.1}$$

On can approximate the left- and right-hand sides of (3.1), for n large, by $ng(e_1 + e_2)$ and $n[g(e_1) + g(e_2)]$ (up to $o(n)$ terms). Then (3.1) will certainly hold if $g(e_1 + e_2) < g(e_1) + g(e_2)$ or equivalently if $g(e_1/2 + e_2/2) < (1/2)[g(e_1) + g(e_2)]$. This follows, for instance, if $\mathcal{B}$ is strictly convex.

3.1. Uniqueness: Licea-Newman. In addition to the above existence statements, one can ask for upper bounds on the number of infinite geodesics. A natural question is whether many geodesics exist in a given (fixed, deterministic) direction θ. At a heuristic level, this possibility seems unlikely on $\mathbb{Z}^2$. Since such geodesics exhibit power-law wandering (see Section 4.2 in [**5**]) in the direction perpendicular to θ, we would expect them to typically cross each other (unless their wandering were somehow synchronized). On the other hand, multiple such crossings would lead to a "loop" of two disjoint geodesics between the same endpoints, a contradiction under our weight assumptions. With this two-dimensional picture in mind, we restrict for the remainder of Section 3.1 to $\mathbb{Z}^2$.

The work of Licea and Newman [**11**] shows a limited form of a "uniqueness" statement for the geodesics in a fixed direction. Namely, when $d = 2$, we have, fixing θ in some full-measure subset of $\mathbb{S}^1$, that there is a.s. at most one distinct geodesic γ such that $\Theta(\gamma) = \{\theta\}$. This will be a partial motivation for our assumption (LimG) introduced below as Assumption 6.1, which allows the construction of a well-behaved "Busemann function".

THEOREM 3.3 (Licea-Newman [**11**]). *Consider FPP on $\mathbb{Z}^2$ under the standard assumptions of Assumption 2.1. There exists a deterministic set $D \subseteq [0, 2\pi)$ of*

Lebesgue measure 2π *such that the following holds. For any* $\theta \in D$,

$$\mathbf{P}\,(\textit{there are distinct } \theta\textit{-directed unigeodesics } \gamma_1, \gamma_2) = 0.$$

Note that the θ appearing in the argument of the probability is deterministic. This is an important distinction; in fact the analogue of Theorem 3.3 with uniqueness simultaneously holding for all $\theta \in [0, 2\pi)$ should not hold. This is because two parallel geodesics should run along directions where there is a "competition interface" (see Theorem 4.7.1 in [**14**]). On the other hand, it is believed that the set D is a purely technical condition — i.e., that the statement of the theorem should hold with $D = [0, 2\pi)$. See Section 7.4, where we discuss improvements to Theorem 3.3.

The proof of Theorem 3.3 rests on the following weaker statement, which we give as a lemma.

LEMMA 3.4. *There exists a deterministic set* $D \subseteq [0, 2\pi)$ *of Lebesgue measure* 2π *such that the following holds. For any* $\theta \in D$,

$$\mathbf{P}\,(\textit{there are distinct } \theta\textit{-directed unigeodesics } \gamma_1, \gamma_2 \textit{ starting from } 0) = 0.$$

PROOF. Consider the random set

$$\Phi = \{\theta \in [0, 2\pi) : \text{there are distinct } \theta - \text{directed unigeodesics starting from } 0\}.$$

The main goal of the argument is to show that Φ is a.s. countable by giving an explicit algorithm for indexing Φ.

Consider a direction $\theta \in \Phi$ and choose two distinct θ-directed unigeodesics γ_1, γ_2 starting from 0. Note that γ_1 and γ_2 must bifurcate at some vertex $z \in \mathbb{Z}^2$. Consider the concatenation of the portions of γ_1 and γ_2 from z onward as a doubly-infinite self-avoiding path Γ. The path Γ divides the plane into two components (that is, $\mathbb{R}^2 \backslash \Gamma$ has two connected components), one of which is asymptotically contained in small sectors around the direction θ. We call this the region "between" γ_1 and γ_2. Moreover, one of the paths γ_1, γ_2 is asymptotically counterclockwise of the other. Without loss of generality, we assume it is γ_1.

Suppose that the first edge of γ_1 after z is (z, y_1), and similarly let (z, y_2) denote the corresponding edge of γ_2. We define algorithms producing two distinct θ-directed unigeodesics $\alpha_1(z, y_1, \mathcal{T}(0))$ and $\alpha_2(z, y_2, \mathcal{T}(0))$. The first edge (considered as directed for purposes of the algorithm) of α_1 is $f_1 = (z, y_1)$. To find the next edge, consider the set of all outgoing directed edges in $\mathcal{T}(0)$ which contain and are directed away from y_1, and which are part of some infinite geodesic in $\mathcal{T}(0)$. Out of all such edges, choose the clockwise-most; this is the second edge f_2 of $\alpha_1(z, y_1, \mathcal{T}(0))$. From the "far" endpoint of f_2 (i.e., the endpoint other than y_1), we repeat the previous step.

The sequence of edges so produced is an infinite path in $\mathcal{T}(0)$ and hence an infinite geodesic. The construction guarantees that α_1 contains only edges of γ_1 and edges in the region between γ_1 and γ_2. Indeed, if α_1 ever branches off of γ_1 and then re-intersects γ_1 or γ_2, there would be a loop in $\mathcal{T}(0)$, which is forbidden by Lemma 3.2. Therefore, α_1 is also θ-directed. Similarly, α_2 (defined via a similar, counter-clockwise search algorithm) is also θ-directed. By a similar reasoning to the above, α_1 and α_2 cannot intersect, since this would again produce a loop.

In particular, on the event Ω_u, there exists some triple of vertices (z, y_1, y_2) for which the corresponding α_1 and α_2 geodesics are distinct and θ-directed. This gives that Φ is a.s. countable.

We now conclude the proof. Since Φ is a.s. countable (and letting Leb denote Lebesgue measure),

$$\mathbb{E}\int_0^{2\pi} \mathbf{1}_\Phi(u)\mathrm{d}u = \mathbb{E}\mathrm{Leb}(\Phi) = 0 \ . \tag{3.2}$$

Interchanging the order of integration in (3.2), we get that $\mathbf{P}(u \in \Phi) = 0$ for (Lebesgue-) almost every u in $[0, 2\pi)$. □

The proof of Theorem 3.3 now proceeds by showing that the existence of parallel geodesics would imply that certain geodesic structures appear, then arguing that translation-invariance of the model prohibits these structures from existing. The way translation-invariance enters in the proof of Licea-Newman is via a "lack of space" argument due originally to Burton and Keane [**4**]. We replace this step with an argument originally appearing in [**6**], which uses a "mass transport" argument as encapsulated in the following lemma.

LEMMA 3.5. *Let $f(\cdot,\cdot) : \mathbb{Z}^2 \times \mathbb{Z}^2 \to [0,\infty]$. Suppose that f is translation-invariant in the sense that $f(x+z, y+z) = f(x,y)$ for all x, y, $z \in \mathbb{Z}^2$. Then*

$$\sum_{y\in\mathbb{Z}^2} f(x,y) = \sum_{y\in\mathbb{Z}^2} f(y,x) \quad \textit{for all } x \in \mathbb{Z}^2 \ . \tag{3.3}$$

PROOF. Note that $\sum_{y\in\mathbb{Z}^2} f(x,y) = \sum_{z\in\mathbb{Z}^2} f(x, x+z)$. Moreover, translation-invariance gives $f(x, x+z) = f(x-z, x)$. Thus,

$$\sum_{y\in\mathbb{Z}^2} f(x,y) = \sum_{z\in\mathbb{Z}^2} f(x-z, x) = \sum_{y\in\mathbb{Z}^2} f(y,x) \ .$$ □

Lemma 3.5 can be used as follows. Suppose we define a "mass transport rule" assigning a (random) nonnegative number $m(x,y)$ to each ordered pair $(x,y) \in \mathbb{Z}^2 \times \mathbb{Z}^2$. This is thought of as an amount of mass moved from vertex x to vertex y. If $\mathbb{E}m(x+z, y+z) = \mathbb{E}m(x,y)$ for all vertices x, y, and z, then we can apply the lemma with $f = \mathbb{E}m$. In our application of this technique, we argue that if Theorem 3.3 did not hold, then one could define such an m in a way that $\sum_y m(x,y) \leqslant 1$ for all x, but such that $\sum_y \mathbb{E}m(y,x) = \infty$ for all x. This contradicts the equality (3.3).

SKETCH OF PROOF OF THEOREM 3.3. The proof is easiest in the case that the edge weights have unbounded support — that is, $\mathbf{P}(t_e > \lambda) > 0$ for all $\lambda > 0$. We will give the proof assuming this in conjunction with our standing assumptions; we direct the reader to the original paper for the modifications that must be made if the weights have bounded support.

The set D of the theorem is the same as the D from Lemma 3.4. For simplicity of notation, we assume that $e_1 \in D$ and show the statement for $\theta = e_1$. Let F' be the event that, from each $x \in \mathbb{Z}^2$, there is at most one unigeodesic in $\mathcal{T}(x)$ having direction e_1; set $F = F' \cap \Omega_u$. It follows from Lemma 3.4 (under the assumption $e_1 \in D$) that $\mathbf{P}(F) = 1$.

The proof proceeds by contradiction. Assume that with positive probability there exist two distinct e_1-directed unigeodesics. By considering these geodesics after their last intersection, we may assume they are disjoint. Since they are e_1-directed, they eventually intersect each line $L_n = \{x \in \mathbb{R}^2 : x \cdot e_2 = n\}$ for n large. They also eventually pass each L_n and never return; in particular, the geodesics have a last intersection point with each L_n for n large. Thus, there is some n and some pair $x, y \in L_n$ such that $\mathbf{P}(E_n(x,y)) > 0$, where

$$E_n(x,y) = \{\text{there exist disjoint } e_1\text{-directed unigeodesics from } x,\ y \\ \text{intersecting } L_n \text{ only at } x,\ y \text{ respectively}\}\ ,$$

and we write $E(\cdot,\cdot) := E_0(\cdot,\cdot)$.

By translation-invariance of the model, we get $\mathbf{P}(E(0,z_0)) > 0$ for some particular vertex z_0 on the positive e_2-axis. We have found two disjoint geodesics in the right half-plane; we now find a third. Letting $E'(x,y,z)$ (where $x, y, z \in L_0$) be the event that there are disjoint e_1-directed unigeodesics starting at and touching L_0 only at x, y, and z respectively, we claim that $\mathbf{P}(E'(0,z_1,z_2)) > 0$ for some vertices z_1 and z_2 of L_0 satisfying $0 < z_1 \cdot e_2 < z_2 \cdot e_2$.

We first note that we can find an integer $m > 1$ such that

$$\mathbf{P}(E(0,z_0) \cap E(mz_0,(m+1)z_0)) > 0.$$

This follows from the ergodic theorem: we have the almost sure and L^2- convergence

$$\frac{1}{M}\sum_{k=0}^{M-1} \mathbf{1}_{E(kz_0,(k+1)z_0)} \to \mathbf{P}(E(0,z_0))\ .$$

Thus $\mathbf{P}(E(k_1 z_0,(k_1+1)z_0) \cap E(k_2 z_0,(k_2+1)z_0)) > 0$ for some $k_1 < k_2 - 1$ and hence (using translation invariance) $\mathbf{P}(E(0,z_0) \cap E((k_2-k_1)z_0,(k_2-k_1+1)z_0)) > 0$.

On the event F, for each $x \in \mathbb{Z}^2$, let $\Gamma(x)$ denote the unique e_1-directed unigeodesic in $\mathcal{T}(x)$ (if it exists). On F, for any x, y such that $\Gamma(x) \cap \Gamma(y) \neq \varnothing$, the geodesics $\Gamma(x)$ and $\Gamma(y)$ are not distinct (they coalesce). Indeed, if $\Gamma(x)$ and $\Gamma(y)$ intersected and subsequently separated, there would be some $z \in \Gamma(x) \cap \Gamma(y)$ such that $\mathcal{T}(z)$ has two distinct e_1-directed unigeodesics, an impossibility on F.

Take m as above. We now show that $\mathbf{P}(E'(0,z_0,(m+1)z_0)) > 0$, which follows from

$$E'(0,z_0,(m+1)z_0) \supseteq E(0,z_0) \cap E(mz_0,(m+1)z_0) \cap F\ . \tag{3.4}$$

To see (3.4) holds, we argue that on the event on the right-hand side, $\Gamma(0)$ cannot intersect $\Gamma(z_0), \Gamma(mz_0)$, or $\Gamma((m+1)z_0)$. The fact that $\Gamma(0)$ cannot intersect $\Gamma(mz_0)$ is a consequence of the occurrence of $E(0,z_0)$. If $\Gamma(z_0)$ did not intersect $\Gamma(mz_0)$, then it would be clear that $\Gamma(0)$ cannot intersect $\Gamma(mz_0)$ either, since it cannot intersect L_0 or $\Gamma(z_0)$. On the other hand, if $\Gamma(mz_0)$ and $\Gamma(z_0)$ intersect, then they must merge; this also prevents $\Gamma(0)$ from intersecting $\Gamma(mz_0)$ (since it would then merge with $\Gamma(z_0)$). Similar arguments hold for the other geodesics; out of $\Gamma(0)$, $\Gamma(z_0)$, $\Gamma(mz_0)$, and $\Gamma((m+1)z_0)$, only $\Gamma(z_0)$ and $\Gamma(mz_0)$ may merge. This establishes (3.4); see Figure 3.1.

We now define the event $E'' = E''(0,z_0,(m+1)z_0)$ to be the subevent of $E'(0,z_0,(m+1)z_0)$ on which no geodesic $\Gamma(y)$ intersects $\Gamma(z_0)$, for any y satisfying either

- $y \cdot e_1 < 0$ or
- $y \cdot e_1 = 0$ with either $y \cdot e_2 > (m+1)z_0 \cdot e_2$ or $y \cdot e_2 < 0$.

We argue $\mathbf{P}(E'') > 0$. Consider the set $\mathfrak{E}$ of edges touching L_0 between 0 and $(m+1)z_0$:

$$\mathfrak{E} = \{\{x - e_1, x\} : x \cdot e_1 = 0,\ 0 \leqslant x \cdot e_2 \leqslant (m+1)z_0 \cdot e_2\}\ .$$

FIGURE 3.1. The event $E'(0, z_0, (m+1)z_0)$ is implied by $E(0, z_0) \cap E(mz_0, (m+1)z_0)$ and uniqueness of e_1-directed geodesics. Although $\Gamma(z_0)$ and $\Gamma(mz_0)$ may coalesce (as depicted), $\Gamma(0)$ cannot intersect $\Gamma((m+1)z_0)$. $\Gamma(z_0)$ is drawn dashed to represent its role as part of a barrier insulating $\Gamma((m+1)z_0)$ from $\Gamma(0)$; the vertical axis (representing L_0) completes the barrier. The edge set $\mathfrak{E}$ is depicted as the shaded gray rectangle.

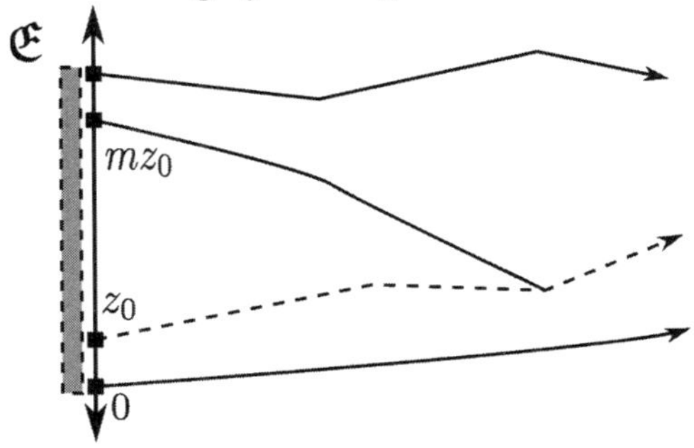

Let A_λ denote the event that, for each $f \in \mathfrak{E}$, there is a path γ connecting the endpoints of f which contains no edge of $\mathfrak{E}$ and which satisfies $T(\gamma) < \lambda$. Clearly $\mathbf{P}(A_\lambda) \to 1$ as $\lambda \to \infty$, so we can fix a value of $\lambda > 0$ such that $\mathbf{P}(A_\lambda \cap E'(0, z_0, (m+1)z_0)) > 0$.

Now we note that $A_\lambda \cap E'$ is increasing in the edge weights in $\mathfrak{E}$. In other words, given an outcome $(t_e) \in A_\lambda \cap E'$, if $(t'_e) \in F$ is an outcome satisfying $t'_e = t_e$ for $e \notin \mathfrak{E}$ and $t'_e \geqslant t_e$ for $e \in \mathfrak{E}$, then $(t'_e) \in A_\lambda \cap E'$ as well. The point is that if we increase edge weights which do not lie on some path Γ which was a geodesic in the original edge-weight configuration, then Γ is also a geodesic in the new configuration.

We use this observation to show $\mathbf{P}(E'') > 0$ by "modifying" outcomes in $A_\lambda \cap E'$ so that each $f \in \mathfrak{E}$ has $t_f > \lambda$ and so that the resulting outcome remains in $A_\lambda \cap E'$. Implementing this modification in a way which is guaranteed to produce a measurable set requires some care; see [**11**, Lemma 3.1]. The resulting event can be shown to have positive probability by the fact that $\mathbf{P}(t_e > \lambda) > 0$ for every λ.

The outcomes produced by the modification are in E'' because the geodesic $\Gamma(z_0)$ is "shielded" by $\mathfrak{E}$ and the surrounding geodesics. Indeed, if a $\Gamma(y)$ as in the definition of E'' intersected $\Gamma(z_0)$, it must first either cross $\Gamma(0)$, $\Gamma((m+1)z_0)$, or an edge of $\mathfrak{E}$. The first two possibilities are ruled out by the definition of E' and the second is ruled out by the fact that A_λ occurs but each edge of $\mathfrak{E}$ has weight $> \lambda$, so no geodesic can use an edge of $\mathfrak{E}$.

We now use the conclusion that $\mathbf{P}(E'') > 0$ to derive a contradiction. Let $\prec$ denote the lexicographic ordering where $x \prec y$ if $x \cdot e_1 \leqslant y \cdot e_1$ or if $x \cdot e_1 = y \cdot e_1$ and $x \cdot e_2 \leqslant y \cdot e_2$. Define the function $m(x, y)$ for $x, y \in \mathbb{Z}^2$ as follows:

$$m(x,y) := \begin{cases} 1, & x \in \Gamma(y) \text{ and } x \notin \Gamma(z) \text{ for any } z \neq y \text{ satisfying } z \prec y; \\ 0, & \text{otherwise.} \end{cases}$$

In other words, $m(x, y) = 1$ if y is the $\prec$-minimal element of $\mathbb{Z}^2$ such that $x \in \Gamma(y)$. Note that $\sum_y m(x, y) \leqslant 1$ almost surely for any x, since there can be at most one such minimal ancestor for any x.

On the other hand, on the event E', there is some $x \in L_0$ between 0 and $(m+1)z_0$ such that $\sum_y m(y,x) = \infty$. Either $x = z_0$, or x is some other vertex of L_0 between 0 and $(m+1)z_0$ having $z_0 \in \Gamma(x)$. Using $\mathbf{P}(E') > 0$ and translation invariance, $\sum_y \mathbb{E}m(y,0) = \infty$, and using the almost sure upper bound above gives $\sum_y \mathbb{E}m(0,y) \leqslant 1$. Setting $f(x,y) = \mathbb{E}m(x,y)$, translation-invariance of the model gives $f(x+z, y+z) = f(x,y)$. Applying Lemma 3.5, these estimates give a contradiction and complete the proof of the theorem. □

4. Showing $\mathcal{N} \geqslant 2$

The first unconditional result showing that $\mathbf{P}(\mathcal{N} \geqslant 2) > 0$ is due to Häggström and Pemantle [**9**] in the case that $d = 2$ and F is an exponential distribution. We will not discuss the details of their proof, since it is very specialized to that setting. We shall instead show a more general result due to both Garet-Marchand [**8**] and Hoffman [**10**], which applies to all $\mathbb{Z}^d$:

THEOREM 4.1 (Garet-Marchand [**8**], Hoffman [**10**]). *Under our standing assumptions,* $\mathbf{P}(\mathcal{N} \geqslant 2) > 0$.

We will give an adaptation of the proof of Garet-Marchand in this section, with an eye towards subsequently introducing the Busemann function methods of Hoffman.

Our main goal is to show that with positive probability there exist sequences $m_k, n_k \to \infty$ such that $\Gamma_0 := \lim_k G(0, -n_k e_1)$ and $\Gamma_1 := \lim_k G(e_1, m_k e_1)$ are distinct. We will actually show something more general, which illustrates an "infection model" principle for finding multiple geodesics; see Lemma 4.2 below. Let us define, for each $z \in \mathbb{Z}^d$, $B_z(x,y) := T(x,z) - T(y,z)$. Note that $B_z(x,y) = -B_z(y,x)$.

We shall see we can find distinct unigeodesics Γ_0 and Γ_1 if we can show $B_{ne_1}(0, e_1)$ is "positively biased" for large positive n. In fact, it is easier to show this with e_1 replaced by ke_1, because the control we get over B will be asymptotic.

LEMMA 4.2. *Fix a finite set of vertices $\{x_1, \ldots, x_k\} \subseteq \mathbb{Z}^d$. Let us define G_i, $1 \leqslant i \leqslant k$ by*

$$G_i := \{z \in \mathbb{Z}^d : B_z(x_i, x_j) < 0 \quad \textit{for all } j \neq i\}$$

— i.e., G_i is the set of vertices closest to x_i in the T-metric. On the event that $|G_i| = \infty$ for all $1 \leqslant i \leqslant k$, we have $\mathcal{N} \geqslant k$.

COROLLARY 4.3. *Assume that for some $\ell > 0$,*

$$\mathbf{P}\left(\left\{\limsup_{n\to\infty} B_{ne_1}(0, \ell e_1) > 0\right\} \cap \left\{\limsup_{n\to\infty} B_{-ne_1}(\ell e_1, 0) > 0\right\}\right) > 0\,. \tag{4.1}$$

Then $\mathbf{P}(\mathcal{N} \geqslant 2) > 0$.

PROOF. Letting $x_1 = 0$, $x_2 = \ell e_1$, on the event in (4.1) we have $|G_i| = \infty$, $i = 1, 2$. The result now follows from Lemma 4.2 . □

PROOF OF LEMMA 4.2. As a preliminary, we observe that G_i and G_j are disjoint by construction, and that with probability one $x_i \in G_i$. We will show that for each i there is an infinite path Γ_i of vertices of G_i, having x_i as its initial vertex, which is actually a unigeodesic.

For each i, choose a sequence $(z_k(i))_k$ of distinct elements of G_i such that $\Gamma_i := \lim_k G(x_i, z_k(i))$ exists; that this is possible follows via the diagonal argument of the proof of Proposition 2.7. We claim that each vertex of Γ_i is an element of G_i. If there were some $z \in \Gamma_i$ with $z \notin G_i$, then for some j, $T(x_i, z) \geqslant T(x_j, z)$. But since $z \in G(x_i, z_k(i))$ for some k, we have $T(x_i, z_k(i)) = T(x_i, z) + T(z, z_k(i))$.

On the other hand, subadditivity gives

$$T(x_j, z_k(i)) \leqslant T(x_j, z) + T(z, z_k(i)) \leqslant T(x_i, z) + T(z, z_k(i)) = T(x_i, z_k(i)) ,$$

so $z_k(i) \notin G_i$. This is a contradiction. □

The structure of Lemma 4.2 is based on viewing the FPP model as defining a competition model (G_i is the set of sites "infected" by a disease with initial patient located at site x_i; sites can be infected only by a single type of disease). This perspective on the geodesic question appeared in the original paper of Häggström-Pemantle. The proof of Lemma 4.2 amounts to showing that each infection spreads along FPP geodesics.

Proof of Theorem 4.1. We use contradiction to show that hypothesis (4.1) of Corollary 4.3 holds. So choose ℓ so large that $\mathbf{E}T(0, \ell e_1) < 3\ell g(e_1)/2$ (this choice will become clear in a moment), and defining

$$A_0 := \left\{\limsup_{n\to\infty} B_{ne_1}(0, \ell e_1) > 0\right\} \quad \text{and} \quad A_\ell := \left\{\limsup_{n\to\infty} B_{-ne_1}(\ell e_1, 0) > 0\right\} ,$$

assume that $\mathbf{P}(A_0 \cap A_\ell) = 0$. We begin with a simple bound: note that the triangle inequality implies that $T(0, ne_1) \leqslant T(0, \ell e_1) + T(\ell e_1, ne_1)$, so $B_{ne_1}(0, \ell e_1) \leqslant T(0, \ell e_1)$ for all n. (A similar argument gives $B_{ne_1}(0, \ell e_1) \geqslant -T(0, \ell e_1)$ and in particular that $B_{ne_1}(0, \ell e_1)$ is integrable).

We can of course bound $B_{ne_1}(0, \ell e_1)$ by 0 on the event A_0^c, and give a similar bound for $B_{-ne_1}(\ell e_1, 0)$. In particular,

$$\begin{aligned}\limsup_{n\to\infty} B_{ne_1}(0, \ell e_1) + \limsup_{m\to\infty} B_{-me_1}(\ell e_1, 0) &\leqslant T(0, \ell e_1)\mathbf{1}_{A_0} + T(\ell e_1, 0)\mathbf{1}_{A_\ell} \\ &= T(0, \ell e_1)[\mathbf{1}_{A_0} + \mathbf{1}_{A_\ell}] .\end{aligned} \tag{4.2}$$

Since $\mathbf{P}(A_0 \cap A_\ell) = 0$, we have $\mathbf{1}_{A_0} + \mathbf{1}_{A_\ell} \leqslant 1$ almost surely. Using this and our choice of ℓ, we take expectations in (4.2) to see

$$\mathbf{E}\left[\limsup_{n\to\infty} B_{ne_1}(0, \ell e_1) + \limsup_{m\to\infty} B_{-me_1}(\ell e_1, 0)\right] \leqslant \mathbf{E}T(0, \ell e_1) < 3\ell g(e_1)/2 . \tag{4.3}$$

So far we have not done anything to show that there is anything wrong with what has been derived. The key step is to show that $B_{ne_1}(0, \ell e_1)$ is typically of order $(\ell - \varepsilon)g(e_1)$ for a large density of n. Combined with a corresponding bound for the other term, we show that the left-hand side of (4.3) is of order $(2\ell - \varepsilon)g(e_1)$, in contradiction to the bound on the right-hand side. This contradicts the assumption that $\mathbf{P}(A_0 \cap A_\ell) = 0$.

The estimate described above is shown following an averaging trick; a more sophisticated version of this will appear in the proof of Theorem 7.9 below. Considering multiples of ℓe_1, we consider terms of the form

$$B_{K\ell e_1}(0, \ell e_1) = T(0, K\ell e_1) - T(\ell e_1, K\ell e_1). \tag{4.4}$$

Note that the model is invariant under lattice shifts, so

$$\begin{aligned}\mathbf{E}B_{K\ell e_1}(0, \ell e_1) &= \mathbf{E}T(0, K\ell e_1) - \mathbf{E}T(\ell e_1, K\ell e_1) \\ &= \mathbf{E}T(0, K\ell e_1) - \mathbf{E}T(0, (K-1)\ell e_1) .\end{aligned} \tag{4.5}$$

This shifts the perspective of the infection process and allows us to apply shape theorem results.

Applying (4.5) to terms of the type (4.4) gives

$$\mathbf{E}\sum_{K=1}^{n} B_{K\ell e_1}(0,\ell e_1) = \mathbf{E}T(0,n\ell e_1)\ .$$

Dividing by n and using the shape theorem gives that

$$\lim_{n\to\infty}\frac{1}{n}\sum_{K=1}^{n}\mathbf{E}B_{K\ell e_1}(0,\ell e_1) = g(\ell e_1) = \ell g(e_1). \tag{4.6}$$

In particular, (4.6) gives the bound $\limsup_{n\to\infty}\mathbf{E}B_{ne_1}(0,\ell e_1)\geqslant \ell g(e_1)$, with a similar bound for $\limsup_{m\to\infty}B_{-me_1}(\ell e_1,0)$. Since $|B_z(0,\ell e_1)|\leqslant T(0,\ell e_1)$, it follows (e.g. by Fatou's lemma) that

$$\mathbf{E}\limsup_{n\to\infty}B_{ne_1}(0,\ell e_1)\geqslant \limsup_{n\to\infty}\mathbf{E}B_{ne_1}(0,\ell e_1)\geqslant \ell g(e_1)\ .$$

Applying this in (4.3) (with the corresponding bound for the other term) completes the proof. □

5. Hoffman's method and Busemann functions

One can think of the function $B_{ne_1}(x,y)$ from the preceding section as a sort of "relative distance to infinity in direction e_1" between points x and y, for n very large. If one could show that the limit $B:=\lim_n B_{ne_1}$ existed, one could work directly with the object B in Garet-Marchand's proof, which would make certain results more natural. For instance, a part of their proof amounts to showing that $\mathbf{E}B_{ne_1}(0,e_1)\approx g(e_1)$ for a large density of but potentially not all n — given existence of B, we could hope instead to show $\mathbf{E}B(0,e_1)=g(e_1)$ and avoid working with subsequences.

We will come back to this perspective in the e_1-direction and other fixed directions later, though the idea of working directly with a limiting difference of passage times will be a key point of this section. Our main goal here will be to outline Hoffman's approach to Theorem 4.1. The key idea is to notice that there is a natural object similar to B_{ne_1} for which limits are easily shown to exist and which has similar implications for geodesics. This object is called a **Busemann function** and was originally introduced by Busemann in the study of (nonrandom) metric geometry.

DEFINITION 5.1. Consider a fixed realization of edge weights and suppose $\gamma=(z_1,z_2,\ldots)$ is a unigeodesic in this configuration. The Busemann function $B_\gamma(\cdot,\cdot)$ is defined by

$$B_\gamma(x,y):=\lim_{n\to\infty}\left[T(x,z_n)-T(y,z_n)\right]\ . \tag{5.1}$$

CLAIM 5.2. *The limit (5.1) exists for any configuration of non-negative edge weights, each pair x, y, and each unigeodesic γ.*

PROOF. Note that the triangle inequality gives that the terms on the right-hand side of (5.1) are bounded in magnitude by $T(x,y)$. For instance, $T(x,z_n)\leqslant T(x,y)+T(y,z_n)$ gives $T(x,z_n)-T(y,z_n)\leqslant T(x,y)$. This implies that if we can show the monotonicity (in n) of these terms for a fixed pair x, y, we will have established existence of the limit.

We first consider the case $y = z_1$. By geodesicity of γ, we have for each $1 < i < n$ that $T(z_1, z_n) = T(z_1, z_i) + T(z_i, z_n)$. On the other hand, subadditivity of course yields $T(x, z_n) \leqslant T(x, z_i) + T(z_i, z_n)$. Thus,

$$T(x, z_n) - T(z_1, z_n) \leqslant T(x, z_i) + T(z_i, z_n) - [T(z_1, z_i) + T(z_i, z_n)]$$
$$= T(x, z_i) - T(z_1, z_i) .$$

Therefore, the terms on the right-hand side of (5.1) are nonincreasing in n, which (combined with the boundedness already established) shows existence of the limit.

Now we consider the case $y \neq z_1$. We can rewrite the typical term on the right-hand side of (5.1) as

$$T(x, z_n) - T(y, z_n) = [T(x, z_n) - T(z_1, z_n)] - [T(y, z_n) - T(z_1, z_n)] . \tag{5.2}$$

Since the limits of each of the bracketed terms on the right-hand side of (5.2) exist, so does the limit of the left-hand side. □

Being a difference of passage times, B_γ inherits several properties which are key to its analysis.

LEMMA 5.3. *On the event Ω_u, the following hold for each unigeodesic $\gamma = (z_1, z_2, \ldots)$ and all vertices x, y, $z \in \mathbb{Z}^d$:*

1. $|B_\gamma(x, y)| \leqslant T(x, y)$;
2. $B_\gamma(x, y) + B_\gamma(y, z) = B_\gamma(x, z)$;
3. $B_\gamma(y, x) = -B_\gamma(x, y)$;
4. *If $x = z_i$ and $y = z_j$ with $i < j$ (so x appears before y in the unigeodesic), then $B_\gamma(x, y) = T(x, y)$.*
5. *If $\tilde{\gamma}$ is another unigeodesic which is not distinct from γ, then $B_\gamma = B_{\tilde{\gamma}}$.*

PROOF. Property (1) already was shown during the proof of Claim 5.2 to establish boundedness of the sequence appearing in (5.1). Properties (2)–(3) follow easily from the fact that B_γ is defined as a difference: for instance, to see (3), write

$$T(y, z_n) - T(x, z_n) = -[T(x, z_n) - T(y, z_n)] \tag{5.3}$$

and take the limit of (5.3) as $n \to \infty$.

To see (4), note that since γ is a geodesic, we have for $n > j$:

$$T(x, z_n) = T(z_i, z_n) = T(z_i, z_j) + T(z_j, z_n) = T(x, y) + T(y, z_n) .$$

Subtracting $T(y, z_n)$ from both sides and taking limits establishes (4).

Property (5) is where we use Ω_u. Let $\tilde{\gamma} = (z_1', z_2', \ldots)$. By Proposition 2.8, we see that the unigeodesics coalesce, so there are some i and j such that $z_i = z_j'$, $z_{i+1} = z_{j+1}', \ldots$. In particular, the limit appearing in (5.1) is unchanged if we replace z_n by z_n'. □

We now are almost equipped to give Hoffman's version of a proof of Theorem 4.1. Since the argument involves applying an ergodic theorem, we take a moment here to define the operators which shift configurations by integer vectors.

DEFINITION 5.4. Let $z \in \mathbb{Z}^d$, and let ω be a realization of the edge weights (t_e). We define the shift θ_z to be the operator which acts on ω, producing a new configuration $\theta_z\omega$, as follows: $\theta_z\omega$ is the realization of edge weights (t_e'), where $t_{\{x,y\}}' = t_{\{x+z, y+z\}}$.

PROOF OF THEOREM 4.1. Assume that $\mathbf{P}(\mathcal{N} = 1) = 1$. Using Proposition 2.7, we see that we are guaranteed the existence of a unigeodesic γ beginning from 0. This γ must be a.s. unique, in the strong sense that there cannot exist an infinite path from 0 which is a unigeodesic, other than γ. Indeed, if there were another unigeodesic $\widetilde{\gamma}$ starting from 0, then by the non-distinctness assumption there must be some $0 \neq z \in \gamma \cap \widetilde{\gamma}$ beyond which the two geodesics are the same. But by a.s. uniqueness of finite geodesics, the subpaths of both γ and $\widetilde{\gamma}$ from 0 to z must be identical, so these two unigeodesics must be exactly the same path.

The translation-invariance of the model gives that we can apply Proposition 2.7 to find a unigeodesic $\Gamma(x)$ from each x in $\mathbb{Z}^d$. Since $\mathcal{N} = 1$, Proposition 2.8 gives us that each $\Gamma(x)$ must coalesce with γ. Let us define $B = B_\gamma$, where γ is as in the preceding paragraph. The proof of Proposition 2.7 in fact shows that every subsequential limit of finite geodesics from any initial vertex x must produce $\Gamma(x)$, which coalesces with γ — so B is in some sense a quite explicit function of the edge weights, and it is not hard to see that this implies the measurability of B. The property (1) of Lemma 5.3 and the integrability of T gives that $B(x, y)$ has finite first moment for all x, y.

In fact, B is translation-covariant, in the following sense. Writing explicitly the dependence of each object on the configuration ω, our assumption that $\mathcal{N} = 1$ gives that, almost surely, $\gamma[\theta_z\omega] = \Gamma(z)[\omega] - z$. This implies the translation-covariance of B: $B(x, y)[\theta_z\omega] = B(z + x, z + y)[\omega]$. We also note that reflections and rotations which fix 0 and leave $\mathbb{Z}^d$ invariant rotate / reflect γ, but that B must be invariant in distribution under these operations (by the invariance of the model and the construction of B). So $\mathbf{E}B(x, y) = 0$ for all x and y.

Fix $0 \neq x \in \mathbb{Z}^d$. Using additivity (property (2) of Lemma 5.3), shift-covariance, and the ergodic theorem, we have for almost every ω:

$$\begin{aligned} \frac{B(0, nx)[\omega]}{n} &= \frac{1}{n}\sum_{i=1}^{n} B((i-1)x, ix)[\omega] \\ &= \frac{1}{n}\sum_{i=1}^{n} B(0, x)[\theta_x^{i-1}\omega] \xrightarrow{n\to\infty} \mathbf{E}B(0, x) = 0\ . \end{aligned} \tag{5.4}$$

Thus, B grows sublinearly in each direction with probability one. We claim a stronger statement: analogous to the usual shape theorem, we can make this sublinearity hold simultaneously in all directions with probability one:

$$\mathbf{P}\left(\lim_{|x|\to\infty} \frac{|B(0, x)|}{|x|} = \limsup_{|x|\to\infty} \frac{|B(0, x)|}{|x|} = 0\right) = 1\ . \tag{5.5}$$

The proof of (5.5) from (5.4) is (*mutatis mutandis*) identical to the proof of the shape theorem (see Theorem 3.1 of [**5**]), so we only recall the main idea here. Fix some configuration ω such that (5.4) holds for each $x \in \mathbb{Z}^d$, and assume for the sake of contradiction that there were some sequence z_k along which $|B(0, z_k)| > \varepsilon|z_k|$ uniformly. Without loss of generality, we may assume that $z_k/|z_k| \to z$ for some z. We can find some $x \in \mathbb{Z}^d$ with $x/|x|$ close to z. For each k large, let n_k minimize $|n_k x - z_k|$; by careful choice of x, we have $|n_k x - z_k| \ll \min\{n_k|x|, |z_k|\}$. In particular, since $|B(0, n_k x) - B(0, z_k)| \leqslant T(n_k x, z_k)$ (and since this passage time is order $|n_k x - z_k|$), we see that $|B(0, n_k x)|$ is large relative to $n_k|x|$. This is in contradiction to the fact that $B(0, nx)/n|x| \to 0$ on ω.

On the other hand, we can see that sublinearity of B is absurd by considering property (4) of Lemma 5.3. Letting $\gamma = (y_1, y_2, \ldots)$, we have $B(y_i, y_j) = T(y_i, y_j)$ for $i < j$. For j large, we have by the shape theorem that $T(0, y_j) \approx g(y_j)$. To be precise, fix $\delta > 0$ such that $g(z) > \delta$ for all z with $|z| = 1$. This is possible by the boundedness of the limit shape — i.e., the fact that passage times asymptotically grow linearly in the Euclidean distance. The shape theorem then implies that with probability one,

$$\limsup_{j\to\infty} \frac{B(0, y_j)}{|y_j|} = \limsup_{j\to\infty} \frac{T(0, y_j)}{|y_j|} \geqslant \delta > 0 .$$

This is in contradiction to (5.5), showing that our assumption that $\mathbf{P}(\mathcal{N} = 1) = 1$ is false. $\square$

6. Directedness and Busemann functions

There is a clear similarity between the Busemann function B_γ and the object B_{ne_1} considered in the proof of Garet-Marchand. We will take some time to develop this idea here under a strong assumption (different from (Curve) and (Expm)). We will not push these ideas as far as we could, since we will in the next section give a framework for getting around these sorts of strong assumptions.

Suppose we wish to avoid assuming (Curve), but still believe the results of Theorem 3.1 and (for at least some dimensions d) Theorem 3.3 should hold. With this as our guidepost, it seems perhaps reasonable to replace assumption (Curve) with the following:

ASSUMPTION 6.1 (LimG). For each $x \in \mathbb{Z}^d$, the geodesics $G(x, ne_1)$ have a limit $\Gamma(x)$. Moreover, these geodesics all coalesce.

One immediate consequence of Assumption (LimG) is that the Busemann function $B(x,y) := \lim_n [T(x, ne_1) - T(y, ne_1)]$ exists, and is covariant with respect to translations by e_1, similarly to the Busemann function in Hoffman's argument. If we continue taking Theorem 3.1 as a goal, one could be led to believe that $\Gamma(x)$ should be directed: $\Theta(\Gamma(x)) = \{e_1\}$, or at least that $\Theta(\Gamma(x)) \neq \mathbb{S}^{d-1}$.

We will not try to prove anything as strong as directedness, but instead just the following much weaker claim. In what follows, let S_δ denote the sector of aperture δ around $-e_1$:

$$S_\delta = \{z \in \mathbb{S}^{d-1} : |z + e_1| < \delta\} .$$

THEOREM 6.2. *Assume* (LimG) (*along with the standard assumptions of Assumption* 2.1). *Then there is some δ such that $\Theta(\Gamma(x)) \cap S_\delta = \varnothing$ almost surely.*

The strategy of the theorem's proof is easy to outline given what we have seen. As in the argument of Garet-Marchand, we can use an "averaging trick" to show that $\mathbf{E}B(0, -me_1)$ grows linearly as $-mg(e_1)$ for large m. In fact, the translation-covariance of B gives that the growth of B occurs almost surely. On the other hand, property 4 of Lemma 5.3 holds for B, giving that $B(0,\cdot)$ behaves as $T(0,\cdot)$ along $\Gamma(0)$. These asymptotics would conflict if $-me_1$ were on $\Gamma(0)$, and in fact exclude the geodesic coming within some conical region of the axis.

PROOF OF THEOREM 6.2. Let us strengthen the assumptions even further to include boundedness: $\mathbf{P}(t_e \leqslant M) = 1$ for some finite M; we prove the theorem in this setting for $\Gamma(0)$. Assume for the sake of contradiction that the statement of the theorem fails when $\delta = (g(e_1)/16M\sqrt{d})$. Consider an outcome ω on which $\Theta(\Gamma(0)) \cap S_\delta \neq \varnothing$ (we will also need to assume that ω lies in various probability one events on which, e.g., $t_e \leqslant M$ for all e; this will become clear in the course of the proof).

The translation-covariance of B gives, as in (5.4),

$$\lim_{m\to\infty} B(0,-me_1)/m = \mathbf{E}B(0,-e_1) \quad \text{a.s. and in } L^1 . \tag{6.1}$$

To compute $\mathbf{E}B$, let us as before define $B_n(x,y) := T(x,ne_1) - T(y,ne_1)$. Note that the shape theorem (in fact, the subadditivity used to establish the shape theorem) shows that $\mathbf{E}T(0,ne_1)/n \to g(e_1)$. We use the Garet-Marchand averaging trick:

$$\begin{aligned}\frac{\mathbf{E}T(0,ne_1)}{n} &= \frac{1}{n}\Big([\mathbf{E}T(0,ne_1) - \mathbf{E}T(0,(n-1)e_1)] \\ &\qquad + [\mathbf{E}T(0,(n-1)e_1) - \mathbf{E}T(0,(n-2)e_1)] + \ldots + \mathbf{E}T(0,e_1)\Big) \\ &= \frac{1}{n}\left(\mathbf{E}B_n(0,e_1) + \mathbf{E}B_{n-1}(0,e_1) + \ldots + \mathbf{E}T(0,e_1)\right) .\end{aligned}$$

Since $B_n(0,e_1) \to B(0,e_1)$ almost surely and since $|B_n(0,e_1)| \leqslant T(0,e_1) \leqslant M$ almost surely, the typical term above converges to $\mathbf{E}B(0,e_1)$. On the other hand, the left-hand side converges to $g(e_1)$, so we see

$$\mathbf{E}B(0,-e_1) = -\mathbf{E}B(0,e_1) = -g(e_1) . \tag{6.2}$$

Combining (6.2) with (6.1), we see that $B(0,-me_1)/m \to -g(e_1)$ almost surely. We now move toward a contradiction similarly to Hoffman's argument. Write $\Gamma(0) = (z_1, z_2, \ldots)$. As in Lemma 5.3, we have $B(0,z_i) = T(0,z_i)$ for all i. In particular, for all i, $B(0,z_i) \geqslant 0$. Let $m_i = \lfloor |z_i| \rfloor$.

$$\begin{aligned}|B(0,z_i) - B(0,-m_ie_1)| &= |B(-m_ie_1,z_i)| \leqslant T(-m_ie_1,z_i) \\ &\leqslant M\|m_ie_1 + z_i\|_1 \\ &\leqslant 2M\sqrt{d}(|z_i|)|(z_i/|z_i|) + e_1| ,\end{aligned} \tag{6.3}$$

when $|z_i| > 1$.

On the other hand, as $i \to \infty$, we have

$$\lim_{i\to\infty} B(0,-m_ie_1)/|z_i| = \lim_{i\to\infty} B(0,-m_ie_1)/m_i = -g(e_1) .$$

If i is sufficiently large, then $|z_i/|z_i| + e_1| \leqslant 2\delta = (g(e_1)/8M\sqrt{d})$ and $B(0,-m_ie_1) \leqslant -|z_i|g(e_1)/2$. Then the above implies along with (6.3) that

$$B(0,z_i)/|z_i| \leqslant -g(e_1)/2 + (2M\sqrt{d})(g(e_1)/8M\sqrt{d}) \leqslant -g(e_1)/4.$$

This contradicts the fact that $B(0,z_i) \geqslant 0$. □

7. Busemann subsequential limits and general directedness statements

We have shown that averaging properties of Busemann functions can be used to control directedness properties of infinite geodesics, and have some idea of how to implement this strategy in practice. Unfortunately, without assumption (LimG), we are lacking a Busemann function and corresponding geodesics on which to run this program. It is obviously reasonable to want to try to construct geodesics without making any unproven assumptions. Our main goal in the remainder of the article is to run a more sophisticated version of last section's argument which allows us to circumvent (LimG).

Fix some $\zeta \in \partial\mathcal{B}$. In all our work in this section, we replace (LimG) with the following assumption:

ASSUMPTION 7.1 (Dif). $\partial\mathcal{B}$ is differentiable at ζ.

Recall the meaning of this statement is that there is a unique supporting hyperplane H for $\mathcal{B}$ at ζ; see Section 3.1.3 in [**5**]. There is a unique ρ such that $H = \{x : x \cdot \rho = 1\}$. Recall that $\mathcal{B}$ is convex by the shape theorem, so (Dif) is a much weaker assumption than it seems at first glance.

Given a ζ as in (Dif), we can of course define the "sector of contact" of H with $\partial\mathcal{B}$:

$$S = H \cap \partial\mathcal{B} .$$

The corresponding set of angles is

$$\Theta_S := \{z \in \mathbb{S}^{d-1} : z/g(z) \in S\} . \tag{7.1}$$

Our main theorem is that, under (Dif), we can produce a geodesic which is directed in a sector no wider than Θ_S.

THEOREM 7.2 ([**6**]). *Assume the standard assumptions and* (Dif). *Then with probability one, there is an infinite geodesic γ from* 0 *which is directed in Θ_S, in the sense that $\Theta(\gamma) \subseteq \Theta_S$.*

In particular, if ζ is an exposed point (i.e., if $S = \{\zeta\}$), then we can produce a directed geodesic. The statement of the theorem should be interpreted carefully: the argument we give here does not give a construction of such a γ that is measurable with respect to the edge weights. Rather, it shows that

$$\mathbf{P}(\text{there exists an infinite geodesic directed in } \Theta_S) = 1$$

by extracting an appropriate (t_e)-measurable event having probability one from a larger probability space.

In our construction, we build a limiting Busemann function B and corresponding limiting geodesics via a particularly chosen limiting procedure. In our analysis of the asymptotics of this Busemman function, we also make a technical improvement on the methods of the last section. Note that our contradiction there came from the observation B_γ grows (at least remaining positive) along its base geodesic γ and becomes negative at a linear rate along the $-e_1$ axis. In fact, we can sharpen these observations by considering the growth of B in a global sense. The main goal here will be to provide a version of (5.5) which gives a "shape theorem" for a particular Busemman function which characterizes completely the linear-order growth.

We note here that the presentation of Theorem 7.2 and its proof are influenced by the versions appearing in [**2**]; in particular, unlike the original paper [**6**], we make clear the generalization to $d > 2$.

7.1. Construction of geodesics, Busemann functions. Our construction benefits greatly from considering point-to-set geodesics. For a subset $S \subseteq \mathbb{R}^d$ and $x \in \mathbb{Z}^d$, let $T(x, S) = \inf_{y \in S} T(x, y)$. Recall the extension of T to points of $\mathbb{R}^d$ — we set $T(x, y) = T(x, [y])$, where $[y]$ is the unique point of $\mathbb{Z}^d$ such that $y \in [y]+[0, 1)^d$]. A point-to-set geodesic from x to S is a path γ from x to some $[y]$ with $y \in S$, such that $T(\gamma) = T(x, S)$.

For each real number α, let $H_\alpha = \{y : y \cdot \rho = \alpha\}$. It is not hard to see following the proof of Lemma 2.4 that geodesics to H_α exist and are unique; we let $\Gamma_\alpha(x)$ denote the geodesic from $x \in \mathbb{Z}^d$ to H_α, and $B_\alpha(x,y) = T(x,H_\alpha) - T(y,H_\alpha)$. We would be very happy if each $\Gamma_\alpha(x)$ and $B_\alpha(x,y)$ had limits as $\alpha \to \infty$, as we could then work with the limiting objects. Because it is not clear how to show this, we will take a particular sort of subsequential limit instead. Particular desiderata which guide the choice of limiting procedure are that the limits have appropriate translation-invariance properties (so we can do versions of averaging tricks and apply ergodic theorems) and that they preserve the relationship between the limiting analogues of Γ_α and B_α (so we can say something about geodesics from Busemann asymptotics).

We consider the edge weight configuration $\omega = (t_e)_e$ to live on the canonical probability space $\Omega_1 := [0, \infty)^{\mathcal{E}^d}$. We will need to consider an enlarged version of this space. Let $\vec{\mathcal{E}}^d$ denote the set of **directed** edges (i.e., an ordered pair (x,y) is in $\vec{\mathcal{E}}^d$ if $\{x,y\} \in \mathcal{E}^d$). We let $\widetilde{\Omega} := \Omega_1 \times \Omega_2 \times \Omega_3$, where $\Omega_2 = \mathbb{R}^{\mathbb{Z}^d}$ and $\Omega_3 = \{0,1\}^{\vec{\mathcal{E}}^d}$. For each α, we will push forward our original measure $\mathbf{P}$ on Ω_1 to a measure μ_α on $\widetilde{\Omega}$.

For $(x,y) = \vec{e} \in \vec{\mathcal{E}}^d$, let $e = \{x, y\}$ be the undirected version of $\vec{e}$ and define the random variable

$$\eta_\alpha(\vec{e}) := \begin{cases} 1 & \text{if } T(x,H_\alpha) = T(y,H_\alpha) + t_e \\ 0 & \text{otherwise.} \end{cases}$$

For each α , define the map $\Phi_\alpha : \Omega_1 \to \widetilde{\Omega}$ as follows:

$$\Phi_\alpha(\omega) = (\omega, B_\alpha, \eta_\alpha) \ .$$

For each α, we define the probability measure μ_α (on $\widetilde{\Omega}$ with the Borel sigma-algebra) to be the push-forward of $\mathbf{P}$ by the map Φ_α.

μ_α keeps track of the joint distribution of the edge weights, Busemann functions, and edges in geodesics. The geodesics are kept track of by the edge variables η_α; for instance, we have almost surely that $\eta_\alpha((x,y)) = 1$ if and only if (x,y) is in $\Gamma_\alpha(x)$. Indeed, if $(x,y) \in \Gamma_\alpha(x)$, then $\eta_\alpha((x,y)) = 1$ by definition of a geodesic; conversely, if $\eta_\alpha((x,y)) = 1$, then by concatenating (x,y) with $\Gamma_\alpha(y)$ we produce a path γ satisfying $T(\gamma) = T(x,H_\alpha)$ (which must be the unique geodesic).

Recall the translation operators from Definition 5.4. We extend them to $\widetilde{\Omega}$ as follows. For $z \in \mathbb{Z}^d$, consider a typical point $((t_e),(B(x,y)),(\eta_\alpha(\vec{e}))$ of $\widetilde{\Omega}$; then

$$\theta_z((t_e),(B(x,y)),(\eta(\vec{e}))) = ((t_{e-z}),(B(x-z,y-z)),(\eta(\vec{e}-z))) \ .$$

LEMMA 7.3. *If $z \in \mathbb{Z}^d$ and $A \subseteq \widetilde{\Omega}$ is an event, then $\mu_\alpha \circ \theta_z^{-1} = \mu_{\alpha-\rho\cdot z}$.*

PROOF. Let us just demonstrate via the event $\{\eta((x,y)) = 1\}$. We have

$$\begin{aligned} &\mu_\alpha \circ \theta_z^{-1}(\eta((x,y)) = 1) \\ &\quad = \mu_\alpha(\eta((x+z,y+z)) = 1) = \mathbf{P}((x+z,y+z) \in \Gamma(x+z,H_\alpha)) \\ &\quad = \mathbf{P}((x,y) \in \Gamma(x,H_{\alpha-z\cdot\rho})) = \mu_{\alpha-\rho\cdot z}(\eta((x,y)) = 1) \ . \end{aligned}$$

Here we have used the invariance of $\mathbf{P}$ under shifts, and the fact that $H_\alpha + z = H_{\alpha+\rho\cdot z}$. □

The following properties of B_α are proved via very similar arguments to those of Lemma 5.3, so we omit their proofs.

LEMMA 7.4. *The following hold* $\mathbf{P}$*-a.s. for each* $x, y, z \in \mathbb{Z}^d$*:*

(1) $|B_\alpha(x,y)| \leqslant T(x,y)$;
(2) $B_\alpha(x,z) = B_\alpha(x,y) + B_\alpha(y,z)$;
(3) $B_\alpha(y,x) = -B_\alpha(x,y)$;
(4) *If* $y \in \Gamma_\alpha(x)$, *then* $B_\alpha(x,y) = T(x,y)$.

In particular, if we replace B_α *and* η_α *by typical points* B *and* η *of* Ω_2 *and* Ω_3*, the analogues of the above hold* μ_α*-a.s.*

When we consider a limit of μ_α, since we do not know that Γ_α's converge, there is no clear way to reconstruct an infinite geodesic corresponding to "the $\alpha = \infty$ version of Γ_α". We wish to use the η variables above to read off an infinite geodesic from a configuration on $\widetilde{\Omega}$. We will be greatly helped here by the following graphical construction.

DEFINITION 7.5. For each $\eta \in \Omega_3$, define a directed graph $\mathbb{G}$ with vertex set $\mathbb{Z}^d$ and directed edge set as follows: (x,y) is an edge of $\mathbb{G}$ if and only if $\eta((x,y)) = 1$. When the configuration η is understood, we write $x \to y$ if there is a directed path in $\mathbb{G}(\eta)$ from x to y.

LEMMA 7.6. *The following hold for* μ_α*-a.e. configuration of* $\widetilde{\Omega}$*.*

(1) *For each* x*, there is a directed path (possibly equal to* (x)*) from* x *to* H_α*.*
(2) *If* γ *is a directed path in* $\mathbb{G}$*, then* γ *is a geodesic.*
(3) *If* $x \to y$*, then* $B(x,y) = T(x,y)$*.*

PROOF. Item (1) is clear by considering the fact that the edges $\vec{e}$ of $\Gamma_\alpha(x)$ have $\eta_\alpha(\vec{e}) = 1$. Item (3) follows from the definition of B_α and property (4) of Lemma 7.4.

Consider an outcome such that geodesics exist and are unique. To prove property (2), let γ start at x and traverse the directed edges $\vec{e}_1 \ldots, \vec{e}_n$ in order. Let $K \leqslant n$ be the maximal index such that the path γ_K which traverses (in order) $\vec{e}_1 \ldots, \vec{e}_K$ is a subpath of $\Gamma_\alpha(x)$. We will show $K = n$.

By the observation that $\eta_\alpha((y,z)) = 1$ if and only if $(y,z) \in \Gamma_\alpha(y)$, we see that $\vec{e}_1$ is the first vertex of $\Gamma_\alpha(x)$ and so $K \geqslant 1$. We now show that if $K < n$, then γ_{K+1} is also a subpath of $\Gamma_\alpha(x)$. Note that there is some path γ' which extends γ_K to a geodesic from x to H_α. Letting $\vec{e}_K = (y,z)$, the subpath of γ' from z to H_α must be the unique geodesic from z to H_α. On the other hand, since $\vec{e}_{K+1}$ is in $\mathbb{G}$ (and is thus in $\Gamma_\alpha(Z)$, we have that the edge of γ' following $\vec{e}_K$ must be $\vec{e}_{K+1}$. $\square$

We average the μ_α's to produce a new measure μ_n^* on $\widetilde{\Omega}$: for $n = 1, 2, \ldots$, set

$$\mu_n^* = \frac{1}{n} \int_0^n \mu_\alpha \, d\alpha \, .$$

There is a technical argument required for this definition: namely, we need to show that for each measurable A, the map $\alpha \mapsto \mu_\alpha(A)$ is measurable (so the integral above makes sense). We refer the interested reader to Appendix A of [**6**].

LEMMA 7.7. *There is a subsequence* (n_k) *and a measure* μ *on* $\widetilde{\Omega}$ *such that*

$$\lim_{k \to \infty} \mu_{n_k}^* = \mu \quad (\text{weakly}).$$

PROOF. The distribution $\mathbf{P}$ on Ω_1 is easily seen to be tight, in the usual sense that for each $\varepsilon > 0$, we can find a compact measurable $K_\varepsilon \subseteq \Omega_1$ such that

$\mathbf{P}(\Omega_1 \backslash K_\varepsilon) < \varepsilon$. By property (1) of Lemma 7.4, we see that the sequence (μ_n^*) is also tight. Prokhorov's theorem now gives the existence of a subsequential weak limit. □

Choose some μ as in the statement of Lemma 7.7. This will be the object we use to construct the geodesic of Theorem 7.2. The bulk of this construction will be done in the next subsection, by analyzing the graphs $\mathbb{G}$ and Busemann functions B sampled from μ. For now, we just give the main reason for using the averaging procedure which constructed μ (and not, say, choosing μ just as a limit of μ_α).

LEMMA 7.8. *μ is translation-invariant: for any $z \in \mathbb{Z}^d$ and any event $A \subseteq \widetilde{\Omega}$,*

$$\mu \circ \theta_z^{-1}(A) = \mu^*(A) .$$

PROOF SKETCH. Let n be a positive integer. We can write (using Lemma 7.3)

$$\mu_n^* \circ \theta_z^{-1}(A) = \frac{1}{n} \int_{-z\cdot\rho}^{n-z\cdot\rho} \mu_\alpha(A) d\alpha .$$

In particular,

$$\left|\mu_n^* \circ \theta_z^{-1}(A) - \mu_n^*(A)\right| \leqslant \frac{1}{n} \left| \int_{n-z\cdot\rho}^{n} \mu_\alpha(A)\, d\alpha + \int_{-z\cdot\rho}^{0} \mu_\alpha(A)\, d\alpha \right| ,$$

which tends to zero in n for any fixed A. The result now follows by approximating μ by μ_n^* for n large. □

7.2. Asymptotics for samples from μ^*. We will now study the asymptotics of a sample $B \in \Omega_2$ from the marginal of μ. Our goal is to replicate the previous "averaging" arguments but in a strong sense. Our first step, as before, is to control the expectation.

THEOREM 7.9. *For any $x, y \in \mathbb{Z}^d$, we have*

$$\mathbf{E}_\mu B(x, y) = \rho \cdot (y - x) .$$

We need the following lemma:

LEMMA 7.10. *Almost surely and in L^1,*

$$\lim_{\alpha \to \infty} T(0, H_\alpha)/\alpha = 1 .$$

This lemma follows from the shape theorem in a fairly straightforward manner, so we omit the proof.

PROOF OF THEOREM 7.9. Let n be a positive integer. Write

$$\begin{aligned}
\mathbf{E}_{\mu_n^*} B(-x, 0) &= \frac{1}{n}\left[\int_0^n \mathbf{E}T(-x, H_\alpha)\, d\alpha - \int_0^n \mathbf{E}T(0, H_\alpha)\, d\alpha\right] \\
&= \frac{1}{n}\left[\int_0^n \mathbf{E}T(0, H_{\alpha + x\cdot\rho})\, d\alpha - \int_0^n \mathbf{E}T(0, H_\alpha)\, d\alpha\right] \\
&= \frac{1}{n}\left[\int_n^{n+x\cdot\rho} \mathbf{E}T(0, H_\alpha)\, d\alpha - \int_0^{x\cdot\rho} \mathbf{E}T(0, H_\alpha)\, d\alpha\right]. \qquad (7.2)
\end{aligned}$$

The second equality comes by translation invariance.

Taking limits in (7.2), the second term goes to zero with n. To deal with the first term, note that Lemma 7.10 implies that for each $\alpha \in [0, x \cdot \rho]$,

$$\lim_{n\to\infty} \frac{\mathbf{E}T(0, H_{\alpha+n})}{n} = \lim_{n\to\infty} \frac{\mathbf{E}T(0, H_{\alpha+n})}{n+\alpha}\frac{n+\alpha}{n} = 1 .$$

Applying this in (7.2) (with a dominated convergence theorem argument to take the limit under the integral) gives $\mathbf{E}_{\mu_n^*} B(-x, 0) \to \rho \cdot x$.

It remains only to show that $\mathbf{E}_{\mu_{n_k}^*} B(-x, 0) \to \mathbf{E}_\mu B(-x, 0)$ (since the expectation at arguments x, y now follows by translation-invariance). For $R > 0$, defining the continuous truncation

$$B_R(-x, 0) = B(-x, 0)\mathbf{1}_{|B(-x,0)|\leqslant R} + R\,\mathrm{sign}(B(-x, 0))\mathbf{1}_{|B(-x,0)>R|} ,$$

we have by continuity and boundedness that

$$\mathbf{E}_{\mu_{n_k}^*} B_R(-x, 0) \to \mathbf{E}_\mu B_R(-x, 0) .$$

The claim now follows by taking $R \to \infty$. Indeed, we have that $B(-x, 0)^2 \leqslant T(0, -x)^2$, which has a finite second moment (bounded uniformly in n by some constant C). The Cauchy-Schwarz inequality implies

$$\begin{aligned}\mathbf{E}_{\mu_n^*}|B(-x, 0)|\mathbf{1}_{|B(-x,0)|>R} &\leqslant \left(\mathbf{E}_{\mu_n^*}|B(-x, 0)|^2\right)^{1/2} \mu_n^*(|B(-x, 0)| > R)^{1/2}\\ &\leqslant C^{1/2}\mathbf{P}(T(0, -x) > R)^{1/2} ,\end{aligned}$$

where we again use $|B(-x, 0)| \leqslant T(-x, 0)$. The probability above goes to zero uniformly in n as $R \to \infty$, so we have established

$$\lim_{R\to\infty} \limsup_{n\to\infty} \mathbf{E}_{\mu_n^*}|B(-x, 0)|\mathbf{1}_{|B(-x,0)|>R} = 0 . \qquad \square$$

As we said earlier, we want to prove a shape theorem for B, establishing the a.s. leading-order behavior of $B(0, x)$ under μ. The preceding lemma suggests a target for this shape theorem: for $|x|$ large, one expects $|B(0, x) - \rho \cdot x| \ll |x|$. Unfortunately, a result like this does not immediately follow in general without information about $\mathcal{B}$, and it is here we use assumption (Dif).

A brief explanation of the problem is as follows. We want to use the ergodic theorem, as in the proof of Garet-Marchand's result, to establish that $B(0, nx)/n \to \rho \cdot x$ for each fixed x; if we could do this, we could patch together a global result by continuity. Unfortunately, the ergodic theorem does not give convergence of $B(0, nx)/n$ to a deterministic limit, because the measure μ is not guaranteed to be ergodic. So the best we can hope for *a priori* is convergence to some functional whose mean is ρ.

The above obstacle is the main reason we assumed (Dif) in the first place. We will see that the differentiability gives us the ability to say the random limiting functional is in fact identical to ρ, almost surely.

THEOREM 7.11. *There exists a random vector ϖ on $\widetilde{\Omega}$ such that*

$$\mu\left(\limsup_{|z|\to\infty} \frac{|B(0, z) - \varpi \cdot z|}{|z|} = 0\right) = 1 .$$

Moreover, $\mathbf{E}_\mu \varpi = \rho$.

We give a sketch of the main ideas of the proof, highlighting the issues with ergodicity.

PROOF SKETCH. We first show that for $e = e_1, e_2, \ldots, e_d$, we have

$$\lim_{n\to\infty} B(0, ne)/n =: \varpi(e) \quad \text{exists } \mu\text{-a.s.} \tag{7.3}$$

To establish (7.3), we write $n^{-1}B(0, ne) = n^{-1}\sum_{j=1}^{n} B((j-1)e, je)$, which follows from (2) in Lemma 7.4 after some work to pass this property through the limit which produces μ. We then rewrite this sum as a sum of copies of $B(0, e)$ evaluated in shifted environments as usual, and apply the ergodic theorem. Since μ is not ergodic but rather merely translation-invariant, we are only guaranteed that the limit exists and defines some random variable: it need not be constant.

We now define $\varpi = (\varpi(e_1), \ldots, \varpi(e_d))$; ϖ is invariant under translations of the realization $((t_e), (B), (\eta))$. Our next step is to see that for each fixed $x = (x(1), x(2), \ldots, x(d)) \in \mathbb{Z}^d$, we have $B(0, nx)/n \to x \cdot \varpi$. This follows by writing

$$B(0, nx) = B(0, nx(1)e_1) + B(nx(1)e_1, nx(1)e_1 + nx(2)e_2) + \ldots$$

and using the previous convergence result (and invariance of ϖ) to approximate each of these terms by $nx(1)\varpi(1)$, $nx(2)\varpi(2)$, etc.

This gives us convergence in fixed directions. To give the global convergence as in the statement of the theorem, we follow the proof of the shape theorem (just as we did in the proof of (5.5)). The form of the mean of ϖ is a consequence of Theorem 7.9. $\square$

As promised, we claimed that we can in fact show that ϖ is ρ under our assumptions. We conclude this subsection by giving the argument.

LEMMA 7.12. *μ-a.s., the hyperplane $H_\varpi := \{x : x \cdot \varpi = 1\}$ is a supporting hyperplane for $\partial\mathcal{B}$ at ζ. In particular, under (Dif), we have $\varpi = \rho$ almost surely.*

PROOF. Note that almost surely, for any fixed $x \in \mathcal{B}$

$$\varpi \cdot x = \lim_{n\to\infty} \frac{B(0, nx)}{n} \leqslant \lim_{n\to\infty} \frac{T(0, nx)}{n} = g(x) \leqslant 1 \ .$$

In particular, $\mathcal{B}$ a.s. lies on one side of H_ϖ. On the other hand,

$$\mathbf{E}_\mu \varpi \cdot \zeta = \rho \cdot \zeta = 1 \ .$$

Thus $\varpi \cdot \zeta = 1$ almost surely, and H_ϖ is a supporting hyperplane. $\square$

7.3. Directedness. We now prove Theorem 7.2. We show that for μ-almost every element of $\widetilde{\Omega}$, there is an infinite path from 0 in $\mathbb{G}(\eta)$ (which must be a geodesic, by the limiting version of Lemma 7.6 (2)) and which has the required directedness.

PROOF OF THEOREM 7.2. It is not hard to show a limiting version of Property (1) of Lemma 7.6 which says that with μ-probability one, there is an infinite path in $\mathbb{G}$ from 0. As mentioned just prior, this path is also easily seen to be a geodesic. So the main argument is to show

$$\text{a.s., for each infinite path } \gamma \text{ in } \mathbb{G},\ \Theta(\gamma) \subseteq \Theta_S \ . \tag{7.4}$$

Let $\gamma = (0 = x_0, x_1, \ldots)$, and let (x_{n_k}) be a subsequence such that $x_{n_k}/|x_{n_k}| \to \theta$. We show $\theta \in \Theta_S$. Applying Theorem 7.11 (with $\varpi = \rho$) gives

$$\lim_k B(0, x_k)/|x_k| = \rho \cdot \theta \ .$$

On the other hand, this limit also equals $\lim_k T(0, x_k)/|x_k| = g(\theta)$. In particular, $\theta/g(\theta)$ is on $\partial\mathcal{B}$ and in the set H, so it is in S. $\square$

7.4. Sector-directed geodesics and uniqueness. We return in this section to the case of $\mathbb{Z}^2$. It is first worth discussing the result of Theorem 7.2 and reflecting on the directedness conjectures of the model. Since it is widely believed that the boundary of the limit shape should be uniformly curved and everywhere differentiable (at least for typical edge weight distributions), one expects that there should exist directed geodesics in each direction. On the other hand, Theorem 7.2 (as well as Theorem 3.1) would require strong and currently unknown information about $\mathcal{B}$ in order to prove the existence of even a single geodesic having asymptotic direction e_1.

In fact, Theorem 7.2 also applies to the setting where the joint distribution of the t_e's is not i.i.d., but rather only assumed translation-ergodic. In this generality, one cannot expect all of the usual FPP conjectures to hold: for instance, we know any "reasonable" set is attainable as a limit shape (see Theorem 3.8 in [**5**]). In the case of an ergodic weight distribution having a polygonal $\mathcal{B}$, Theorem 7.2 guarantees only the existence of some $\mathcal{B}$-dependent finite number of geodesics and does not rule out their wandering across entire sectors. In recent work by Brito and Hoffman [**3**], an ergodic weight distribution exhibiting this kind of counterintuitive behavior is constructed: in their example, there exist exactly four infinite geodesics, and each wanders across an entire quadrant of the lattice. See also the recent work by Alexander and Berger [**1**], where an ergodic distribution is constructed having an octogon as its limit shape, but having geodesics directed only toward the axis directions of the lattice.

With these ideas in mind, we return to the problem raised in Section 3.1. Recall that Licea and Newman showed that for "typical" directions θ — i.e., all θ in some set $D \subseteq [0, 2\pi)$ — there is a.s. a unique θ-directed geodesic. We will describe work from [**7**] which extends these uniqueness statements using Busemann function techniques. In keeping with the above discussion, these results give not uniqueness of geodesics having some single direction θ, but rather uniqueness of geodesics directed within sectors (as produced by Theorem 7.2). The result also requires further assumptions on the limit shape; we comment on these after the statement of the theorem.

THEOREM 7.13 ([**7**]). *Consider FPP on $\mathbb{Z}^2$ under the standard assumptions of Assumption 2.1. Assume that $\partial\mathcal{B}$ is differentiable everywhere. Let $\zeta \in \partial\mathcal{B}$; let H be the line tangent to $\mathcal{B}$ at ζ, let S be the sector of of contact between H and $\mathcal{B}$, and let Θ_S be the corresponding set of angles (see (7.1) above). Then there is a.s. a unique geodesic γ_x from each x such that γ_x is directed in Θ_S (i.e., $\Theta(\gamma_x) \subseteq \Theta_S$). Moreover, these geodesics a.s. coalesce: $|\gamma_x \triangle \gamma_y| < \infty$ for all x, y.*

In fact, the theorem requires only a weaker assumption of differentiability at ζ and the endpoints of S. Under the additional assumption that every point of $\partial\mathcal{B}$ is also an exposed point (so that $S = \{\zeta\}$ in the theorem), the statement gives us an existence and uniqueness statements of geodesics that are directed in the sense of Theorem 3.1. We describe here the outline of the proof of Theorem 7.13 with many of the more graph-theoretic arguments truncated, preferring to emphasize the role of Busemann functions in the analysis.

SKETCH OF PROOF OF THEOREM 7.13. One of the hardest parts of the argument is dealing with the undirectedness of the model. This will be dealt with by considering both FPP on $\mathbb{Z}^2$ and on the right half-plane $\mathbb{H} = \{(a, b) \in \mathbb{Z}^2 : a \geqslant 0\}$;

the edges of $\mathbb{H}$ are the edges of $\mathbb{Z}^2$ having both endpoints in $\mathbb{H}$. The models on the two graphs can be coupled in the natural way by using the same edge weights. The main reason for introducing $\mathbb{H}$ is that two geodesics cannot change their relative top / bottom ordering without crossing each other, and geodesic intersections yield a useful monotonicity property for the corresponding Busemann functions.

For simplicity, we consider just the sector in the e_1-direction (i.e., ζ is taken to be a multiple of e_1); as before, this allows us to simplify notation without sacrificing any important parts of the argument. Note that our assumption implies $\Theta_S \subseteq (-\pi/4, \pi/4)$. Let $G_H(x, y)$ denote the geodesic between $x, y \in \mathbb{H}$ for the FPP model on $\mathbb{H}$.

Step 1: Finding "trapping" geodesics. We first establish the following two statements:

(1) There exist sequences $(\Theta_B^n)_{n=1}^{\infty}$ and $(\Theta_T^n)_{n=1}^{\infty}$ of sets of angles in the interval $(-\pi/2, \pi/2)$ such that $\sup \Theta_T^{n+1} < \inf \Theta_T^n$ and $\inf \Theta_B^{n+1} > \sup \Theta_B^n$ such that $(\inf \Theta_B^n)$ converges to the bottom (i.e. clockwise) endpoint of Θ_S and $(\sup \Theta_T^n)$ converges to the top (counterclockwise) endpoint of Θ_S, such that the following holds. For each n, there a.s. exist geodesics in $\mathcal{T}(0)$ directed in Θ_T^n and Θ_B^n.

(2) The statement of Item 1 holds on $\mathbb{H}$, where the geodesics mentioned there are geodesics in $\mathbb{H}$ for the FPP model defined on $\mathbb{H}$.

The first item above follows immediately from Theorem 7.2, using the fact that $\partial\mathcal{B}$ is differentiable at the endpoints of S.

To establish the result on the half-plane, we use the result on $\mathbb{Z}^2$. The geodesic in $\mathcal{T}(0)$ from item (1) directed in Θ_T^n eventually crosses each vertical line; from this we derive that there is a.s. a positive density of sites along the e_2-axis having $\mathbb{H}$-geodesics directed in Θ_T^n. A similar statement holds for Θ_T^{n+1}. Let ke_2 and $-\ell e_2$ $(k, \ell > 0)$ be two vertices of the e_2-axis having such Θ_T^n and Θ_T^{n+1}-directed (respectively) $\mathbb{H}$-geodesics. Denote these by Γ^n and Γ^{n+1}; the directedness of these geodesics moreover allows us to choose them so that $\Gamma^n \cap \Gamma^{n+1} = \varnothing$.

Now choose some sequence $(v_n) \subseteq \mathbb{H}$ such that $v_n/|v_n| \to \theta$ for some $\theta \in (\sup \Theta_T^{n+1}, \inf \Theta_T^n)$. Let Γ be some subsequential limit of the sequence $(G_H(0, v_n))$. Uniqueness of passage times forces Γ to be contained in the region containing 0 and delimited by Γ^n, Γ^{n+1}, and the segment of the e_2-axis between $-\ell e_2$ and ke_2. Thus Γ is directed in $[\sup \Theta_T^{n+1}, \inf \Theta_T^n]$, and continuing in this way we can find an ordered sequence of sectors converging to the top endpoint of Θ_S and half-plane geodesics from 0 directed in these sectors; a similar argument establishes the analogous statement for bottom endpoint sectors.

Step 2: Existence of extremal geodesics. We define subtrees of $\mathcal{T}(0)$ and its half-plane analogue consisting of S-directed geodesics, then find topmost and bottommost geodesics in them. More generally, for any x, let $\mathcal{G}(x)$ and $\mathcal{G}_H(x)$ be the sets of Θ_S-directed $\mathbb{Z}^2$- and $\mathbb{H}$-geodesics starting at x, respectively. We define an ordering $\prec$ on each of these sets as follows: $\Gamma \prec \Gamma'$ if the topmost intersection of Γ with $\{y : y \cdot e_1 = n\}$ is below the corresponding topmost intersection of Γ', for all large n. This $\prec$ turns out to a.s. define a total ordering, due to the directedness of these geodesics (so they eventually pass each vertical line) and the uniqueness of passage times (which prevents crossing and then re-crossing).

Given this ordering, one shows that there a.s. exist a unique maximal element $\Gamma_T(x) \in \mathcal{G}(x)$ and minimal element $\Gamma_B(x) \in \mathcal{G}(x)$, with analogous statements for

$\mathbb{H}$ (we denote the $\mathbb{H}$ geodesics by Γ_T^H and Γ_B^H). To do this, one constructs $\Gamma_T(x)$ explicitly as the limit of $G(x, x_T(n))$, where x_T is the element of $\mathcal{G}(x) \cap \{y : y \cdot e_1 = n\}$ having largest e_2-coordinate. One argues that any such subsequential limit must be S-directed (being trapped by the geodesics from Item 1 via uniqueness of passage times) but asymptotically lie above any other S-directed geodesic by virtue of the topmost position of $x_T(n)$ and the fact that geodesics in $\mathcal{G}(x)$ cannot backtrack too far (to interchange orderings).

Step 3: Coalescence of extremal geodesics. One argues that the family $\{\Gamma_T(x)\}_x$ is a coalescing family of geodesics, with analogous statements for the $\Gamma_B(x)$'s and their half-plane analogues. The $\mathbb{Z}^2$ coalescence statements follow from a version of the Licea-Newman argument (see Section 3.1 above). The limiting procedure used to produce topmost geodesics shows that a subsegment of a topmost geodesic is also a topmost geodesic from its initial vertex. Since there is a unique Γ_T from each x, if there were multiple distinct Γ_T geodesics, one could use them and an edge-modification argument to construct geodesic structures which would be forbidden by translation-invariance.

The coalescence result on $\mathbb{H}$ follows from the coalescence result on $\mathbb{Z}^2$; we describe the case that the geodesics start on the e_2-axis. Since the $\mathbb{Z}^2$ geodesics $\Gamma^T(x)$ pass every vertical line, we can find a positive density of sites along the e_2-axis whose topmost $\mathbb{Z}^2$ geodesics stay in $\mathbb{H}$. In such a case, it turns out that $\Gamma^T(x)$ is also the topmost S-directed geodesic from x in $\mathbb{H}$. Now the fact that these geodesics coalesce also forces every other Γ_T^H geodesic from axial vertices to coalesce with this family. The key observation forcing the trapping is again that subsegments of topmost geodesics are topmost geodesics, so the aforementioned topmost geodesics cannot cross.

Given these coalescing families, we can define Busemann functions B_T and B_B and their half-plane analogues B_T^H and B_B^H. If $x \in \mathbb{Z}^2$ is any vertex and $\Gamma_T(x) = (x = x_1, x_2, \ldots)$, we set

$$B_T(y, z) = \lim_{n \to \infty} \left[T(y, x_n) - T(z, x_n)\right] .$$

By coalescence, similarly to item (5) of Lemma 5.3, the definition above does not depend on the choice of x. We make similar definitions for the other Busemann functions.

Step 4: Limiting behavior of Busemann functions. One follows an argument similar to that used in the proof of Theorem 7.11 to establish shape theorems for B_T and B_B. These say that there exists a (deterministic, since B_T depends only on the i.i.d. environment) vector ρ_T such that, for each $\delta > 0$,

$$\mathbf{P}\left(\limsup_{|x| \to \infty} \left|\frac{B_T(0, x) - \rho_T \cdot x}{|x|}\right| > \delta\right) = 0 .$$

Again a similar statement holds with B_B and some ρ_B. Just as in the preceding Lemma 7.12, we see that $\{y : \rho_T \cdot y = 1\}$ must be the unique tangent line to the limit shape at ζ and similarly for ρ_B, giving $\rho_T = \rho_B$.

Since the tangent line at ζ is vertical, $\rho \cdot e_2 = 0$. Note that $B_T^H(0, ke_2)/k \to \mathbf{E}B_T^H(0, e_2)$ by the ergodic theorem. One can now show that $B_T^H(0, ke_2) = o(k)$ using the fact that $\Gamma_T(\ell e_2) = \Gamma_T^H(\ell e_2)$ for infinitely many ℓ and then applying the above shape theorem. An analogous statement holds for B_B^H, so defining $\Delta(x, y) = B_T^H(x, y) - B_B^H(x, y)$, we have $\mathbf{E}\Delta(0, e_2) = 0$.

Step 5: Deriving a contradiction. We assume that $\Gamma_B(0) \neq \Gamma_T(0)$ with positive probability to derive a contradiction. We show that $\Delta(0, e_2)$ has a definite sign, which combined with the previous step gives $\Delta(0, e_2) = 0$ a.s. On the other hand, this is impossible if $\Gamma_B(0)$ and $\Gamma_T(0)$ are not identical almost surely.

To see the "definite sign" claim, we note that the coalescence of top and bottom geodesics, along with the fact that geodesics from axis points in $\mathbb{H}$ must cross to change their top-bottom ordering, gives that $\Gamma_T^H(0)$ and $\Gamma_B^H(e_2)$ must share a vertex z. We will use the vertex z to "switch paths" and bound the Busemann functions via subadditivity. Let a_T and a_B be vertices along $\Gamma_T^H(0)$ and $\Gamma_B^H(0)$, respectively, which come after both z and the coalescence points with $\Gamma_T^H(e_2)$ and $\Gamma_B^H(e_2)$. We note that

$$B_T(0, e_2) = T(0, a_T) - T(e_2, a_T); \quad B_B(0, e_2) = T(0, a_B) - T(e_2, a_B)\ . \tag{7.5}$$

We now compute

$$\begin{aligned} T(0, a_B) + T(e_2, a_T) &\leqslant T(0, z) + T(z, a_B) + T(e_2, z) + T(z, a_T) \\ &= T(0, a_T) + T(e_2, a_B)\ . \end{aligned}$$

The inequality above follows from subadditivity of T; the equality comes from the geodesics $\Gamma_T^H(0)$ and $\Gamma_B^H(e_2)$ (so that, for instance, $T(0, z) + T(z, a_T) = T(0, a_T)$). Subtracting the terms involving e_2 from both sides of the above and using (7.5) gives

$$B_B(0, e_2) \leqslant B_T(0, e_2)\ , \quad \text{so} \quad \Delta(0, e_2) \geqslant 0 \quad \text{a.s.} \tag{7.6}$$

Combining (7.6) with the fact that $\mathbf{E}\Delta(0, e_2) = 0$ gives that

$$\Delta(0, e_2) = 0 \quad \text{a.s.} \tag{7.7}$$

We contradict (7.7) with the following estimate:

LEMMA 7.14. *If* $\mathbf{P}(\Gamma_B(0) \neq \Gamma_T(0)) > 0$, *then* $\mathbf{P}(\Delta(0, e_2) > 0) > 0$.

Note that together, Lemma 7.14 and (7.7) show that $\mathbf{P}(\Gamma_B(0) = \Gamma_T(0)) = 1$. Since every unigeodesic from 0 directed in Θ_S asymptotically lies between $\Gamma_B(0)$ and $\Gamma_T(0)$ under the order $\prec$, we see that $\Gamma_B(0) = \Gamma_T(0)$ a.s. implies that there is a.s. at most one Θ_S-directed geodesic from 0.

PROOF OF LEMMA 7.14. Suppose for the sake of contradiction that (under the assumption of the lemma) $\mathbf{P}(\Delta(0, e_2) = 0) = 1$. Since B_T and B_B are additive (compare to the second property of Lemma 5.3), we have

$$\mathbf{P}(\Delta(0, ke_2) = 0) = 1 \quad \text{for all integers } k \geqslant 1\ . \tag{7.8}$$

We show that (7.8) contradicts the a.s. uniqueness of passage times.

Choosing b_1 sufficiently large, we can find a $\delta > 0$ and an edge $f \in [0, b_1] \times [-b_1, b_1]$ such that with probability at least δ, $f \in \Gamma_T^H(0) \Delta \Gamma_B^H(0)$ (we assume that $f \in \Gamma_T^H(0)$, the other case being similar). By directedness of the half-plane geodesics in Θ_S, we can choose a $k > b_1$ such that

$$\mathbf{P}\left((\Gamma_T^H(ke_2) \cup \Gamma_B^H(ke_2)) \cap ([0, b_1] \times [-b_1, b_1]) = \varnothing \right) > 1 - \delta/2\ .$$

In particular, on the above event, f is not in $\Gamma_T^H(ke_2)$ or $\Gamma_B^H(ke_2)$.

We have shown that with probability at least $\delta/2$, the edge f is in $\Gamma_T^H(0)$ but not $\Gamma_B^H(0)$, $\Gamma_T^H(ke_2)$, or $\Gamma_B^H(ke_2)$, and furthermore $\Delta(0, ke_2) = 0$. Recalling the coalescence properties of these half-plane geodesics, let $a_T \in \Gamma_T^H(0) \cap \Gamma_T^H(ke_2)$ and similarly for a_B; we have

$$\begin{aligned}\Delta(0, ke_2) &= B_T(0, ke_2) - B_B(0, ke_2)\\ &= T(0, a_T) - T(ke_2, a_T) - T(0, a_B) + T(ke_2, a_B) = 0 \ .\end{aligned}$$

Only one of the terms above involves t_f. This contradicts the almost sure uniqueness of passage times from Lemma 2.4. ☐

Theorem 7.13 is proved. ☐

References

[1] K. Alexander and Q. Berger. Geodesics toward corners in first passage percolation. Arxiv preprint: arXiv:1709.09072.

[2] Antonio Auffinger, Michael Damron, and Jack Hanson, *50 years of first-passage percolation*, University Lecture Series, vol. 68, American Mathematical Society, Providence, RI, 2017. MR3729447

[3] G. Brito and C. Hoffman. Private communication.

[4] R. M. Burton and M. Keane, *Density and uniqueness in percolation*, Comm. Math. Phys. **121** (1989), no. 3, 501–505. MR990777

[5] Michael Damron, *Random growth models: shape and convergence rate*, Proc. Sympos. Appl. Math. **75** (2018).

[6] Michael Damron and Jack Hanson, *Busemann functions and infinite geodesics in two-dimensional first-passage percolation*, Comm. Math. Phys. **325** (2014), no. 3, 917–963, DOI 10.1007/s00220-013-1875-y. MR3152744

[7] Michael Damron and Jack Hanson, *Bigeodesics in first-passage percolation*, Comm. Math. Phys. **349** (2017), no. 2, 753–776, DOI 10.1007/s00220-016-2743-3. MR3595369

[8] Olivier Garet and Régine Marchand, *Coexistence in two-type first-passage percolation models*, Ann. Appl. Probab. **15** (2005), no. 1A, 298–330, DOI 10.1214/105051604000000503. MR2115045

[9] Olle Häggström and Robin Pemantle, *First passage percolation and a model for competing spatial growth*, J. Appl. Probab. **35** (1998), no. 3, 683–692. MR1659548

[10] Christopher Hoffman, *Coexistence for Richardson type competing spatial growth models*, Ann. Appl. Probab. **15** (2005), no. 1B, 739–747, DOI 10.1214/105051604000000729. MR2114988

[11] Cristina Licea and Charles M. Newman, *Geodesics in two-dimensional first-passage percolation*, Ann. Probab. **24** (1996), no. 1, 399–410, DOI 10.1214/aop/1042644722. MR1387641

[12] Charles M. Newman, *A surface view of first-passage percolation*, Proceedings of the International Congress of Mathematicians, Vol. 1, 2 (Zürich, 1994), Birkhäuser, Basel, 1995, pp. 1017–1023. MR1404001

[13] Charles M. Newman and Marcelo S. T. Piza, *Divergence of shape fluctuations in two dimensions*, Ann. Probab. **23** (1995), no. 3, 977–1005. MR1349159

[14] Firas Rassoul-Agha, *Busemann functions, geodesics, and the competition interface for directed last-passage percolation*, Proc. Sympos. Appl. Math. **75** (2018).

Department of Mathematics, City College of New York, New York 10031
Email address: jhanson@ccny.cuny.edu

Proceedings of Symposia in Applied Mathematics
Volume **75**, 2018
https://doi.org/10.1090/psapm/075/00658

Fluctuations in first-passage percolation

Philippe Sosoe

ABSTRACT. We present a survey of techniques to obtain upper bounds for the variance of the passage time in first-passage percolation. The methods discussed are a combination of tools from the theory of concentration of measure, some of which we briefly review. These are combined with variations on an idea of Benjamini-Kalai-Schramm to obtain a logarithmic improvement over the linear bound implied by the Efron-Stein/Poincare inequality, for general edge-weight distributions.

CONTENTS

1. Introduction

We look at first-passage percolation in $\mathbb{Z}^d$, with i.i.d. weights

$$t_e, \quad e \in \mathcal{E}(\mathbb{Z}^d).$$

Here $\mathcal{E}(\mathbb{Z}^d)$ is the set of edges between points of $\mathbb{Z}^d$ that differ by 1 in one coordinate. The common distribution of the weights is denoted by μ, a probability measure on $[0, \infty)$. We will assume throughout that

$$\mu(\{0\}) < p_c(d), \tag{1.1}$$

where $p_c(d)$ is the critical probability for independent Bernoulli percolation on the edges.

Let $\mathbf{e}_1 = (1, \dots, 0)$ be the first standard basis vector. The quantity we are interested in is the passage time:

$$T_n := T(0, n\mathbf{e}_1) = \inf_{\gamma: 0 \to n\mathbf{e}_1} \sum_{e \in \gamma} t_e, \tag{1.2}$$

2010 *Mathematics Subject Classification.* 60K35, 60K37, 82B43.

where the infimum is taken over the collection of lattice paths

$$\gamma = (0 = x_0, \dots, n\mathbf{e}_1 = x_N), \quad \|x_i - x_{i+1}\|_1 = 1.$$

1.1. First order. Under our assumptions, the leading order behavior of T_n is linear. This follows from the subadditive ergodic theorem, which is presented in [**8**]. Here we just record the result for future reference.

THEOREM 1.1. *Let the edge weights be i.i.d. with distribution μ satisfying* (1.1), *as well as*

$$\mathbb{E}\min(Y_1, \dots, Y_{2d}) < \infty,$$

where Y_i, $i = 1, \dots, 2d$ are independent and have distribution μ. Then there exists a constant $\nu > 0$ such that

$$\lim_{n\to\infty} \frac{1}{n} T(0, n\mathbf{e}_1) = \nu, \tag{1.3}$$

almost surely and in L^1.

1.2. Fluctuations. The objective in this article will be to investigate the fluctuations of T_n around its mean $\mathbf{E}T_n$. The simplest measure of the magnitude of these fluctuations is the variance

$$\mathbf{Var}(T_n) = \mathbf{E}(T_n - \mathbf{E}T_n)^2.$$

It is expected that

$$\mathbf{Var}(T_n) \approx n^{2\chi} \tag{1.4}$$

for some dimension dependent exponent $\chi = \chi(d)$. We leave the exact nature of the approximation (1.4) unspecified. In some exactly solvable models, the correct value of χ is known and we have asymptotics of the form

$$\frac{\mathbf{Var}(T_n)}{n^{2\chi}} \to \sigma, \tag{1.5}$$

for some $\sigma > 0$ (see the section on predictions below). We will be interested in upper and lower bounds on the variance.

1.3. Predictions. The validity of the approximation (1.4) is widely assumed in the physics literature. Moreover, the exponents $\chi(d)$ are believed to be universal among a large class of growth models, some of which are more complicated than first-passge percolation. The models considered in the physics literature include ballistic aggregation, domain walls in two-dimension random-bond Ising models, and directed polymers in random potentials [**14, 18**].

Physicists generally agree that $\chi(d) < 1$ and that χ decreases with d, but there are differring predictions for the value of χ [**18, 21, 22**]. A possibility which is investigated in several works on random growth models is the existence of an upper critical dimension d_0, such that $\chi(d) = 0$ for $d \geqslant d_0$. See the discussion in [**22**].

The most precise predictions are available in dimension $d = 2$, where it is expected [**14, 15, 17, 18**] that

$$\chi = \frac{1}{3}.$$

This prediction has been verified rigorously in a number of exactly solvable growth models. We cite one of the early results below. Articles [**24**] and [**7**] will explore solvable models in greater depth. In any case, these predictions are far from being proved in the first-passage percolation model that concerns us here.

Another important prediction from physics, the universal scaling relation, links the order of the fluctuations χ to another exponent ξ, the transversal fluctuation exponent. Discussing this here would take us too far afield, so we refer to [**8**] for a statement of the scaling relation and discussion of its derivation under suitable assumptions.

1.4. Known results. The best lower bound for χ in general dimension is due to H. Kesten in [**20**], and is

$$\mathbf{Var}(T_n) \geqslant C, \tag{1.6}$$

which translates to

$$\chi(d) \geqslant 0.$$

Note that if the model has an upper critical dimension in the sense discussed above, or simply $\chi(d) \to 0$, this is the best that can be hoped for without any assumptions on d, although it is likely a poor bound in low dimensions. C. Newman and M. Piza obtain the following improvement on (1.6) in dimensions 2:

THEOREM 1.2 (Newman, Piza [**23**]). *Let $d = 2$. Suppose $\mathbf{E}t_e^2 < \infty$, $\mathbf{Var}(t_e) > 0$ and that* (1.1) *holds. Then we have the lower bound*

$$\mathbf{Var}(T_n) \geqslant C \log n$$

for some $C > 0$.

Note that the previous result still does not imply that $\chi(2) > 0$. As for upper bounds, the best known result is the following, obtained in [**9**] by M. Damron, J. Hanson and myself, following work of I. Benjamin, G. Kalai, and O. Schramm [**4**], as well as M. Benaim and R. Rossignol [**3**]:

THEOREM 1.3 (Damron, Hanson, Sosoe [**9**]). *Suppose*

$$\mathbf{E}\Big[t_e^2 \log_2^+ t_e\Big] < \infty \tag{1.7}$$

and (1.1) *holds. Then there is a constant $C > 0$ such that*

$$\mathbf{Var}(T_n) \leqslant C\frac{n}{\log n}. \tag{1.8}$$

Most of the remainder of this article will be devoted to giving a sketch of the proof of Theorem 1.3. Under stronger assumptions on the moments, this bound on the variance can be supplemented with concentration results:

THEOREM 1.4 (Damron, Hanson, Sosoe [**10**]). *Under the same assumptions as Theorem* 1.3, *there exist constants $C, c > 0$, such that for every $\lambda \geqslant 0$:*

$$\mathbf{P}(T_n - \mathbf{E}T_n \leqslant -\lambda\sqrt{n/\log n}) \leqslant Ce^{-c\lambda}. \tag{1.9}$$

If $\mathbf{E}e^{\alpha t_e} < \infty$, then

$$\mathbf{P}(|T_n - \mathbf{E}T_n| \geqslant \lambda\sqrt{n/\log n}) \leqslant Ce^{-c\lambda}. \tag{1.10}$$

We discuss concentration in the final section.

The upper and lower bounds in (1.6) and (1.8) imply

$$0 \leqslant \chi(d) \leqslant 1/2,$$

in all dimensions, which is quite far from the values found based on arguments from statistical physics and simulations. We close this section by mentioning a remarkable result due to Kurt Johansson in an exactly solvable model of last passage percolation which confirms the predictions of physicists in dimension 2.

In the model considered by Johansson, the weights t_v are placed at the vertices of the square $([0,n]\times[0,n])\cap\mathbb{Z}^2$ in the first quadrant. Allowable paths from $(0,0)$ to (n,n) are constrained to have non-decreasing coordinates, and the *last passage time*

$$T_n = \max_{\pi:(0,0)\to(n,n)} \sum_{v\in\pi} t_v, \tag{1.11}$$

where the maximum is taken over paths

$$\pi = \{x_0 = (0,0),\dots,x_n = (n,n)\},$$

such that $x_{i+1} = x_i + (1,0)$ or $x_{i+1} = x_i + (0,1)$.

THEOREM 1.5 (Johansson [**16**]). *Let A be an $n\times n$ matrix with complex Gaussian entries of mean zero and variance 1. Let λ_n be the largest eigenvalue of AA^*. If the edge weights have geometric distribution with mean* 1*,*

$$\mathbf{P}(T_n \leqslant t) = \mathbf{P}(\lambda_n \leqslant t), \tag{1.12}$$

for all $t\geqslant 0$.

The moments of the quantity

$$Z_n = \frac{\lambda_n - 4n}{2^{4/3}n^{1/3}}$$

converges to the GUE *Tracy-Widom distribution.* It follows immediately from this result that $\chi = 1/3$.

2. Kesten's linear bound for the variance

In this part of the article, we take a first step towards the derivation of Theorem 1.3, by giving a proof of Kesten's result from [**20**] that the variance is at most linear in any dimension (i.e. $\chi(d)\leqslant 1/2$):

THEOREM 2.1 (Kesten [**20**]). *Assume $\mathbf{E}t_e^2<\infty$ and that* (1.1) *holds. Then*

$$\mathbf{Var}(T_n) \leqslant Cn. \tag{2.1}$$

Here, as later in this article, we use the symbol C to denote a constant independent of n, whose value may change between occurrences. In [**20**], Theorem 2.1 is supplemented by a concentration result on scale n in case μ has exponential moments. We discuss concentration in the final part of the chaper.

Our main goal in this article is to make explicit the ideas behind the proofs of the results. We give proofs when they are deemed instructive, but we do not always provide all the details. The reader can find these in the relevant papers. In particular, in several places below we deal with infinite sums whose convergence can readily be justified by suitable approximations, without giving the details of such justifications.

A common theme in the derivation of the theorem and the refinements leading to (1.8) is the application of standard tools in concentration inequalities to the random variable T_n, viewed as a function of the independent edge weight variables t_e, combined with some simple inputs from the model. We summarize the two key properties of the model we use in the next section. We will see that we use only a limited amount of information about the model. This makes it obvious that the techniques we present here are unlikely to get us close to the values of the exponents predicted by physicists.

For Theorem 2.1, the general concentration tool we use and apply to T_n is the *Efron-Stein inequality*:

LEMMA 2.2. *Let $f : \mathbb{R}^{\mathbb{N}} \to \mathbb{R}$ be a measurable function, and $X_1, X_2, \ldots$ be a collection of independent random variable. Let also X_i', $i \geqslant 0$, be an independent copy of X_i. Assume the random variable $f(X_1, X_2, \ldots)$ has finite second moment. Then*

$$\mathbf{Var}(f(X_1, X_2, \ldots)) \leqslant \frac{1}{2}\sum_{i=1}^{\infty} \mathbf{E}\Big[f(X_1, \ldots, X_i, \ldots) - f(X_1, \ldots, X_i', \ldots)\Big]^2. \tag{2.2}$$

PROOF. Since there is no risk of confusion, we will denote the random variable

$$f(X_1, X_2, \ldots)$$

simply by f. We define a sequence of σ-algebras by

$$\begin{aligned} \mathcal{F}_0 &= \varnothing \\ \mathcal{F}_i &= \sigma(X_1, \ldots, X_i), \end{aligned}$$

for $i \geqslant 0$. Consider the martingale increments

$$\Delta_i = \mathbf{E}[f \mid \mathcal{F}_i] - \mathbf{E}[f \mid \mathcal{F}_{i-1}].$$

Then, by orthogonality of the Δ_i, we have

$$\mathbf{E}[\mathbf{E}[f \mid \mathcal{F}_n]^2] - \mathbf{E}[f]^2 = \mathbf{E}[\mathbf{E}[f \mid \mathcal{F}_n]^2] - \mathbf{E}[f \mid \mathcal{F}_0]^2 = \sum_{i=1}^{n} \mathbf{E}[\Delta_i^2]. \tag{2.3}$$

Denote by $\mathbf{E}_i$ integration over X_i:

$$\mathbf{E}_i f = \int f(X_1, \ldots, x_i, \ldots)\, \mathbf{P}_{X_i}(\mathrm{d}x_i).$$

Here we have denoted by $\mathbf{P}_{X_i}$ the distribution of X_i on $\mathbb{R}$. Notice that

$$\Delta_i = \mathbf{E}[f \mid \mathcal{F}_i] - \mathbf{E}[f \mid \mathcal{F}_{i-1}] = \mathbf{E}[f - \mathbf{E}_i[f] \mid \mathcal{F}_i].$$

By Jensen's inequality, we have

$$\mathbf{E}\Delta_i^2 = \mathbf{E}[|\mathbf{E}[f - \mathbf{E}_i[f] \mid \mathcal{F}_i]|^2] \leqslant \mathbf{E}[|f - \mathbf{E}_i f|^2] = \mathbf{E}\mathbf{Var}_i(f).$$

where

$$\mathbf{Var}_i(f) = \int f^2(X_1, \ldots, x_i, \ldots)\, \mathbf{P}_{X_i}(\mathrm{d}x_i) - \left(\int f(X_1, \ldots, x_i, \ldots)\, \mathbf{P}_{X_i}(\mathrm{d}x_i)\right)^2.$$

We can rewrite the latter quantity as

$$\mathbf{Var}_i(f) = \frac{1}{2}\int\int (f(X_1, \ldots, x_i, \ldots) - f(X_1, \ldots, x_i', \ldots))^2\, \mathbf{P}_{X_i}(\mathrm{d}x_i)\mathbf{P}_{X_i'}(\mathrm{d}x_i').$$

Returning to (2.3), this gives

$$\begin{aligned} \mathbf{E}[\mathbf{E}[f \mid \mathcal{F}_n]^2] - \mathbf{E}[f]^2 &\leqslant \frac{1}{2}\sum_{i=1}^{n} \mathbf{E}\Big[f(X_1, \ldots, X_i, \ldots) - f(X_1, \ldots, X_i', \ldots)\Big]^2 \\ &\leqslant \frac{1}{2}\sum_{i=1}^{\infty} \mathbf{E}\Big[f(X_1, \ldots, X_i, \ldots) - f(X_1, \ldots, X_i', \ldots)\Big]^2. \end{aligned}$$

By the assumption of finite second moment, the Doob martingale $\mathbf{E}[f \mid \mathcal{F}_n]$ converges to f in L^2, and we obtain the result. □

2.1. Two ingredients from FPP. The two pieces of information about first passage percolation we use to pass from Lemma 2.2 to Theorem 2.1 are

(1) A linearization of the passage time. That is, we need to know how the random variable T_n changes when we modify the value of a single edge, conditional on all the others. This means an estimate for

$$T_n(t, (t_{e'})_{e' \neq e}) - T_n(s, (t_{e'})_{e' \neq e}).$$

(2) An estimate showing that *geodesic paths* in first-passage percolation have typical size of order n. Geodesic paths are minimizing paths γ in (1.2). Under our assumption 1.1 [**1, 19**], such minimizing paths exist almost surely, although they may not be unique if μ has atoms.

To state more precisely the results we will need, define G_n to be the intersection of all geodesics from the origin of $\mathbb{Z}^d$ to $n\mathbf{e}_1$. Then, concerning the point 2. above, we have

LEMMA 2.3 (Kesten). *There is a constant C such that, for all n*

$$\mathbf{E}[\#G_n^2] \leqslant C\mathbf{E}T_n^2, \tag{2.4}$$

where $\#G_n$ denotes the cardinality of G_n.

To understand why this implies that geodesics are one dimensional, note that by simply considering the path

$$(0, \dots, 0), \mathbf{e}_1, 2\mathbf{e}_1, \dots, n\mathbf{e}_1,$$

we have

$$\mathbf{E}T_n^2 \leqslant n\mathbf{E}t_e^2 + n(n-1)(\mathbf{E}t_e)^2 \leqslant Cn^2.$$

With (2.4), this gives

$$\mathbf{E}\#G_n \leqslant (\mathbf{E}(\#G_n)^2)^{1/2} \leqslant Cn. \tag{2.5}$$

Note that (2.4) is easily seen to be true if the edge weight distribution is supported on $[a, \infty)$ for some $a > 0$. Indeed, in that case we always have

$$\#G_n \leqslant \sum_{e \in \gamma} t_e \leqslant \frac{1}{a} T_n,$$

where γ is any minimizing path. This simple argument in the case when the distribution is supported on values bounded below suggests that to prove Lemma 2.3, we must control the proportion of very small edge weights along geodesics. That is the purpose of the following result.

LEMMA 2.4 (Kesten). *For a lattice path γ, we define*

$$T(\gamma) = \sum_{e \in \gamma} t_e.$$

There exist a, c such that for all $n \geqslant 0$,

$$\mathbf{P}(\exists \text{ self-avoiding } \gamma \text{ containing } 0 \text{ with } \# \geqslant n \text{ but } T(\gamma) < an) \leqslant e^{-cn}.$$

Given the lemma, we obtain Lemma 2.3 by considering

$$Z_n = (\#G_n)\mathbf{1}_{T_n < a\#G_n}.$$

Using Lemma 2.4, we have

$$\mathbf{P}(Z_n \geqslant n) \leqslant e^{-Cn},$$

so

$$\mathbf{E}[\#G_n^2] \leqslant a^{-2}\mathbf{E}T_n^2 + \mathbf{E}Z_n^2 \leqslant C\mathbf{E}T_n^2.$$

For the first "ingredient" above, we have the following concerning the effect of changing one edge weight:

LEMMA 2.5. *If $t \geqslant s$, then*

$$0 \leqslant T_n(t, (t_{e'})_{e' \neq e}) - T_n(s, (t_{e'})_{e' \neq e}) \leqslant (t - s)\mathbf{1}_{e \in G_n}(s, (t_{e'})_{e' \neq e}),$$

where the indicator function is 1 on the event if the edge e is in the intersection of all geodesics from 0 to $n\mathbf{e}_1$ in the configuration where the edge e has weight s.

PROOF. Unless the edge e is in the intersection of all geodesic paths when $t_e = s$, raising the weight of that edge has no effect on the passage time. If $e \in G_n$ for $t_e = s$, then the passage time changes at most linearly in t_e. □

2.2. Finishing the proof of Theorem 2.1. Applying Lemma 2.5 with $f = T_n$ and $X_i = t_{e_i}$ for some enumeration

$$e_1, e_2, \ldots$$

of the edges, we see that the quantity to control is

$$\sum_{e \in \mathcal{E}(\mathbb{Z}^d)} \mathbf{E}\Big[T_n(t_e, (t_f)_{f \neq e}) - T_n(t'_e, (t_f)_{f \neq e})\Big]^2.$$

To simplify notation, we represent the vector $(t_f)_{f \neq e}$ by t_{e^c}. Since t_e and $t_{e'}$ have the same distribution, we have

$$\mathbf{E}[T_n(t_e, t_{e^c}) - T_n(t'_e, t_{e^c})]^2 = 2\mathbf{E}\Big[(T_n(t_e, t_{e^c}) - T_n(t'_e, t_{e^c}))^2 \mathbf{1}_{t'_e > t_e}\Big].$$

The insertion of the indicator function $\mathbf{1}_{t_e > t'_e}$ allows us to apply Lemma 2.5 to bound

$$\begin{aligned} \mathbf{E}[(T_n(t_e, t_{e^c}) - T_n(t'_e, t_{e^c}))^2 \mathbf{1}_{t_e > t'_e}] &\leqslant \mathbf{E}[(t'_e - t_e)_+^2 \mathbf{1}_{e \in G_n(t_e, t_e^c)}] \\ &\leqslant \mathbf{E}[(t'_e)^2 \mathbf{1}_{e \in G_n(t_e, t_e^c)}] \\ &= \mathbf{E}(t'_e)^2 \mathbf{P}(e \in G_n). \end{aligned} \tag{2.6}$$

In the final step, we used independence between t'_e and (t_e, t_e^c). Since $\mathbf{E}t_e^2 < \infty$, we have now shown that

$$\mathbf{Var}(T_n) \leqslant C \sum_{e \in \mathcal{E}(\mathbb{Z}^d)} \mathbf{P}(e \in G_n) = C\mathbf{E}[\#G_n].$$

We can now use our second ingredient, the linear bound for the size of G_n, to conclude:

$$\mathbf{Var}(T_n) \leqslant C\mathbf{E}\#G_n \leqslant Cn. \tag{2.7}$$

2.3. Summary. Let us summarize the steps that led us to Theorem 2.1. The improvement (1.8) will be obtained by following the same strategy, replacing each step with a slight refinement.

(1) We used a martingale decomposition to isolate the contribution of the individual edges. (The Efron-Stein inequality, Lemma 2.2).
(2) Using the linearization in Lemma 2.5 ("Ingredient 1"), we find that the variance $\mathbf{Var}(T_n)$ is bounded by a sum, where each summand has the form $\mathbf{E}[X\mathbf{1}_{e\in G_n}]$, where X can be decorrelated from the indicator.
(3) The sum over e of $\mathbf{P}(e \in G_n)$ contributes Cn ("Ingredient 2").

3. The Benjamini-Kalai-Schramm strategy

In their paper [**4**], Benjamini, Kalai and Schramm (BKS) showed how to obtain the sublinear bound (1.8) for an edge weight distribution of Bernoulli type

$$\mu = \frac{1}{2}\delta_a + \frac{1}{2}\delta_b, \tag{3.1}$$

where $0 < a < b$.

Theorem 3.1 (Benjamini, Kalai, Schramm). *If μ has the form (3.1), then there is a constant $C > 0$ such that*

$$\mathbf{Var}(T_n) \leqslant C\frac{n}{\log n}. \tag{3.2}$$

Benjamini-Kalai-Schramm obtained this result by replacing the Efron-Stein inequality 2.2 by the following result of M. Talagrand [**25**]:

Theorem 3.2. *Consider a function $f : \{-1,1\}^n \to \mathbb{R}$. Let $X = (X_1,\dots,X_n)$ be uniformly distributed over the hypercube $\{-1,1\}^n$. Let*

$$X^i = (X_1,\dots,-X_i,\dots X_n).$$

Then, for some constant $C > 0$:

$$\mathbf{Var}(f) \leqslant C\sum_{i=1}^n \frac{\mathbf{E}[f(X)-f(X^i)]^2}{1+\log\frac{\|f(X)-f(X^i)\|_2}{\|f(X)-f(X^i)\|_1}}. \tag{3.3}$$

The L^1 and L^2 norms in (3.3) are taken with respect to the uniform measure on $\{-1,1\}^n$.

Note that without the logarithm term in the denominator, (3.3) is the same as the right side of the of the Efron-Stein inequality (2.2). Indeed, if $g : \{-1,1\} \to \mathbb{R}$; and x, x' are uniformly distributed on $\{-1,1\}$, then

$$\mathbf{E}|g(x)-g(-x)|^2 = \mathbf{E}|g(-1)-g(1)|^2 = 2\mathbf{E}|g(x)-g(x')|^2. \tag{3.4}$$

We will not give the proof of Theorem 3.2, referring instead to [**4, 25**]. The inequality is specific to the uniform distribution on the hypercube. We derive a related inequality below when we discuss the proof of (1.8) for general edge weights.

3.1. Small influences.

$$\mathbf{E}\#G_n \leqslant Cn.$$

In dimensions $d \geqslant 2$, this means geodesics are a "thin" set compared to the ambient space. Writing $\#G_n = \sum_e \mathbf{1}_{e\in G_n}$ and exchanging $\mathbf{E}$ and the sum over e, (2.5) implies

$$\sum_{e\in\mathcal{E}(\mathbb{Z}^2)} \mathbf{P}(e \in G_n) \leqslant Cn.$$

Because of (2.3), we expect about n^d edges to contribute to the sum on the left. This suggests that, on average, $\mathbf{P}(e \in G_n)$ is of order n^{1-d}: each individual has only a small probability of having any effect on the passage time T_n. In this case, we say the corresponding variable t_e has *small influence* [**12**].

3.2. FPP on the torus. C. Garban and J. Steif introduced a model where it is easy to show that edges have small influences in a strong sense. Instead of $\mathbb{Z}^d$, consider the torus $\mathbb{T}_n^d = (\mathbb{Z}/n\mathbb{Z})^d$, and the passage time

$$T_{\mathbb{T}_n^d} = \inf_{\gamma\in C_m} \sum_{e\in\gamma} t_e, \tag{3.5}$$

where C_m is the set of closed lattice paths in $\mathbb{T}_n^d$ that have winding number one in the first coordinate direction.

The advantage of this model is that, because of the symmetries of the torus, the probability that an edge $e \in \mathbb{T}_n^d$ is in the intersection $G_n(\mathbb{T}_n^d)$ of all minimizing paths in (3.5) is independent of e:

$$\mathbf{P}(e \in G_n(\mathbb{T}_n^d)) = \mathbf{P}(e' \in G_n(\mathbb{T}_n^d)) \tag{3.6}$$

for all $e, e' \in \mathbb{T}_n^d$. On the other hand, if the edge weight distribution μ has the form (3.1), then almost surely,

$$\#G_n(\mathbb{T}_n^d) \leqslant \frac{1}{a} T_{\mathbb{T}_n^d} \leqslant \frac{b}{a} n.$$

Together with (3.6), this gives

$$\begin{aligned} n^d \mathbf{P}(e \in G_n(\mathbb{T}_n^d)) &= \sum_{e'\in\mathcal{E}(\mathbb{T}_n^d)} \mathbf{P}(e' \in G_n(\mathbb{T}_n^d)) \\ &= \mathbf{E}\#G_n(\mathbb{T}_n^d) \\ &\leqslant \frac{b}{a} n. \end{aligned}$$

From this we obtain the influence bound:

$$\mathbf{P}(e \in G_n(\mathbb{T}_n^d)) \leqslant Cn^{-c}, \tag{3.7}$$

where $c > 0$ if $d \geqslant 2$, uniformly in e.

Let us see how (3.7) allows us to apply (3.2) to prove the analog of the Benjamini-Kalai-Schramm result, Theorem 3.1 on $\mathbb{T}_n^d$. We take $f(X) = T_{\mathbb{T}_n^d(X)}$, where

$$t_{e_i} = \frac{X_i - 1}{2} a + \frac{X_i + 1}{2} \cdot b,$$

for some enumeration $e_1, e_2, \ldots, e_{n^d}$ of the edges of $\mathbb{T}_n^d$.

To exploit the small influence bound, note that

$$|T_{\mathbb{T}_n^d}(X) - T_{\mathbb{T}_n^d}(X^i)| \leqslant |T_{\mathbb{T}_n^d}(X) - T_{\mathbb{T}_n^d}(X^i)| \mathbf{1}_{e_i\in G_n(\mathbb{T}_n^d)}, \tag{3.8}$$

because the left side is only non-zero if the edge e_i is in $G_n(\mathbb{T}_n^d)$. By Cauchy-Schwarz, (3.8), (3.7):

$$\begin{aligned}\|T_{\mathbb{T}_n^d}(X) - T_{\mathbb{T}_n^d}(X^i)\|_1 &\leqslant \mathbf{P}(e \in G_n(\mathbb{T}_n^d))^{1/2}\|T_{\mathbb{T}_n^d}(X) - T_{\mathbb{T}_n^d}(X^i)\|_2 \\ &\leqslant Cn^{-c/2}\|T_{\mathbb{T}_n^d}(X) - T_{\mathbb{T}_n^d}(X^i)\|_2.\end{aligned}$$

Thus, for $i = 1, \ldots, n^d$:

$$\frac{\|T_{\mathbb{T}_n^d}(X) - T_{\mathbb{T}_n^d}(X^i)\|_2}{\|T_{\mathbb{T}_n^d}(X) - T_{\mathbb{T}_n^d}(X^i)\|_1} \geqslant Cn^{c/2},$$

for some constant C. Inserting this into (3.3) and using $\mathbf{E}\#G_n(\mathbb{T}_n^d) \leqslant Cn$, we find

$$\begin{aligned}\mathbf{Var}(T_{\mathbb{T}_n^d}) &\leqslant \frac{C}{\log n} \sum_{e \in \mathcal{E}(\mathbb{T}_n^d)} \mathbf{P}(e \in G_n(\mathbb{T}_n^d)) \\ &\leqslant C\frac{n}{\log n}.\end{aligned}$$

which concludes the proof of sublinear variance on the torus.

3.3. Averaging. The difficulty in reproducing the argument in the previous section for the passage time T_n in $\mathbb{Z}^d$ is that no inequality like (3.7) can be true of edges that are close to the origin or $n\mathbf{e}_1$. For instance, if $\mathbf{e}_i = (0, \ldots, 1, \ldots, 0)$, $i = 1, \ldots, d$ are the standard basis vectors, then

$$1 \leqslant \sum_{i=1}^{d} (\mathbf{P}(\mathbf{e}_i \in G_n) + \mathbf{P}(-\mathbf{e}_i \in G_n)),$$

so one of the $2d$ probabilities on the right is bounded below by $1/2d$.

Benjamini-Kalai-Schramm resolved this difficulty by considering an averaged version of the passage time for which the edge variables do have small influence. Here we will use a slightly different average. Let $T(x, y)$ be the passage time between x and y:

$$T(x, y) = \inf_{\gamma: x \to y} \sum_{e \in \gamma} t_e.$$

We let

$$m = \lceil n^{1/4} \rceil, \tag{3.9}$$

and define the averaged passage time

$$F_n = \frac{1}{\#B_m} \sum_{z \in B_m} T(z, z + n\mathbf{e}_1), \tag{3.10}$$

where

$$B_m = \{x \in \mathbb{Z}^d : \|x\|_1 \leqslant m\}.$$

The next approximation result shows that for the purposes of showing (1.8), we can consider F_n:

LEMMA 3.3. *Suppose $\mathbf{E}t_e^2 < \infty$. For $d \geqslant 2$, there is a constant such that for all $n \geqslant 1$,*

$$|\mathbf{Var}(T_n) - \mathbf{Var}(F_n)| \leqslant Cn^{3/4}. \tag{3.11}$$

PROOF. First, write

$$|\mathbf{Var}T(0,n\mathbf{e}_1)-\mathbf{Var}F_n|\leqslant\|T(0,n\mathbf{e}_1)-F_n\|_2(\mathbf{Var}^{1/2}(T_n)+\mathbf{Var}^{1/2}(F_n)).$$

By subadditivity,

$$|T(0,n\mathbf{e}_1)-T(z,z+n\mathbf{e}_1)|\leqslant T(0,z)+T(n\mathbf{e}_1,z+n\mathbf{e}_1),$$

so from (3.10) and Minkowski's inequality,

$$\|F_n-T(0,n\mathbf{e}_1)\|_2\leqslant\frac{2}{\#B_m}\sum_{\|z\|_1\leqslant m}\|T(0,z)\|_2\leqslant 2\max_{\|z\|_1\leqslant m}\|T(0,z)\|_2.$$

Since $\mathbf{E}t_e^2<\infty$ and the passage time from the origin to z can be dominated by a sum of $O(\|z\|_1)$ i.i.d. μ-distributed variables, we have

$$\|T(0,z)\|_2\leqslant C\|z\|_1,$$

for some constant $C>0$, so using (2.6) on $\mathbf{Var}T_n$, $\mathbf{Var}F_n$, we have

$$|\mathbf{Var}T_n-\mathbf{Var}F_n|\leqslant C\|F_n-T(0,x)\|_2\cdot n^{1/2}\leqslant Cn^{3/4},$$

by our choice of m (3.9). □

3.4. Entropy. To control the averaged passage time (3.10), we will use a substitute for Theorem 3.2, due to F. Falik and G. Samorodnitsky [**11**], based on an entropy inequality.

Let $X\in L^1(\mu)$ be a nonnegative random variable. The entropy of X relative to μ is defined by

$$Ent_\mu X=\mathbf{E}\Big[X\log\frac{X}{\mathbf{E}X}\Big]. \tag{3.12}$$

By Jensen's inequality, $Ent_\mu X\geqslant 0$ for any X.

LEMMA 3.4 (Falik, Samorodnitsky). *Let $(\Omega,\mathbf{P})$ be a probability space and $f\in L^2(\Omega,\mathbf{P})$ be a function of independent random variables $X_1,X_2,\dots$. Let*

$$\mathcal{F}_0=\varnothing,\mathcal{F}_i=\sigma(X_1,\dots,X_i),$$

and

$$\Delta_i f=\mathbf{E}[f\mid\mathcal{F}_i]-\mathbf{E}[f\mid\mathcal{F}_{i-1}] \tag{3.13}$$

denote the martingale increments. We have the inequality

$$\mathbf{Var}(f)\cdot\log\frac{\mathbf{Var}(f)}{\sum_{i=1}^\infty(\mathbf{E}|\Delta_i f|)^2}\leqslant\sum_{i=1}^\infty Ent_\mu(\Delta_i f)^2 \tag{3.14}$$

Before giving the proof of Lemma 3.4, we need one more auxiliary result:

PROPOSITION 3.5 (Falik-Samorodnitsky). *If $X\geqslant 0$ almost surely,*

$$Ent_\mu(X^2)\geqslant\mathbf{E}_\mu X^2\log\frac{\mathbf{E}_\mu X^2}{(\mathbf{E}_\mu X)^2}\ . \tag{3.15}$$

PROOF. Since both sides are homogeneous under scaling, we assume $\mathbf{E}f^2=1$, so the inequality to prove is

$$-\log(\mathbf{E}f)^2\leqslant\mathbf{E}f^2\log f^2,$$

or, as $f\geqslant 0$,

$$0\leqslant\mathbf{E}f^2\log(f\mathbf{E}f).$$

Next, use the elementary inequality

$$1 - x \leqslant \log x^{-1}$$

with $x = (f\mathbf{E}f)^{-1}$ to find

$$0 = \mathbf{E}f^2 - 1 = \mathbf{E}\left[f^2(1 - 1/(f\mathbf{E}f)); f > 0\right] \leqslant \mathbf{E}f^2 \log(f\mathbf{E}f). \qquad \square$$

We can now give the proof of (3.14):

PROOF OF LEMMA 3.4. We apply (3.15) to each of the $\Delta_i f$:

$$\begin{aligned}\sum_{i=1}^{\infty} Ent_\mu(\Delta_i f)^2 &\geqslant \sum_{i=1}^{\infty} \mathbf{E}[(\Delta_i f)^2] \log \frac{\mathbf{E}[(\Delta_i f)^2]}{(\mathbf{E}|\Delta_i f|)^2} \\ &= -\mathbf{Var}(f) \sum_{i=1}^{\infty} \frac{\mathbf{E}(\Delta_i f)^2}{\mathbf{Var}(f)} \log \frac{(\mathbf{E}|\Delta_i f|)^2}{\mathbf{E}[(\Delta_i f)^2]}.\end{aligned}$$

Since $f \in L^2$, $\mathbf{Var}(f) = \sum_{i=1}^{\infty} \mathbf{E}[(\Delta_i f)^2]$, we apply Jensen's inequality to the function $-\log x$ to obtain the lower bound

$$-\mathbf{Var}(f) \log \sum_{i=1}^{\infty} \frac{\mathbf{E}[(\Delta_i f)^2]}{\mathbf{Var}(f)} \frac{(\mathbf{E}|\Delta_i f|)^2}{\mathbf{E}[(\Delta_i f)^2]}. \qquad \square$$

3.5. BKS in $\mathbb{Z}^d$. Let us see how (3.14) applied to $f = F_n$ is used to obtain the Benjamini-Kalai-Schramm result in $\mathbb{Z}^d$, and then extended to more general edge weights. The sigma algebra $\mathcal{F}_i$ is $\sigma(t_{e_1}, \dots, t_{e_i})$ for some enumeration of the edges of $\mathbb{Z}^d$. The proof has two essentially distinct parts

(1) Showing that the factor

$$\log \frac{\mathbf{Var}(f)}{\sum_{i=1}^{\infty} (\mathbf{E}|\Delta_i f|)^2}$$

on the left of (3.14) gives a logarithmic improvement. Here we use a more refined version of our "Ingredient 2", the one-dimensionality of geodesics, in the form of Lemma 3.6.

(2) Proving the bound

$$\sum_{i=1}^{\infty} Ent_\mu |\Delta_i F_n|^2 \leqslant Cn. \tag{3.16}$$

For Bernoulli edge weights, or more generally when a *logarithmic Sobolev inequality* is available, this can be simply estimated by $\sum_{e \in \mathcal{E}(\mathbb{Z}^d)} \mathbf{P}(e \in G_n)$. In the case of general edge weights μ, we represent μ-distributed random variables as suitable images of Bernoulli random variables, which leads to a more complicated calculation presented in the next part of this article.

To see 1. above, note that we may assume

$$\mathbf{Var}(f) \geqslant n^{7/8}, \tag{3.17}$$

otherwise there is nothing left to prove.

Then, write as in the proof of Kesten's linear variance bound

$$\begin{aligned}\mathbf{E}|\Delta_i F_n| &\leqslant \frac{2}{\#B_m} \sum_{z \in B_m} \mathbf{E}|\Delta_i T(z, z + n\mathbf{e}_1)| \\ &\leqslant \frac{C}{\#B_m} \sum_{z \in B_m} \mathbf{P}(e_i \in G_n(z, z + n\mathbf{e}_1)),\end{aligned} \tag{3.18}$$

where $G_n(x, y)$ is the intersection of all geodesics from x to y. By translation invariance, the sum in the last step is

$$\begin{aligned}\sum_{z \in B_m} \mathbf{P}(e_i \in G_n(z, z + n\mathbf{e}_1)) &= \sum_{z \in B_m} \mathbf{P}(e_i + z \in G_n(0, n\mathbf{e}_1)) \\ &= \mathbf{E}\#(G_n(0, n\mathbf{e}_1) \cap \{e_i + B_m\}).\end{aligned} \tag{3.19}$$

Since any geodesic is typically a "1-dimensional" set, we would expect the last quantity to be of order $\mathrm{diam} B_m$, which is the content of the following

LEMMA 3.6. *Let $\mathcal{G}$ be the set of all finite self-avoiding geodesics from 0 to $n\mathbf{e}_1$. For any finite set E,*

$$\mathbf{E} \max_{\gamma \in \mathcal{G}} \#(E \cap \gamma) \leqslant C \mathrm{diam} E.$$

We prove Lemma 3.6 at the end of this section. Assuming the lemma for now, an argument similar to (2.6), but assuming only $\mathbf{E} t_e < \infty$, gives

$$\mathbf{E}|\Delta_i F_n| \leqslant \frac{C}{\#B_m} \mathrm{diam} B_m \leqslant C n^{(1-d)/4}. \tag{3.20}$$

Using (3.20), we can bound the sum appearing on the left side in (3.14):

$$\sum_i (\mathbf{E}|\Delta_i F_n|)^2 \leqslant C n^{(1-d)/4} \sum_i \frac{1}{\#B_m} \sum_{z \in B_m} \mathbf{P}(e_i \in G_n(z, z + n\mathbf{e}_1)) \leqslant C n^{(5-d)/4}. \tag{3.21}$$

Here we have used translation invariance and $\sum_i \mathbf{P}(e_i \in G_n) = \mathbf{E}[\#G_n]$. From (3.21), (3.17), we now have, for $d \geqslant 2$,

$$\mathbf{Var}(F_n) \log n \leqslant C \sum_{i=1}^{\infty} Ent_\mu |\Delta_i F_n|^2.$$

Theorem 1.3 would now be proved if we had the estimate (3.16) for general μ.

3.6. Log-Sobolev. When μ is of the form (3.1), the left side of (3.16) can be estimated by the right side of (2.2). This is due to the following discrete *logarithmic Sobolev inequality*, due to Bonami [**5**] and Gross [**13**]:

PROPOSITION 3.7. *Let $f : \{0, 1\} \to \mathbb{R}$ and $\mu = \frac{1}{2}(\delta_0 + \delta_1)$. Then*

$$Ent_\mu f^2 \leqslant (1/2)|f(0) - f(1)|^2. \tag{3.22}$$

The proof of this result is elementary and can be found in [**6**].

At this point we will need the following "tensorization property" of entropy.

PROPOSITION 3.8. *Let f be a non-negative random variable on a product probability space*

$$\left(\prod_{i=1}^{\infty} \Omega_i, \mathcal{F}, \mu = \prod_{i=1}^{\infty} \mu_i \right),$$

where $\mathcal{F} = \bigvee_{i=1}^{\infty} \mathcal{G}_i$ and each triple $(\Omega_i, \mathcal{G}_i, \mu_i)$ is a probability space. Then

$$EntX \leqslant \sum_{i=1}^{\infty} \mathbf{E} Ent_i X \ , \tag{3.23}$$

where $Ent_i f$ is the entropy of $f(\omega) = f(\omega_1, \ldots, \omega_i, \ldots))$ with respect to μ_i, as a function of the i-th coordinate (with all other values fixed).

We refer to [**10**, Theorem 2.3] for a proof using the variational characterization of entropy:

$$Ent f = \sup\{\mathbf{E} fg : \mathbf{E}_\mu e^g \leqslant 1\}. \tag{3.24}$$

Using (3.22), we can finish the derivation of Theorem 3.1. Indeed, we then have, first by Proposition 3.8, and then by (3.22)

$$\begin{aligned}\sum_{i=1}^{\infty} Ent_\mu |\Delta_i F_n|^2 &\leqslant \sum_{i=1}^{\infty}\sum_{k=1}^{\infty} \mathbf{E} Ent_k |\Delta_i F_n|^2 \\ &\leqslant C \sum_{i=1}^{\infty}\sum_{k=1}^{\infty} \mathbf{E}\Big[|\Delta_i F_n(a, t_{e_k^c}) - \Delta_i F_n(b, t_{e_k^c})|^2\Big]\end{aligned} \tag{3.25}$$

Now we use

$$\begin{aligned}&(\Delta_i F_n)(a, t_{e_k^c}) - (\Delta_i F_n)(b, t_{e_k^c}) \\ &= \begin{cases} 0, & i < k \\ \mathbf{E}[F_n(a, t_{e_k^c}) - F_n(b, t_{e_k^c}) \mid \mathcal{F}_i], & i = k \\ \mathbf{E}[F_n(a, t_{e_k^c}) - F_n(b, t_{e_k^c}) \mid \mathcal{F}_i] & \\ \qquad - \mathbf{E}[F_n(a, t_{e_k^c}) - F_n(b, t_{e_k^c}) \mid \mathcal{F}_{i-1}], & i > k \end{cases}\end{aligned} \tag{3.26}$$

and orthogonality of martingale differences to find

$$\begin{aligned}\sum_{i=1}^{\infty}\sum_{k=1}^{\infty} &\mathbf{E}\Big[|\Delta_i F_n(a, t_{e_k^c}) - \Delta_i F_n(b, t_{e_k^c}|^2\Big] \\ &\leqslant \sum_{k=1}^{\infty} \mathbf{E}\Big[|F_n(a, t_{e_k^c}) - F_n(b, t_{e_k^c})|^2\Big].\end{aligned}$$

At this point, we can proceed as in (2.6), to estimate

$$\sum_{k=1}^{\infty} \mathbf{E}\Big[|F_n(a, t_{e_k^c}) - F_n(b, t_{e_k^c})|^2\Big] \leqslant \mathbf{E}[\#G_n(0, n\mathbf{e}1)] \leqslant Cn.$$

This completes our derivation of Theorem 3.1, except for the proof of Lemma 3.6.

Let us note that if instead of (3.1), the edge weight distribution μ is absolutely continuous and satisfies the logarithmic Sobolev inequality

$$Ent_\mu f \leqslant C_{LS} \|f'\|^2_{L^2(\mu)} \tag{3.27}$$

for all smooth $f : \mathbb{R} \to \mathbb{R}$, then the same argument we have given for Bernoulli also gives Theorem 1.3 for μ. Indeed, applying (3.27) in each variable as in (3.25), we find

$$\begin{aligned}\sum_{i=1}^{\infty} Ent_{\mu}|\Delta_i F_n|^2 &\leqslant C \sum_{k=1}^{\infty}\sum_{i=1}^{\infty} \mathbf{E}|\partial_{t_{e_k}}(\Delta_i F_n)|^2. \\ &= C \sum_{e\in\mathcal{E}(\mathbb{Z}^d)}^{\infty} \mathbf{E}|\partial_{t_e} F_n|^2,\end{aligned}$$

where we have used orthogonality of martingale increments and

$$\partial_{t_{e_k}}(\Delta_i F_n) = \begin{cases} 0, & i < k, \\ \mathbf{E}[\partial_{t_{e_k}} F_n \mid \mathcal{F}_i], & i = k, \\ \mathbf{E}[\partial_{t_{e_k}} F_n \mid \mathcal{F}_i] - \mathbf{E}[\partial_{t_{e_k}} F_n \mid \mathcal{F}_{i-1}], & i > k. \end{cases}$$

Then, use

$$\partial_{t_e} T(0,x) = \mathbf{1}_{\{e\in G_n(0,x)\}},$$

which holds Lebesgue almost surely (see Lemma 4.2), and conclude as in (2.7).

The class of distributions for which (3.27) holds includes, for example, any absolutely continuous distribution with bounded density on an interval [**2**, Proposition 5.1.6]. The case of general μ is more delicate, and will be discussed in the next part of this article.

3.7. Geodesics are one-dimensional. Recall from (3.18)-(3.21) that what allowed us to obtain a logarithmic improvement over the linear bound in Kesten's result Theorem 2.1 is the "one-dimensionality" of geodesics. This is expressed by the estimate in Lemma 3.6. This section we show how to obtain this result from Kesten's geodesic length estimate, Lemma 2.4.

PROOF OF LEMMA 3.6. Choose a, c from Theorem 2.4.

If $\#(E\cap\gamma) \geqslant \lambda$ for some $\gamma\in\mathcal{G}$, then we may find the first and last intersections (say y and z respectively) of γ with V, the set of endpoints of edges in E. The portion of γ from y to z is then a geodesic with at least λ edges. This means

$$\mathbf{P}(\#(E\cap\gamma) \geqslant \lambda \text{ for some } \gamma\in\mathcal{G}) \leqslant (\#V)\exp(-c\lambda) + \mathbf{P}\left(\max_{y,z\in V} T(y,z) \geqslant a\lambda\right).$$

Therefore

$$\begin{aligned}\mathbf{E}\max_{\gamma\in\mathcal{G}} \#(E\cap\gamma) \leqslant \operatorname{diam}(E) &+ \sum_{\lambda=\operatorname{diam}(E)}^{\infty} (\#V)\exp(-c\lambda) \\ &+ \sum_{\lambda=\operatorname{diam}(E)}^{\infty} \mathbf{P}\left(\max_{y,z\in V} T(y,z) \geqslant a\lambda\right).\end{aligned}$$

By the inequality $\operatorname{diam}(E) \geqslant C(\#V)^{1/d}$, the middle term is bounded uniformly in E, so we get the upper bound

$$C\operatorname{diam}(E) + \frac{1}{a}\mathbf{E}\max_{y,z\in V} T(y,z).$$

By Lemma 3.9 below, this is bounded by $C\operatorname{diam}(E)$. □

We conclude the proof with the following result, which we used at the end of the previous derivation.

LEMMA 3.9. *If* $\mathbf{E}t_e^2 < \infty$, *there exists* C *such that for all finite subsets* S *of* $\mathbb{Z}^d$,

$$\mathbf{E}\left[\max_{x,y\in S} T(x,y)\right] \leqslant C\text{diam } S\ .$$

PROOF. Given $x, y \in S$, we first obtain an estimate for the tail of the distribution of $T(x,y)$. For this, note that we can build $2d$ disjoint (deterministic) paths from x to y of length $N\|x-y\|_1)$ for some integer N. This means that $\tau(y,z)$ is bounded above by the minimum of $2d$ variables $T_1, \ldots, T_{2d}$, the collection being i.i.d. and each variable distributed as the sum of $N\text{diam}(S)$ i.i.d. variables t_e, so

$$\mathbf{P}(T(x,y) \geqslant \lambda) \leqslant \prod_{i=1}^{2d} \mathbf{P}(T_i \geqslant \lambda) \leqslant \left[\frac{N\text{diam}(S)\mathbf{Var}(t_e)}{(\lambda - N\text{diam}(S)\mathbf{E}t_e)^2}\right]^{2d}.$$

Therefore if we fix some $x_0 \in S$, for $M = 2N\mathbf{E}t_e$,

$$\begin{aligned}\sum_{\lambda=M\text{diam}(S)}^{\infty} &\lambda \max_{y\in S} \mathbf{P}(T(x_0,y) \geqslant \lambda) \\ &\leqslant (2M\text{diam}(S)\mathbf{Var}t_e)^{2d} \sum_{\lambda=M\text{diam}(S)}^{\infty} \lambda^{1-4d} = C(\text{diam } S)^{2-2d}\ .\end{aligned} \tag{3.28}$$

The result now follows by subadditivity,

$$\begin{aligned}\mathbf{E}\left[\max_{x,y\in S} T(x,y)\right] \leqslant 2\mathbf{E}\left[\max_{y\in S} T(x_0,y)\right] &\leqslant 2M\text{diam } S \\ &\quad + 2\#S \sum_{\lambda=M\text{diam}(S)}^{\infty} \max_{y\in S} \mathbf{P}(T(x_0,y) \geqslant \lambda) \\ &\leqslant C\text{diam } S\ ,\end{aligned}$$

where in the second step we have used

$$\mathbf{P}(\max_{y\in S} T(x_0,y) \geqslant \lambda) \leqslant (\#S) \max_{y\in S} \mathbf{P}(T(x_0,y) \geqslant \lambda),$$

and in the final step we have used (3.28). □

4. General edge-weight distributions

To prove Theorem 1.3 for general edge weight distributions we must find an alternative to the log-Sobolev inequality that we used in the case of Bernoulli weights to obtain the bound

$$\sum_{i=1}^{\infty} Ent_i|\Delta_i F_n|^2 \leqslant Cn.$$

The proof rests on two main ideas:

(1) Representing each of the edge weights t_e as the pushforward of a Bernoulli sequence $(\omega_{e,j})_{j\geqslant 1}$. This representation allows to use the discrete log-Sobolev inequality (3.22) to bound the entropy by a double sum over e and j of discrete derivatives corresponding to the two values of $\omega_{e,j}$. After a computation involving a variant of our "Ingredient 1" (see Lemma 2.5), it is found that

$$\sum_{i=1}^{\infty} Ent_i|\Delta_i F_n|^2 \leqslant C\mathbf{E} \sum_{e\in G_n(0,n\mathbf{e}_1)} (1 - \log F(t_e)), \tag{4.1}$$

where F is the distribution function of μ. Note that without the $\log F$ term, the sum on the right would be $O(n)$ by (2.5).

(2) A lattice animals argument to show that under the assumption (1.7), the quantity (4.1) is of order $O(n)$.

Let $\omega_{e,j}$, $e \in \mathcal{E}(\mathbb{Z}^d)$, $j \geqslant 1$ be a collection of independent Bernouli($\frac{1}{2}$)-distributed random variables on a product probability space $(\Omega_B, \pi = \prod_{e,j} \pi_{e,j})$. We denote

$$\omega_B = \left\{\omega_{e,j} : e \in \mathcal{E}^d,\ j \geqslant 1\right\} .$$

For fixed e, denote by ω_e the vector $(\omega_{e,j})_{j\geqslant 1}$, and define the random variable

$$U_e(\omega_e) = \sum_{j=1}^{\infty} \frac{\omega_{e,j}}{2^j}. \tag{4.2}$$

Under π, for each e, $U_e(\omega_e)$ is uniformly distributed on $[0,1]$, and the collection $\{U_e(\omega_e)\}_{e\in\mathcal{E}^d}$ is independent.

We denote by $F(x)$, $x \geqslant 0$ the distribution function of μ, and by I the infimum of the support:

$$I = \inf\{x : F(x) > 0\}.$$

The right-continuous inverse of F is

$$F^{-1}(y) = \inf\{x : F(x) \geqslant y\}.$$

If U is uniformly distributed on $[0,1]$, we have

$$\mathbb{P}(F^{-1}(U) \leqslant x) = \mathbb{P}(U \leqslant F(x)) = F(x), \tag{4.3}$$

that is, $F^{-1}(U)$ has distribution μ. Letting

$$\varphi_e(\omega_e) = F^{-1}(U_e(\omega_e)), \tag{4.4}$$

the distribution of φ_e under π is μ.

We gather the φ_e into a product map $\varphi := \Omega_B \to \Omega = [0,\infty)^{\mathcal{E}^d}$:

$$\varphi(\omega_B) = \left(\varphi_e(\omega_e) : e \in \mathcal{E}^d\right) .$$

By (4.3), $\pi \circ \varphi^{-1} = \mathbb{P}$.

Functions f (in particular, F_n) on the original space Ω can be written as functions on Ω_B, through the map φ. We will estimate discrete derivatives, so for a function $f : \Omega_B \to \mathbb{R}$, set

$$\left(\nabla_{e,j} f\right)(\omega_B) = f(\omega_B^{e,j,+}) - f(\omega_B^{e,j,-}) , \tag{4.5}$$

where $\omega_B^{e,j,+}$ agrees with ω_B except possibly at $\omega_{e,j}$, where it is 1, and $\omega_B^{e,j,-}$ agrees with ω_B except possibly at $\omega_{e,j}$, where it is 0.

LEMMA 4.1. *We have the following inequality:*

$$\sum_{i=1}^{\infty} Ent(\Delta_i F_n^2) \leqslant \sum_{e,j} \mathbf{E}_\pi \left(\nabla_{e,j} F_n \circ \varphi\right)^2 .$$

PROOF. Once the notation is properly set up, the proof is a straightforward application of the log-Sobolev inequality and tensorization of entropy like what we saw in the case of μ Bernoulli.

We let $G = F_n \circ \varphi$, and denote by $\mathcal{G}_k$ the sigma-algebra generated by $\{\omega_{e_r,j}, r \leqslant k, j \in \mathbb{N}\}$. Then we have

$$\mathbf{E}[F_m \mid \mathcal{F}_i](\varphi(\omega_B)) = \mathbf{E}[G \mid \mathcal{G}_i](\omega_B)$$

π-almost surely. Define

$$W_i = \mathbf{E}_\pi[G \mid \mathcal{G}_i] - \mathbf{E}_\pi[G \mid \mathcal{G}_{i-1}]$$

Using Propostion 3.8, we find

$$\sum_{i=1} Ent_\mu(\Delta_i F_n^2) = \sum_{i=1} Ent_\pi(W_i^2) \leqslant \sum_{k=1}^{\infty} \mathbf{E}_\pi \sum_{e,j} Ent_{\pi_{e,j}}(W_i)^2.$$

Using the log-Sobolev inequality (3.22), the last quantity is bounded by

$$\sum_{i=1}^{\infty} \mathbf{E}_\pi \sum_{e,j} Ent_{\pi_{e,j}}(W_i)^2 \leqslant C \sum_{i=1}^{\infty} \sum_{e,j} \mathbf{E}_\pi[(\nabla_{e,j} W_i)^2].$$

Computing $\nabla_{e,j} W_k$ as in (3.26) and using orthogonality of increments, we can sum over k to find

$$C \sum_{e,j} \mathbf{E}_\pi(\nabla_{e,j} G)^2. \tag{4.6}$$

The lemma is proved. □

4.1. The key estimate. The L^2 norm of the discrete derivatives appearing in (4.6) can be expressed in terms of the variables $U(\omega_e)$ in (4.2) as

$$\begin{aligned}
\mathbf{E}_\pi[(\nabla_{e,j} G)^2] &= \mathbf{E}_\pi(F_n(1, (\omega_{e',i})_{(e',i)\neq(e,j)}) - F_n(0, (\omega_{e',i})_{(e',i)\neq(e,j)}))^2 \\
&= \frac{1}{2}\mathbf{E}_\pi[(F_n(\omega_{e,j}, (\omega_{e',i})_{(e',i)\neq(e,j)}) - F_n(0, (\omega_{e',i})_{(e',i)\neq(e,j)}))^2 \mathbf{1}_{\{\omega_{e,j}=1\}}] \\
&\leqslant \frac{1}{2}\mathbf{E}_\pi[(F_n(U_e(\omega_e), (U_{e'})_{e'\neq e}) - F_n(U_e(\omega_e - 2^{-j}), (U_{e'})_{e'\neq e}))^2 \mathbf{1}_{\{U_e \geqslant 2^{-j}\}}],
\end{aligned}$$

with the obvious identifications of F_n with $F_n \circ \varphi$. Recalling the definition of F_n (3.10), the last quantity is bounded by

$$\begin{aligned}
&\mathbf{E}[(F_n(U_e(\omega_e), (U_{e'})_{e'\neq e}) - F_n(U_e(\omega_e - 2^{-j}), (U_{e'})_{e'\neq e}))^2 \mathbf{1}_{\{U_e \geqslant 2^{-j}\}}] \\
&\qquad \leqslant \frac{1}{\#B_m} \sum_{z \in B_m} \mathbf{E}_\pi[(T_z(U_e(\omega_e)) - T_z(U_e(\omega_e) - 2^{-j}))^2 \mathbf{1}_{\{U_e \geqslant 2^{-j}\}}].
\end{aligned} \tag{4.7}$$

For simplicity of notation, we have omitted the dependence of $T_z := T(z, z + n\mathbf{e}_1)$ on variables $U_{e'}$, $e' \neq e$. We have so far shown that to estimate

$$\sum_{i=1} Ent_\mu(\Delta_i F_n^2),$$

it will suffice to estimate

$$\sum_{e \in \mathcal{E}(\mathbb{Z}^d)} \frac{1}{\#B_m} \sum_{z \in B_m} \sum_{j=1}^{\infty} \mathbf{E}_\pi[(T_z(U_e(\omega_e)) - T_z(U_e(\omega_e) - 2^{-j}))_+^2 \mathbf{1}_{\{U_e \geqslant 2^{-j}\}}]. \tag{4.8}$$

We would like to perform the summation over j and obtain a term similar to the summands in (4.1).

To move further, we need an estimate for

$$T_z(U_e(\omega_e)) - T_z(U_e(\omega_e) - 2^{-j}) \tag{4.9}$$

that is summable in j. For this purpose, we introduce a more refined version of "Ingredient 1" (See section 2.1), concerning the effect on T_n of changing one edge weight. Whereas previously Lemma 2.5 was sufficient, we will need the following more precise result.

LEMMA 4.2. *The random variable*

$$D_{z,e_i} = \sup[\{r \geqslant 0 : t_e = r \text{ and } e \text{ is in a geodesic from } z \text{ to } z + z_i\} \cup \{0\}]$$

is almost surely finite. If $0 \leqslant s \leqslant t$,

$$T_z(t, t_{e_i^c}) - T_z(s, t_{e_i^c}) = \min\{t - s, (D_{z,e} - s)_+\}. \tag{4.10}$$

The content of Lemma 4.2 is clear: the difference in passage times $T_z(t) - T_z(s)$ is linear, as long as t is not so large that the edge e_i is no longer in a geodesic. The logic of the proof is similar to the argument in Lemma 2.5, see [**9**]. Two crucial properties of $D_{z,e}$ are

(1) $D_{z,e}$ depends only on edge weights $t_{e'}$, $e' \neq e$, but not on t_e.
(2) $T_z(U_e(\omega_e))$ is constant on $U_e \in [F(D_{z,e}^-), 1]$ (for fixed values of t_{e^c}.

To perform the summation over j in (4.8), we combine Lemma 4.2 with the following lemma, due to R. Rossignol, which summarizes a less transparent computation which appeared in [**9**].

LEMMA 4.3. *Let* $a, \tau \in [0, 1]$. *Suppose* f *is nonnegative, non-decreasing on* $[0, 1]$, *and that* f *is constant on* $[a, 1]$. *If* $\tau \leqslant 1/2$, *then*

$$\int_\tau^1 (f(x) - f(x - \tau))^2 \,\mathrm{d}x \leqslant \int_0^1 f^2(x)\mathbf{1}_{\{x \geqslant 1-\tau\}} \,\mathrm{d}x. \tag{4.11}$$

Moreover:

(1) *If* $a \leqslant \tau \leqslant \frac{1}{2}$,

$$\int_\tau^1 (f(x) - f(x - \tau))^2 \,\mathrm{d}x \leqslant 2a \int_0^1 f^2(x) \,\mathrm{d}x. \tag{4.12}$$

(2) *If* $\tau \leqslant a \leqslant \frac{1}{2}$,

$$\int_\tau^1 (f(x) - f(x - \tau))^2 \,\mathrm{d}x \leqslant 2\tau \int_0^1 f^2(x) \,\mathrm{d}x. \tag{4.13}$$

We defer the proof to section 4.3 below.

With Lemmas 4.2 and 4.3 in hand, we can now obtain (4.1). Fix an edge e. We apply Lemma 4.3 with

$$\begin{aligned} f(x) &= T_z(x) - T_z(0), \\ \tau &= 2^{-j}, \\ a &= F(D_{z,e}^-). \end{aligned}$$

Writing the expectation over U_e explicitly:

$$\begin{aligned} &\mathbf{E}_{\pi_e}[(T_z(U_e(\omega_e)) - T_z(U_e(\omega_e) - 2^{-j}))^2 \mathbf{1}_{\{U_e \geqslant 2^{-j}\}}] \\ &\quad = \int_{2^{-j}}^1 (T_z(x) - T_z(x - 2^{-j}))^2 \,\mathrm{d}x. \end{aligned}$$

For j such that $F(D_{z,e}^-) < 2^{-j}$ $(a < \tau)$, we use (4.12):

$$\int_{2^{-j}}^1 (T_z(x) - T_z(x - 2^{-j}))^2 \,\mathrm{d}x \leqslant 2F(D_{z,e}^-) \int_0^1 (T_z(x) - T_z(0))^2 \,\mathrm{d}x.$$

Recall that $T_z(x)$ also depends on $\omega_{e'}$, $e' \neq e$. By Lemma 4.2, we have

$$0 \leqslant T_z(x) - T_z(0) \leqslant \min\{D_{z,e}, \varphi_e(\omega_e)\},$$

where φ_e was defined in (4.4), so

$$\int_0^1 (T_z(x)-T_z(0))^2\,\mathrm{d}x \leqslant \mathbf{E}_{\pi_e}[\min\{D_{z,e},\varphi_e(\omega_e)\}]^2.$$

For j such that $2^{-j}\leqslant F(D_{z,e}^-)\leqslant \frac{1}{2}$, ($\tau\leqslant a\leqslant \frac{1}{2}$), we use (4.13):

$$\int_{2^{-j}}^1 (T_z(x)-T_z(x-2^{-j}))^2\,\mathrm{d}x \leqslant 2\cdot 2^{-j}\mathbf{E}_{\pi_e}[\min\{D_{z,e},\varphi_e(\omega_e)\}]^2.$$

We now sum over j in (4.8). For $F(D_{z,e}^-)\leqslant\frac{1}{2}$ we find

$$\begin{aligned}\sum_{j=1}^\infty &\mathbf{E}_{\pi_e}\Big[(T_z(U_e(\omega_e)-T_z(U_e(\omega_e)-2^{-j}))^2\mathbf{1}_{\{U_e\geqslant 2^{-j}\}}\Big]\\ &\leqslant \sum_{j:2^{-j}>F(D_{z,e}^-)}\mathbf{E}_{\pi_e}\Big[\min\{D_{z,e},\varphi_e(\omega_e)\}\Big]^2+\sum_{j:2^{-j}\leqslant F(D_{z,e}^-)}2^{1-j}\mathbf{E}_{\pi_e}\Big[\min\{D_{z,e},\varphi_e(\omega_e)\}\Big]^2\\ &\leqslant 2F(D_{z,e}^-)(2-\log_2 F(D_{z,e}^-))\mathbf{E}_{\pi_e}\Big[\min\{D_{z,e},\varphi_e(\omega_e)\}\Big]^2.\end{aligned}$$

If instead $F(D_{z,e}^-)\geqslant\frac{1}{2}$, we use (4.11) in the sum over j such that $2^{-j}\leqslant F(D_{z,e}^-)$:

$$\begin{aligned}\sum_{j:2^{-j}\leqslant F(D_{z,e}^-)}&\int_0^1[(T_z(x)-T_z(x-2^{-j}))^2\mathbf{1}_{\{x\geqslant 2^{-j}\}}\,\mathrm{d}x\\ &\leqslant\int_0^1\min\{D_{z,e},F^{-1}(x)\}^2\sum_{j:2^{-j}\leqslant F(D_{z,e}^-)}\mathbf{1}_{\{1-x\leqslant 2^{-j}\}}(x)\,\mathrm{d}x\\ &\leqslant\int_0^1\min\{D_{z,e},F^{-1}(x)\}^2\log_2\frac{1}{1-x}\,\mathrm{d}x\\ &\leqslant 2F(D_{z,e}^-)\int_0^1\min\{D_{z,e},F^{-1}(x)\}^2\log_2\frac{1}{1-x}\,\mathrm{d}x.\end{aligned}$$

Combining the previous calculations, we find:

$$\begin{aligned}&\sum_{j=1}^\infty\mathbf{E}_\pi[T_z(U_e(\omega_e)-T_z(U_e(\omega_e)-2^{-j})_+^2\mathbf{1}_{U_e\geqslant 2^{-j}\}}]\\ &\quad\leqslant C\mathbf{E}_{\pi_e}F^{-1}(U_e)^2\left(1+\log_2^+\frac{1}{1-U_e}\right)\mathbf{E}_\pi F(D_{z,e}^-)(1-\log_2 F(D_{z,e}^-).\end{aligned}\tag{4.14}$$

Note the use of independence of $D_{z,e}$ from ω_e. To obtain an expression of the form (4.1), we integrate by parts:

$$\begin{aligned}-F(y^-)\log F(y^-)&=-\int\mathbf{1}_{[I,y)}(x)\log F(y^-)\,\mu(\mathrm{d}x)\\ &\leqslant-\int\mathbf{1}_{[I,y)}(x)\log F(x)\,\mu(\mathrm{d}x).\end{aligned}\tag{4.15}$$

Thus

$$F(D_{z,e}^-)(1-\log_2 F(D_{z,e}^-)=\int_I^{D_{z,e}}(1-\log F(t_e)\,\mu(\mathrm{d}t_e).$$

Finally, the quantity

$$\mathbf{E}_{\pi_e}F^{-1}(U_e)^2\left(1+\log_2\frac{1}{1-U_e}\right)$$

can be estimated by

$$2Ent_\mu t_e^2 + 2\mathbf{E}_\mu t_e^2 \mathbf{E}_{\pi_e} \frac{1}{(1-U_e)^{1/2}}$$

using the variational charaterization of entropy, (3.24). By (1.7), this is a finite constant.

By Lemma 4.2, if $t_e < D_{z_e}$, then $e \in G_n(z, z+n\mathbf{e}_1)$, so

$$\int_I^{D_{z,e}} \sum_{e \in \mathcal{E}(\mathbb{Z}^d} (1 - \log F(t_e)\, \mu(\mathrm{d}t_e) \leqslant \sum_{e \in G_n(z,z+n\mathbf{e}_1)} (1 - \log F(t_e))\, \mu(\mathrm{d}t_e)$$

almost surely. Thus (4.14) gives

$$\sum_{j=1}^{\infty} \mathbf{E}_\pi[(T_z(U_e(\omega_e) - T_z(U_e(\omega_e) - 2^{-j}))_+^2 \mathbf{1}_{\{U_e \geqslant 2^{-j}\}}] \tag{4.16}$$

$$\leqslant \mathbf{E}_\pi \sum_{e \in G_n(z,z+n\mathbf{e}_1)} (1 - \log F(t_e))\, \mu(\mathrm{d}t_e). \tag{4.17}$$

Shifting by z, we have now derived the key estimate in [**9**]:

PROPOSITION 4.4. *Define F_n by (3.10) and $\Delta_i F_n$ as in (3.13). Then there is a constant $C > 0$ such that:*

$$\sum_{i=1}^{\infty} Ent_i |\Delta_i F_n|^2 \leqslant C\mathbf{E} \sum_{e \in G_n} (1 - \log F(t_e)).$$

4.2. Estimating the sum (4.1)**.** To complete the proof of Theorem 1.3, we must show

$$\mathbf{E} \sum_{e \in G_n} (1 - \log F(t_e)) \leqslant Cn.$$

The sum in the expectation has the form

$$Y_n := \sum_{e \in G_n} w_e,$$

where the weights w_e are given by

$$w_e = 1 - \log F(t_e).$$

Recall that if F^{-1} is the right-continuous inverse of F and U is uniform on $[0,1]$, then $F^{-1}(U)$ is distributed like t_e :

$$\mathbf{P}(w \geqslant r) = \mathbf{P}(F(F^{-1}(U)) \leqslant e^{1-r}) \leqslant \mathbf{P}(U \leqslant e^{1-r}) = e^{1-r}. \tag{4.18}$$

Thus, the weights w_e have exponential tails. For a lattice path γ, we define

$$N(\gamma) = \sum_{e \in \gamma} N(\gamma),$$

and

$$N_n := \max_{\substack{\gamma : \#\gamma = n \\ 0 \in \gamma}} N_n(\gamma). \tag{4.19}$$

Clearly,

$$\{N_n > \beta n\} \subset \bigcup_{\gamma : \#\gamma = n, 0 \in \gamma} \{N(\gamma) > \beta n\}.$$

On the other hand, the number of *lattice animals* (connected subsets of $\mathbb{Z}^d$) of size $\leqslant n$ containing the origin is bounded by e^{Cn} for some n, so

$$\mathbf{P}(N_n > \beta n) \leqslant e^{Cn}\mathbf{P}(\sum_{i=1}^n w_i > \beta n).$$

Here w_i, $1 \leqslant i \leqslant n$ are i.i.d. random variables with the common distribution of the w_e. By (4.18), it is straightforward to show that

$$\mathbf{P}(N_n \geqslant \beta n) \leqslant e^{-\beta n/4}$$

for large β. From this tail bound, we find

$$\mathbf{E}N_n^4 = 4\int_0^\infty x^3\mathbf{P}(N_n \geqslant x)\,\mathrm{d}x \leqslant (\beta_0 n)^4 + 4\int_{\beta_0 n}^\infty x^3 e^{-x/4}\,\mathrm{d}x \leqslant Cn^4. \tag{4.20}$$

We can now estimate the expectation

$$\begin{aligned}\mathbf{E}Y_n &= \sum_{j=1}^\infty \mathbf{E}Y_n\mathbf{1}_{2^{j-1}n\leqslant \#G_n<2^j n}\\ &\leqslant \sum_{j=1}^\infty (\mathbf{E}N_{2^j n}^4)^{1/4}\mathbf{P}(2^{j-1}n \leqslant \#G_n < 2^j n)^{3/4}\\ &\leqslant Cn\sum_{j=1}^\infty 2^j\mathbf{P}(2^{j-1}n \leqslant \#G_n < 2^j n)^{3/4}.\end{aligned}$$

we have used Hölder's inequality in the second step. Using Hölder (for sums) again on the final sum in j, we find

$$\begin{aligned}&\sum_{j=1}^\infty 2^{-j/2}2^{3j/2}\mathbf{P}(2^{j-1}n \leqslant \#G_n < 2^j n)^{3/4}\\ &\qquad \leqslant C\left(\sum_{j=1}^\infty 2^{2j}\mathbf{P}(2^{j-1}n \leqslant \#G_n < 2^j n)\right)^{3/4}.\end{aligned}$$

By (2.5), the final sum is bounded by a constant, and Theorem 1.3 is proved.

4.3. Proof of Lemma 4.3. Here we give the derivation of Lemma 4.3. Notice the crucial role of positive association.

PROOF. If $\tau \leqslant 1/2$, then

$$\int_\tau^1 (f(x) - f(x-\tau))^2\,\mathrm{d}x \leqslant \int_\tau^1 (f^2(x) - f^2(x-\tau))\,\mathrm{d}x$$

where we have used $f \geqslant 0$. Next,

$$\begin{aligned}\int_\tau^1 f^2(x) - f^2(x-\tau)\,\mathrm{d}x &= \int_\tau^1 f^2(x)\,\mathrm{d}x - \int_0^{1-\tau} f^2(x)\,\mathrm{d}x\\ &\leqslant \int_{1-\tau}^1 f^2(x)\,\mathrm{d}x\\ &= \int_0^1 f^2(x)\mathbf{1}_{\{x\geqslant 1-\tau\}}\,\mathrm{d}x.\end{aligned}$$

We have used the non-negativity of $f^2(x)$ to drop the integral over $[0,\tau]$. This shows (4.11).

If $a \leqslant \tau \leqslant \frac{1}{2}$, since f is constant over $[a,1]$,

$$\int_\tau^1 (f(x)-f(x-\tau))^2\,\mathrm{d}x = \int_\tau^{a+\tau} (f(x)-f(x-\tau))^2\,\mathrm{d}x.$$

By monotonicity, the right side is no bigger than

$$\int_\tau^{a+\tau} f^2(x)\,\mathrm{d}x = \int_\tau^1 f^2(x)\mathbf{1}_{\{x\leqslant a+\tau\}}\,\mathrm{d}x.$$

The Chebyshev association inequality [**6**, Theorem 2.14] now gives (4.12):

$$\begin{aligned}\int_\tau^1 f^2(x)\mathbf{1}_{\{x\leqslant a+\tau\}}\,\mathrm{d}x &\leqslant \frac{1}{1-\tau}\int_\tau^1 f^2(x)\,\mathrm{d}x\int_\tau^1 \mathbf{1}_{\{x\leqslant a+\tau\}}\,\mathrm{d}x\\ &= \frac{a}{1-\tau}\int_\tau^1 f^2(x)\,\mathrm{d}x\\ &\leqslant 2a\int_0^1 f^2(x)\,\mathrm{d}x.\end{aligned}$$

When $\tau \leqslant a \leqslant \frac{1}{2}$, we again have

$$\int_\tau^1 (f(x)-f(x-\tau))^2\,\mathrm{d}x = \int_\tau^{a+\tau} (f(x)-f(x-\tau))^2\,\mathrm{d}x.$$

Expanding the square and using monotonicity as in the case $\tau \geqslant 1/2$, we find

$$\begin{aligned}\int_\tau^{a+\tau} (f(x)-f(x-\tau))^2\,\mathrm{d}x &\leqslant \int_\tau^{a+\tau} f^2(x)\,\mathrm{d}x - \int_0^a f^2(x)\,\mathrm{d}x\\ &= \int_a^{a+\tau} f^2(x)\,\mathrm{d}x - \int_0^\tau f^2(x)\,\mathrm{d}x\\ &\leqslant \int_a^1 f^2(x)\mathbf{1}_{\{x\leqslant a+\tau\}}\,\mathrm{d}x\\ &\leqslant \frac{\tau}{1-a}\int_0^1 f^2(x)\,\mathrm{d}x\\ &\leqslant 2\tau\int_0^1 f^2(x)\,\mathrm{d}x.\end{aligned}$$

In the second-to-last step, we have used Chebyshev's association inequality. □

5. Concentration

We remark briefly on how to derive the concentration result 1.4. The proof is relies on the following (see [**6**]).

PROPOSITION 5.1. *Suppose $Z \geqslant 0$ is a random variable such that there exist constants $0 < C \leqslant B$ such that*

$$\mathbf{Var}e^{tZ/2} \leqslant Ct^2\mathbf{E}e^{tZ} < \infty$$

for $t \in (0, B^{-1/2})$, then

$$\psi_Z(t) := \mathbf{E}[e^{tZ}] \leqslant -2\log(1-Ct^2), \quad t \in (0, B^{-1/2}).$$

In particular, we have exponential concentration:

$$\mathbf{P}(Z \geqslant \lambda) \leqslant e^{-t\lambda}\mathbf{E}e^{tZ} \leqslant e^{-t\lambda}\frac{1}{(1-Ct^2)^2}.$$

Proposition 5.1 reduces Theorem 1.4 to showing

$$\mathbf{Var}(e^{\lambda F_n/2}) \leqslant K\lambda^2\mathbf{E}e^{\lambda F_n}, \quad |\lambda| < \frac{1}{2\sqrt{K}}. \tag{5.1}$$

with $K = \frac{Cn}{\log n}$. The proof of (5.1) follows a similar strategy to the bound for $\mathbf{Var}F_n$ presented in the previous part of this article, starting from inequality (3.14) with $f = e^{\lambda F_n}$. The derivation of the key estimate for the discrete derivatives is somewhat more involved. We refer to [**10**] for details.

Finally, we note the following result, which improves the exponential concentration result of Kesten [**20**] to Gaussian concentration. It was first derived by M. Talagrand. An alternate proof appears in [**10**].

THEOREM 5.2. *Let $d \geqslant 2$. Assuming* (1.1) *and $\mathbf{E}e^{\alpha t_e} < \infty$ for some $\alpha > 0$, there exist $c, C > 0$ such that*

$$\mathbf{P}(T_n - \mathbf{E}T_n \geqslant t\sqrt{n}) \leqslant e^{-ct^2} \quad t \in (0, C\sqrt{n}).$$

If $\mathbf{E}Y^2 < \infty$, where Y is the minimum of $2d$ i.i.d. copies of t_e, then also

$$\mathbf{P}(T_n - \mathbf{E}T_n \leqslant -t\sqrt{n}) \leqslant e^{-ct^2},$$

for all $t \geqslant 0$.

References

[1] Antonio Auffinger, Michael Damron, and Jack Hanson, *50 years of first-passage percolation*, University Lecture Series, vol. 68, American Mathematical Society, Providence, RI, 2017. MR3729447

[2] Dominique Bakry, Ivan Gentil, and Michel Ledoux, *Analysis and geometry of Markov diffusion operators*, Grundlehren der Mathematischen Wissenschaften [Fundamental Principles of Mathematical Sciences], vol. 348, Springer, Cham, 2014. MR3155209

[3] Michel Benaïm and Raphaël Rossignol, *Exponential concentration for first passage percolation through modified Poincaré inequalities* (English, with English and French summaries), Ann. Inst. Henri Poincaré Probab. Stat. **44** (2008), no. 3, 544–573, DOI 10.1214/07-AIHP124. MR2451057

[4] Itai Benjamini, Gil Kalai, and Oded Schramm, *First passage percolation has sublinear distance variance*, Ann. Probab. **31** (2003), no. 4, 1970–1978, DOI 10.1214/aop/1068646373. MR2016607

[5] Aline Bonami, *Étude des coefficients de Fourier des fonctions de $L^p(G)$* (French, with English summary), Ann. Inst. Fourier (Grenoble) **20** (1970), no. fasc. 2, 335–402 (1971). MR0283496

[6] Stéphane Boucheron, Gábor Lugosi, and Pascal Massart, *Concentration inequalities*, Oxford University Press, Oxford, 2013. A nonasymptotic theory of independence; With a foreword by Michel Ledoux. MR3185193

[7] Ivan Corwin, *Exactly solving the KPZ equation*, Proc. Sympos. Appl. Math. **75** (2018).

[8] Michael Damron, *Random growth models: shape and convergence rate*, Proc. Sympos. Appl. Math. **75** (2018).

[9] Michael Damron, Jack Hanson, and Philippe Sosoe, *Sublinear variance in first-passage percolation for general distributions*, Probab. Theory Related Fields **163** (2015), no. 1-2, 223–258, DOI 10.1007/s00440-014-0591-7. MR3405617

[10] Michael Damron, Jack Hanson, and Philippe Sosoe, *Subdiffusive concentration in first-passage percolation*, Electron. J. Probab. **19** (2014), no. 109, 27, DOI 10.1214/EJP.v19-3680. MR3286463

[11] Dvir Falik and Alex Samorodnitsky, *Edge-isoperimetric inequalities and influences*, Combin. Probab. Comput. **16** (2007), no. 5, 693–712, DOI 10.1017/S0963548306008340. MR2346808

[12] Christophe Garban and J Steif, *Lectures on noise sensitivity and percolation.*

[13] Leonard Gross, *Logarithmic Sobolev inequalities*, Amer. J. Math. **97** (1975), no. 4, 1061–1083, DOI 10.2307/2373688. MR0420249

[14] David A. Huse and Christopher L. Henley, *Pinning and roughening of domain walls in ising systems due to random impurities*, Phys. Rev. Lett. **54** (1985), 2708–2711.

[15] David A. Huse, Christopher L. Henley, and Daniel S. Fisher, *Huse, henley, and fisher respond*, Phys. Rev. Lett. **55** (1985), 2924–2924.

[16] Kurt Johansson, *Shape fluctuations and random matrices*, Comm. Math. Phys. **209** (2000), no. 2, 437–476, DOI 10.1007/s002200050027. MR1737991

[17] Mehran Kardar, Giorgio Parisi, and Yi-Cheng Zhang, *Dynamic scaling of growing interfaces*, Phys. Rev. Lett. **56** (1986), 889–892.

[18] Mehran Kardar and Yi-Cheng Zhang, *Scaling of directed polymers in random media*, Phys. Rev. Lett. **58** (1987), 2087–2090.

[19] Harry Kesten, *Aspects of first passage percolation*, École d'Été de Probabilités de Saint Flour XIV-1984.

[20] Harry Kesten, *On the speed of convergence in first-passage percolation*, Ann. Appl. Probab. **3** (1993), no. 2, 296–338. MR1221154

[21] J. M. Kim and J. M. Kosterlitz, *Kim and kosterlitz reply*, Phys. Rev. Lett. **63** (1989), 1192–1192.

[22] Sang-Woo Kim and Jin Min Kim, *A restricted solid-on-solid model in higher dimensions*, J. Stat. Mech. Theory Exp. **7** (2014), P07005, 9. MR3249152

[23] Charles M. Newman and Marcelo S. T. Piza, *Divergence of shape fluctuations in two dimensions*, Ann. Probab. **23** (1995), no. 3, 977–1005. MR1349159

[24] Timo Seppäläinen, *The corner growth model with exponential weights*, Proc. Sympos. Appl. Math. **75** (2018).

[25] Michel Talagrand, *On Russo's approximate zero-one law*, Ann. Probab. **22** (1994), no. 3, 1576–1587. MR1303654

Center for Mathematical Sciences and Applications, Harvard University, 20 Garden Street, Cambridge, Massachusetts 01451

Current address: Cornell University, 584 Malott Hall, Ithaca, New York 14853

Email address: psosoe@math.cornell.edu

Proceedings of Symposia in Applied Mathematics
Volume **75**, 2018
https://doi.org/10.1090/psapm/075/00659

Busemann functions, geodesics, and the competition interface for directed last-passage percolation

Firas Rassoul-Agha

ABSTRACT. In this survey article we consider the directed last-passage percolation model on the planar square lattice with nearest-neighbor steps and general i.i.d. weights on the vertices, outside of the class of exactly solvable models. We show how stationary cocycles are constructed from queueing fixed points and how these cocycles characterize the limit shape, yield existence of Busemann functions in directions where the shape has some regularity, describe the direction of the competition interface, and answer questions on existence, uniqueness, and coalescence of directional semi-infinite geodesics, and on nonexistence of doubly infinite geodesics.

CONTENTS

1. Introduction

In 1965, Hammersley and Welsh [**33**] introduced a model of a fluid flowing through a porous medium that they called first-passage percolation (FPP). Roughly speaking, the model consists of putting random positive numbers (called weights) on the nearest-neighbor edges of the square lattice $\mathbb{Z}^d$. These numbers describe the time it takes to traverse the various edges. Only nearest-neighbor paths are allowed between lattice points. The (random) distance between two points is then given by

2010 *Mathematics Subject Classification.* 60K35, 60K37, 82B43.

The author was partially supported by NSF grant DMS-1407574.

the passage time – the shortest time it takes to go between the points. A path that realizes the passage time is naturally called a geodesic. Immediate questions arise regarding the structure of the random metric, its balls, and its geodesics (finite, semi-infinite, and bi-infinite).

As it is often the case in probability theory, structure emerges from randomness when we look at the large scale behavior of the model. This means scaling the lattice down as we look at farther and farther sites. When done properly, the random metric on the scaled lattice converges to a deterministic one on $\mathbb{R}^d$. As a result, random balls approach deterministic convex sets as their radii grow and geodesics from the origin to far away points approach straight lines. This raises the next layer of questions regarding the size and statistics of the fluctuations of the random objects from their deterministic limits. See Section 1 of [**18**] for more details and precise formulations.

In a related model, called directed last-passage percolation or just last-passage percolation (LPP) paths are only allowed to take steps in $\{e_1, \dots, e_d\}$. One can think of $e_1 + \cdots + e_d$ as a time direction that cannot be reversed. In fact, in LPP passage time is maximized rather than minimized and the random weights the path collects are usually put on the vertices instead of on the edges, but these are only minor differences from FPP. The crucial difference is that paths are now directed. This creates some simplifications, but also introduces some complications. For example, in LPP it is clear that there exist maximizing paths (still called geodesics by analogy with LPP) from the origin to any point in the first quadrant. There are after all finitely many such paths, as opposed to the situation in FPP. On the other hand, the passage time is no longer symmetric in LPP: there is no path back to the origin, starting from a point in the first quadrant.

Nevertheless, LPP and FPP are expected to a have similar qualitative behavior at large scales. One of the advantages the two-dimensional LPP has over its cousin FPP is a connection to other models, such as queues and particle systems. Furthermore, with certain specific weight distributions, LPP becomes what we call an exactly solvable model. This means that many computations can be carried out all the way down to exact formulas. In this case, one is able to extract very detailed information about the behavior of the model, which then drives or confirms predictions about general LPP and FPP (and related) models.

The focus of this article is on surveying recent developments in connection with the structure of geodesics in LPP. A main tool will be Busemann functions, originally used to study the large-scale geometry of geodesics in Riemannian manifolds.

The next section collects some notation for easier reference. Section 3 introduces the last-passage percolation model and Section 4 discusses its connections to the corner growth model, queues in tandem, a system of interacting particles, and a competition model. In Section 5 we describe the scaling limit of the passage time. Section 6 introduces the Busemann functions and studies their properties. To prove existence of these functions we borrow tools from queuing theory. This is done in Section 7. Busemann functions are then used in Section 8 to construct semi-infinite geodesics and describe their structure. They are also used to study the competition interface, in Section 9. In the last two sections we give a brief account of the recent history of Busemann functions and geodesics in percolation and a preview of the topic of the next two lecture notes [**16, 55**]: fluctuations.

2. Notation

$\mathbb{Z}$ denotes the integer numbers, $\mathbb{Z}_+$ the nonnegative integers, and $\mathbb{N}$ the positive integers. $\mathbb{R}$ denotes the real numbers and $\mathbb{R}_+$ the nonnegative real numbers. $\mathbb{Q}$ denotes the rational numbers. e_1 and e_2 are the canonical basis vectors of $\mathbb{R}^2$. $\mathcal{U} = \{(t, 1-t) = te_1 + (1-t)e_2 : 0 \leqslant t \leqslant 1\}$ and its relative interior is $\mathcal{U}^\circ = \{(t, 1-t) : 0 < t < 1\}$. We will write ξ, ζ, η for elements in $\mathcal{U}$. $x_{i,j} = (x_i, \dots, x_j)$, for $-\infty \leqslant i < j \leqslant \infty$. $x \cdot y$ denotes the scalar product of vectors $x, y \in \mathbb{R}^2$. $|x|_1 = |x_1| + |x_2| = |x \cdot e_1| + |x \cdot e_2|$. $\lfloor a \rfloor$ is the largest integer $\leqslant a$. If a is a vector, then $\lfloor a \rfloor$ is taken coordinatewise. For a real number a we write a^+ for $\max(a, 0)$ and a^- for $-\min(a, 0)$. Thus, $a = a^+ - a^-$. For $x, y \in \mathbb{R}^2$, $y \geqslant x$ and $y \leqslant x$ mean the inequalities hold coordinatewise. We will say that a sequence x_n is asymptotically directed into a subset $A \subset \mathcal{U}$ if all the limit points of x_n/n are inside A A random variable X has an exponential distribution with rate $\theta > 0$ if $P(X > s) = e^{-\theta s}$ for all $s \geqslant 0$. The probability density function of such a variable equals $\theta e^{-\theta s}$ for $s \geqslant 0$ and 0 for $s < 0$. The mean of X then equals $E[X] = \theta \int_0^\infty s e^{-\theta s}\, ds = \theta^{-1}$. Its variance equals

$$E[(X^2 - E[X])^2] = E[X^2] - E[X]^2 = \theta \int_0^\infty s^2 e^{-\theta s}\, ds - \theta^{-2} = \theta^{-2}.$$

A random variable X is said to have a continuous distribution if $P(X = s) = 0$ for all $s \in \mathbb{R}$.

3. Directed last-passage percolation (LPP)

Consider the two-dimensional square lattice $\mathbb{Z}^2$. Put a random real number ω_x at each site $x \in \mathbb{Z}^2$ and let these assignments be independent. In more precise terms, let $\Omega = \mathbb{R}^{\mathbb{Z}^2}$ and endow it with the product topology and the Borel σ-algebra. A generic element in Ω is denoted by $\omega = \{\omega_x : x \in \mathbb{Z}^2\}$ and called an *environment* or a *configuration*. The numbers ω_x are called *weights*. Let μ be a probability measure on $\mathbb{R}$ and let $\mathbb{P}$ be the product probability measure on Ω with all marginals equal to μ: for a finite collection $x_1, \dots, x_n$ and measurable sets $A_1, \dots, A_n \subset \mathbb{R}$

$$\mathbb{P}\big\{\omega : \omega_{x_i} \in A_i : 1 \leqslant i \leqslant n\big\} = \prod_{i=1}^n \mu(A_i).$$

(In particular $\mathbb{P}(a < \omega_x \leqslant b) = \mu((a, b])$ for all $x \in \mathbb{Z}^2$ and $-\infty \leqslant a < b \leqslant \infty$.) We will abbreviate the average weight value as $m_0 = \mathbb{E}[\omega_0]$.

When weights ω_x are nonnegative they can be thought of as times spent at sites x. Thus, the *passage time* of a path $x_0, \dots, x_n \in \mathbb{Z}^2$ is the time it takes to traverse the path and equals $\sum_{i=1}^n \omega_{x_i}$. (We choose not to count the time spent at the very first point of the path.) We will abbreviate $x_{i,j}$ for a sequence $x_i, x_{i+1}, \dots, x_j$, with a similar notation for $x_{i,\infty}$, $x_{-\infty,j}$, and $x_{-\infty,\infty}$.

The model under consideration is directed, which means that we will only consider *up-right* paths, i.e. paths $x_{0,n} \in \mathbb{Z}^2$ with $x_{i+1} - x_i \in \{e_1, e_2\}$ for all $0 \leqslant i < n$. We sometimes also call such paths *admissible*. If $x, y \in \mathbb{Z}^2$ are such that $x \leqslant y$ coordinatewise, then the *last-passage time* between x and y is given by

$$G_{x,y} = \max\Big\{\sum_{i=1}^n \omega_{x_i} : x_{0,n} \text{ up-right, } x_0 = x,\, x_n = y,\, n = |y - x|_1\Big\}. \tag{3.1}$$

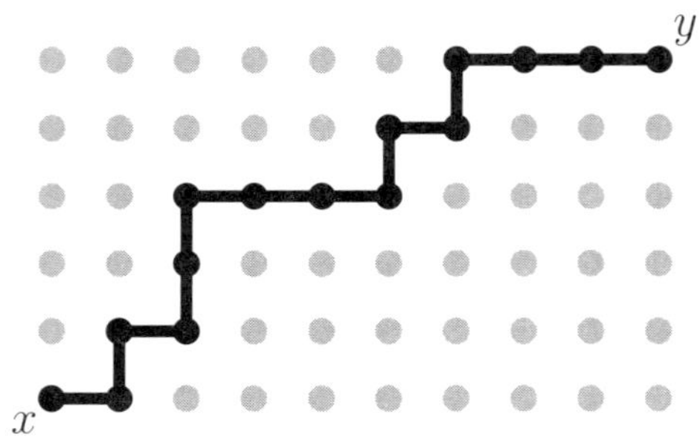

FIGURE 3.1. Illustration of a possible geodesic from x to y.

Here, $|\cdot|_1$ is the ℓ^1 norm on $\mathbb{R}^2$. Let us immediately record an important inequality, called *superadditivity*: for $x \leqslant y \leqslant z$ coordinatewise

$$G_{x,y} + G_{y,z} \leqslant G_{x,z}. \tag{3.2}$$

Paths that maximize in (3.1) are called *geodesics*. See Figure 3.1. The above inequality comes simply from observing that concatenating a geodesic from x to y with one from y to z gives an up-right path from x to z (which may or may not be a geodesic).

In what follows the math will make sense even when ω_x can be negative and we continue to use the term *passage time* even with negative weights.

REMARK 3.1. The above model is related to the standard first-passage site-percolation model (FPP) as follows. Allow admissible paths to take steps in $\{\pm e_1, \pm e_2\}$, i.e. be nearest-neighbor paths. Then, paths from x to y no longer have to have a determined length and the last-passage time is defined by

$$G_{x,y} = \max\Big\{ \sum_{i=1}^{n} \omega_{x_i} : x_{0,n} \text{ nearest-neighbor}, \, x_0 = x, \, x_n = y, \, n \geqslant 1 \Big\}.$$

Replacing ω with $-\omega$ turns the max into a min and the model becomes the standard first-passage percolation model. (Note that now weights must be taken nonnegative for otherwise $G_{x,y}$ will be infinite.) In this case, the inequality in (3.2) is reversed and $G_{x,y}$ defines a (random) metric on $\mathbb{Z}^2$. It is by analogy with this situation that optimizing paths in the directed model are called geodesics.

4. Connections to other models

Say weights ω_x are positive and for simplicity assume they have a continuous distribution, i.e. $\mathbb{P}(\omega_x = s) = 0$ for all $s \in \mathbb{R}$. Then the last-passage percolation model we defined in Section 3 has several other equivalent descriptions.

4.1. Random corner growth model (CGM).

For $x \in \mathbb{Z}^2$ set $G_{x,x} = 0$. An infection starts at the origin and spreads into the first quadrant $\mathbb{Z}^2_+$. The set of infected sites at time $t \geqslant 0$ is

$$\mathcal{B}(t) = \{x \in \mathbb{Z}^2_+ : G_{0,x} \leqslant t\}.$$

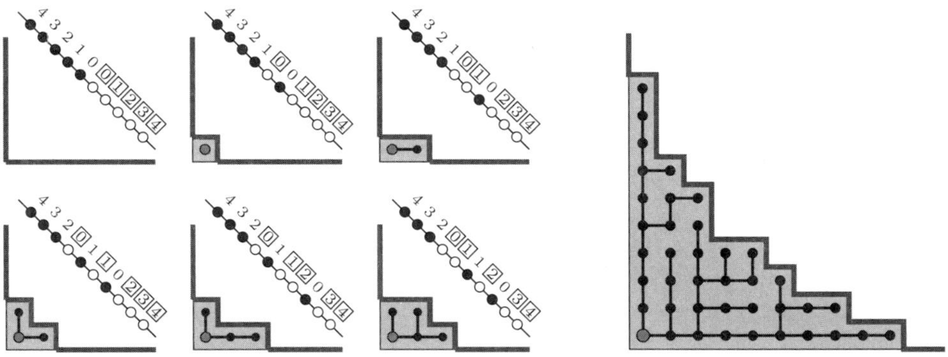

FIGURE 4.1. A possible early evolution of the corner growth model on the first quadrant of the plane. Bullets mark infected lattice points x with $G_{0,x} \leqslant t$. The origin is distinguished with the larger red bullet. The gray region is the fattened set $\mathcal{B}(t)+[-1/2,1/2]^2$. The thick purple down-right path is the height function that is the boundary of the fattened set $\mathcal{H}(t) = (\mathbb{Z}_+^2\backslash\mathcal{B}(t))+[-1/2,1/2]^2$. The bold black edges are the paths of minimal passage time from the origin. They are all directed. The antidiagonals illustrate the mapping of the corner growth model to TASEP. Whenever a point is added to the growing infected cluster, a particle (solid purple circle) switches places with the hole (open circle) to its right. For the queuing picture, boxed numbers indicate the service stations and numbers without a box are the customers. When a customer and a station switch places, the customer has left that station and moved to the back of the line at the next station.

Let us see how $\mathcal{B}(t)$ evolves. Since for a site $x = ke_i$, $k \in \mathbb{N}$ and $i \in \{1,2\}$, we have

$$G_{0,ke_i} = \sum_{j=1}^{k} \omega_{je_i} > G_{0,(k-1)e_i};$$

such a site cannot get infected before $(k-1)e_i$ is infected. Similarly, for a site $x \in \mathbb{N}^2$ we have the induction

$$G_{0,x} = \omega_x + \max(G_{0,x-e_1}, G_{0,x-e_2}). \tag{4.1}$$

Thus, $G_{0,x} > G_{0,x-e_i}$ for both $i \in \{1,2\}$ and a site $x \in \mathbb{N}^2$ cannot be infected until after both $x-e_1$ and $x-e_2$ were infected.

There is a nice description of the evolution of $\mathcal{B}(t)$ using the fattened set of heathy sites

$$\mathcal{H}(t) = (\mathbb{Z}_+^2\backslash\mathcal{B}(t)) + [-1/2,1/2]^2.$$

See the sequence of snapshots in Figure 4.1. Start with $\mathcal{B}(0-) = \varnothing$ and

$$\mathcal{H}(0-) = \mathbb{Z}_+^2 + [-1/2,1/2]^2 = \{x \in \mathbb{R}^2 : x \geqslant -(e_1+e_2)/2\}. \tag{4.2}$$

The boundary of $\mathcal{H}(0-)$ is given by the path $\{h(s) : s \in \mathbb{R}\}$ where

$$h(s) = s^+e_1 + s^-e_2 - (e_1+e_2)/2. \tag{4.3}$$

At time 0 the origin $x=0$ is infected, $\mathcal{B}(0) = \{0\}$, $\mathcal{H}(0) = \mathcal{H}(0-)\backslash[-1/2,1/2)^2$, and the boundary of $\mathcal{H}(0)$ is obtained from that of $\mathcal{H}(0-)$ by flipping the south-west corner located at $-(e_1+e_2)/2$ to a north-east corner, creating two new south-west corners at $(e_1-e_2)/2$ and $(e_2-e_1)/2$.

For concreteness, say $\omega_{e_1} < \omega_{e_2}$. Then e_1 is the next site to become infected, exactly at time $t_1 = \omega_{e_1} > 0$. We have $\mathcal{B}(t) = \{0\}$ for $0 \leqslant t < t_1$ and $\mathcal{B}(t_1) = \{0, e_1\}$. Region $\mathcal{H}$ gets another square taken away: $\mathcal{H}(t_1) = \mathcal{H}(0)\backslash(e_1 + [-1/2, 1/2)^2)$. Its boundary changes by the south-west corner at $(e_1 - e_2)/2$ getting flipped into a north-east corner.

Site (k, ℓ) becomes infected at time $t = G_{0,(k,\ell)}$. Right before this time, the boundary of $\mathcal{H}(t-)$ has a south-west corner at $ke_1 + \ell e_2 - (e_1 + e_2)/2$ that then flips into a north-east corner. Hence the name "corner growth".

The evolution is particularly nice when weights ω_x have an exponential distribution, i.e. when $\mathbb{P}(\omega_x > s) = \mu((s, \infty)) = e^{-\theta s}$ for some $\theta > 0$ and all $s \in \mathbb{R}_+$. Parameter θ is called the rate of the exponential random variable.

In this special case the evolution goes as follows. Given set $\mathcal{B}(t_0)$ at some point in time $t_0 \geqslant 0$ consider the south-west corners on the boundary of $\mathcal{H}(t_0)$. (There are always finitely many such corners.) Assign to these corners independent random variables, exponentially distributed with the same rate θ as weights ω_x. Think of these variables as the time an "alarm clock" goes off at the corner the variable is assigned to. When the first of these clocks rings the corresponding south-west corner gets flipped to a north-east corner. At that point in time, we have a new $\mathcal{H}$ and the procedure is repeated. The mathematics of this evolution is quite clear in the next equivalent description.

4.2. Queues in tandem. The queueing interpretation of LPP in terms of tandem service stations goes as follows. Imagine a queueing system with customers labeled by $\mathbb{Z}_+$ and service stations also labeled by $\mathbb{Z}_+$. The random weight $\omega_{k,\ell}$ is the service time of customer k at station ℓ. Right before time 0 all customers are lined up at service station 0 and customer 0 is first in line and has just been served. At time $t = 0$ customer 0 is first in line at queue 1 and the rest of the customers are still at queue 0 with customer 1 being first in line there, then customer 2, and so on. Service of customers 0 and 1 begins. Customers proceed through the system in order, obeying FIFO (first-in-first-out) discipline, and joining the queue at station $\ell + 1$ as soon as service at station ℓ is complete. Once customer $k \geqslant 0$ is first in line at station $\ell \geqslant 0$, it takes $\omega_{k,\ell}$ time units to perform service. (The only exception is $k = \ell = 0$ where customer 0 advances immediately from queue 0 to queue 1.)

For each $k \geqslant 0$ and $\ell \geqslant 0$, $G_{0,(k,\ell)}$ is the time when customer k departs station ℓ and joins the end of the queue at station $\ell + 1$. To see this observe that on the one hand this is clear when $k = 0$ or $\ell = 0$ and, on the other hand, this departure time satisfies the same recurrence (4.1). Indeed, the departure time of customer k from station ℓ is equal to the service time $\omega_{k,\ell}$ plus the time $G_{0,(k-1,\ell)}$ when customer $k - 1$ departs station ℓ or the time customer k departs station $\ell - 1$, whichever is larger. (If the former time is larger, customer k will have to wait for customer $k - 1$ before service starts at station ℓ. If instead the latter time is the larger of the two, then station ℓ will be empty when customer k arrives and service starts immediately.) In terms of the corner growth model, this is exactly when site (k, ℓ) gets infected. See Figure 4.1. Among the seminal references for these ideas are [**31**, **47**].

When weights are exponentially distributed, the description is again quite transparent. At every point in time there are only finitely many non-empty queues. Assign to the first customer in each of these queues an independent random variable (a clock) with exponential distribution having the same rate as weights ω_x. The customer whose clock rings first has been served and moves on to the end of the next station and then the procedure repeats.

4.3. Totally asymmetric simple exclusion process (TASEP). In this model, a configuration is an assignment of 0s and 1s to the integers $\mathbb{Z}$. More precisely, it is a function $\eta : \mathbb{Z} \to \{0, 1\}$. Think of $\eta_j = 1$ as a particle occupying site $j \in \mathbb{Z}$ and then $\eta_j = 0$ means j is empty (or a hole). Given a configuration η such that $\exists j_0$ with $\eta_j = 0$ for all $j \geqslant j_0$ define a curve (known as a *height function*) $h : \mathbb{R} \to \mathbb{R}^2$ by $h(j_0 - 1/2) = j_0 e_1 - (e_1 + e_2)/2$, $h(j + 1/2) - h(j - 1/2) = (1 - \eta_j)e_1 - \eta_j e_2$ for all $j \in \mathbb{Z}$, and linear interpolation on $\mathbb{R}\backslash(\mathbb{Z} + 1/2)$. (This is well defined, regardless of the choice of j_0.)

Right before time $t = 0$ we start by placing a particle at every site $j \leqslant -1$ and leaving sites $j \geqslant 0$ empty. In other words, $\eta_j(0-) = \mathbb{1}\{j < 0\}$. The corresponding height function is given by (4.3), i.e. it is the boundary of $\mathcal{H}(0-)$ from (4.2). At time $t = 0$ the particle at $j = -1$ jumps to the empty site $j = 0$, leaving site $j = -1$ empty. Now, the corresponding height function is given by the boundary of $\mathcal{H}(0)$.

As t grows particles move around. At any point in time only one particle is allowed to make a move and it can only move one step to its right, if there is no other particle there already. Here is a more precise description of the particle dynamics. Particles move only at times when $\mathcal{B}(t)$ changes (i.e. when new sites are infected). Think of the boundary of $\mathcal{H}(t)$ as a height function. Then, there is a one-to-one correspondence between south-west corners in the boundary of $\mathcal{H}(t)$ and particle-hole pairs (the hole being immediately to the right of the particle). When a south-west corner in the boundary of $\mathcal{H}(t)$ is flipped, the corresponding particle jumps one step to its right, switching positions with the hole that was there. See Figure 4.1.

Comparing this description to the queuing system we see that holes play the role of service stations and particles to the left of a hole play the role of customers in line at that service station.

Once again, when the weights are exponentially distributed the evolution can be described in a Markovian way using exponential clocks. Given a configuration at some time t_0 there are only finitely many particles that have a hole immediately to their right. Each of these particles is given an independent exponential random variable with the same rate as weights ω_x and the particle whose clock rings first moves one position to the right, effectively switching places with the hole that was there. Then the process is repeated. This is one of the most fundamental interacting particle systems. See [**43, 56**] for two of the earliest papers on the model.

Here is a table that summarizes the meaning of $G_{0,(k,\ell)}$ in the models in Sections 4.1-4.3.

Model	$G_{0,(k,\ell)}$ **is the time when:**
CGM	site (k, ℓ) is infected
Queues	customer k clears server ℓ
TASEP	particle k exchanges places with hole ℓ

4.4. The competition interface (CIF). Start two infections at e_1 and at e_2 and let sites get infected as before. Mark sites infected by e_1 purple and those infected by e_2 green. Once a site is infected by one of the two types, it remains like that forever. This partitions the first quadrant $\mathbb{Z}^2_+\backslash\{0\}$ into two regions of infection.

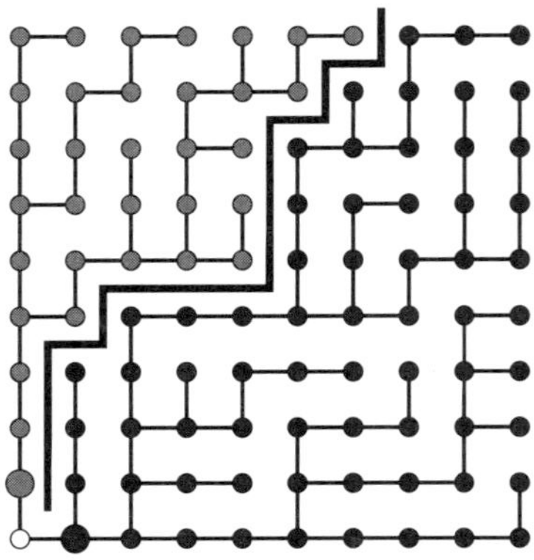

FIGURE 4.2. The full tree of infection in the corner growth model. The origin is the open circle at the bottom left. The solid black line marks the competition interface that separates the two competing infections that grow from points e_1 and e_2 marked with larger circles.

Mathematically, recall induction (4.1). Since we assumed weights to be independent and with a continuous distribution, equality $G_{0,x-e_1} = G_{0,x-e_2}$ happens with zero probability. Thus,

$$G_{0,x} = \omega_x + G_{0,x-e_i}$$

for exactly one of $i \in \{1, 2\}$. This indicates who infected site x.

Another description of the spread of infection comes using geodesics. Again, because weights are independent and have a continuous distribution, there is a unique geodesic between any two distinct sites $x \leqslant y$. As such, the union of all geodesics from 0 to sites $x \in \mathbb{Z}_+^2 \setminus \{0\}$ forms a spanning tree of $\mathbb{Z}_+^2$ that represents the genealogy of the infection. The subtrees rooted at e_1 and e_2 are precisely the vertices infected by these two sites, respectively. They are separated by an up-right path on the dual lattice $\mathbb{Z}^2 + (e_1 + e_2)/2$ called the *competition interface*. See Figure 4.2. Properties of this interface, as well as references for further reading, are in Section 9.

5. The shape function

One of the central questions in probability theory is to describe the order that emerges out of randomness as the number of random inputs into the system grows. For instance, the law of large numbers says that if $\{X_n : n \in \mathbb{N}\}$ are independent random variables that are identically distributed (i.e. random samples from the same population), then the sample or empirical mean $(X_1 + \cdots + X_n)/n$ converges, with probability one, to the (population) mean $E[X_1]$ (provided this mean is well defined). Even though $G_{0,x}$ is not simply a sum of independent random variables (it has a maximum), the law of large numbers raises the question of whether or not $G_{0,x}$ should grow at most linearly with $|x|_1$. This is indeed the case. For $a \in \mathbb{R}$ let $\lfloor a \rfloor$ be the largest integer no greater than a. For $x \in \mathbb{R}^2$ let $\lfloor x \rfloor$ act coordinatewise. Recall that $m_0 = \mathbb{E}[\omega_0]$.

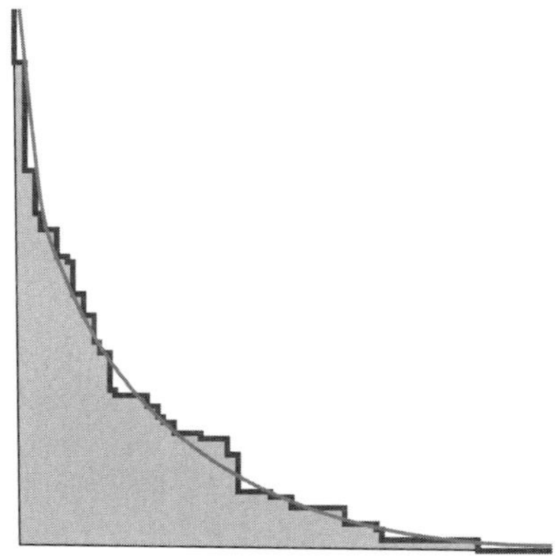

FIGURE 5.1. LPP with exponentially distributed vertex weights with rate 1. The gray region is a simulation of the scaled growing set $t^{-1}(\mathcal{B}(t) + [-1/2, 1/2]^2)$ at time $t = 160$. Its boundary (the thick blue down-right path) approximates the smooth red limit curve $\{x \in \mathbb{R}_+^2 : \sqrt{x \cdot e_1} + \sqrt{x \cdot e_2} = 1\}$, as first proved by Rost in 1981.

THEOREM 5.1. [45] *Assume* $\mathbb{E}[|\omega_0|] < \infty$. *Then for any* $\xi \in \mathbb{R}_+^2$ *the limit*

$$g(\xi) = \lim_{n\to\infty} \frac{G_{0,\lfloor n\xi\rfloor}}{n} \tag{5.1}$$

exists almost surely and (if $g(\xi) < \infty$*) in* L^1. *It is deterministic, 1-homogenous, concave, and satisfies* $g(x_1, x_2) = g(x_2, x_1)$, $x_1, x_2 \in \mathbb{R}_+$, $g(e_1) = g(e_2) = m_0$, *and* $g(\xi) \geqslant m_0\,|\xi|_1$. *If furthermore*

$$\int_0^\infty \sqrt{\mathbb{P}(\omega_0 > s)}\, ds < \infty \tag{5.2}$$

(e.g. if $\mathbb{E}[|\omega_0|^{2+\varepsilon}] < \infty$ *for some* $\varepsilon > 0$*), then* g *is finite and continuous on all of* $\mathbb{R}_+^2$ *and*

$$\lim_{n\to\infty} \max_{x\in\mathbb{Z}_+^2 : |x|_1 = n} \frac{|G_{0,x} - g(x)|}{n} = 0 \quad \text{almost surely.} \tag{5.3}$$

In particular, limit (5.3) says that the fattened set $(\mathcal{B}(t) + [-1/2, 1/2])/t$ converges almost surely, as $t \to \infty$, to the set $\{x \in \mathbb{R}_+^2 : g(x) \leqslant 1\}$. Thus, (5.3) is called a *shape theorem* and g is the *shape function*. See Figure 5.1.

PROOF OF THEOREM 5.1. Let us start by considering the case $\xi \in \mathbb{Z}_+^2$. By superadditivity (3.2) we have for $m \leqslant n$

$$G_{0,m\xi} + G_{m\xi,n\xi} \leqslant G_{0,n\xi}.$$

If we had additivity instead of superadditivity, then we could write

$$G_{0,n\xi} = \sum_{i=0}^{n-1} G_{i\xi,(i+1)\xi}.$$

The summands are *independent and identically distributed* (i.i.d.) and as such $n^{-1}\sum_{i=0}^{n-1} G_{i\xi,(i+1)\xi}$ is the sample mean of random samples of $G_{0,\xi}$. A generalization of the law of large numbers, called the ergodic theorem, tells us then that this sample mean converges to the population mean $\mathbb{E}[G_{0,\xi}]$.

Unfortunately, additivity does not hold. However, it turns out that one can prove a stochastic version of Fekete's subadditive lemma and apply this *subadditive ergodic theorem* to $-G_{0,n\xi}$ to obtain the limit (5.1). A version of this theorem is given in [**18**] as Theorem 3.2. A bit more work is needed for the general case $\xi \in [0,\infty)^2$ and even more work is needed to get more uniform control and prove (5.3). The details are omitted as they are similar to the ones in the proof of the standard first-passage percolation shape theorem, given in [**18**]. See also Proposition 2.1(i) of [**45**].

Symmetry of g follows from that of the lattice and the fact that i.i.d. random variables are exchangeable (i.e. switching them around does not change the joint distribution). Since $G_{0,ne_i} = \sum_{k=0}^{n-1} \omega_{ke_i}$, $i \in \{1,2\}$, (5.1) and the law of large numbers give $g(e_1) = g(e_2) = \mathbb{E}[\omega_0] = m_0$. Consider next the up-right path $x_{0,\lfloor n\xi \rfloor}$ from 0 to $\lfloor n\xi \rfloor$ that first takes $\lfloor n\xi_1 \rfloor$ e_1-steps and then $\lfloor n\xi_2 \rfloor$ e_2-steps. We have

$$n^{-1} G_{0,\lfloor n\xi \rfloor} \geqslant n^{-1} \sum_{i=0}^{\lfloor n\xi_1 \rfloor - 1} \omega_{ie_1} + n^{-1} \sum_{i=0}^{\lfloor n\xi_2 \rfloor - 1} \omega_{\lfloor n\xi_2 \rfloor + ie_2}.$$

By (5.1) the left-hand side converges to $g(\xi)$ almost surely and hence also in probability. By the weak law of large numbers, the two sums on the right-hand side converge in probability to $m_0\xi_1$ and $m_0\xi_2$, respectively. It follows that $g(\xi) \geqslant m_0|\xi|_1$.

Finiteness of $g(\xi)$ comes easily if weights ω_x are bounded above by some constant, i.e. if $\mathbb{P}(\omega_0 \leqslant c) = 1$ for some $c > 0$ then clearly $G_{0,\lfloor n\xi \rfloor} \leqslant cn|\xi|_1$ and thus $g(\xi) \leqslant c|\xi|_1$. More generally, finiteness would follow from the fact that g is concave, homogenous, and continuous on $\mathbb{R}_+^2$ (all the way up to the boundary).

Let us now prove the regularity properties claimed in the theorem. Homogeneity of g comes simply from

$$g(c\xi) = \lim_{n\to\infty} \frac{G_{0,\lfloor nc\xi \rfloor}}{n} = c \lim_{n\to\infty} \frac{G_{0,\lfloor cn\xi \rfloor}}{cn} = cg(\xi), \quad \text{for } c > 0.$$

Then concavity follows from homogeneity and superadditivity: for $\alpha \in (0,1)$

$$\alpha g(\xi) + (1-\alpha) g(\zeta) = g(\alpha\xi) + g((1-\alpha)\zeta)$$

and for $x, y \in \mathbb{R}_+^2$

$$\begin{aligned}
&g(x) + g(y) \\
&= \lim_{n\to\infty} \frac{G_{0,\lfloor nx \rfloor}}{n} + \lim_{n\to\infty} \frac{G_{0,\lfloor ny \rfloor}}{n} && \text{(almost surely)} \\
&= \lim_{n\to\infty} \frac{G_{0,\lfloor nx \rfloor}}{n} + \lim_{n\to\infty} \frac{G_{\lfloor nx \rfloor, \lfloor ny \rfloor + \lfloor nx \rfloor}}{n} && \text{(in probability, due to shift-invariance)} \\
&\leqslant \lim_{n\to\infty} \frac{G_{0,\lfloor ny \rfloor + \lfloor nx \rfloor}}{n} && \text{(by supperadditivity (3.2))} \\
&= g(x+y).
\end{aligned}$$

In the second equality, we used the fact that if one shifts the picture, placing the origin where say z used to be, then the two situations are statistically equivalent: for all $z \in \mathbb{Z}^2$ and $x \geqslant 0$, $G_{z,z+x}$ has the same distribution as $G_{0,x}$. It remains to prove continuity. We will do so under the assumption that weights are bounded in absolute value. This will capture the essence of the argument. The general case requires some extra technical work that we will avoid. The interested reader can find the details in the proof of Proposition 2.2 of [**45**].

Fix an $\varepsilon \in \mathbb{Q} \cap (0, 1/2)$ and an integer k such that $k\varepsilon \in \mathbb{N}$. The above computation gives us

$$g(e_2 + \varepsilon e_1) - g(e_2) \geqslant g(\varepsilon e_1) = \varepsilon g(e_1) = \varepsilon m_0. \tag{5.4}$$

We next prove a similar upper bound.

Each up-right path from 0 to $kne_2 + \varepsilon kne_1$ consists of kn e_2-steps and εkn e_1-steps. Thus, there are $\binom{nk(1+\varepsilon)}{nk\varepsilon}$ such paths. By Stirling's approximation $N! \sim \sqrt{2\pi N}N^N e^{-N}$ we have that

$$\binom{nk(1+\varepsilon)}{nk\varepsilon} \sim \sqrt{\frac{1+\varepsilon}{2\pi nk\varepsilon}}e^{nkh(\varepsilon)},$$

where $h(\varepsilon) = (1+\varepsilon)\log(1+\varepsilon) - \varepsilon\log\varepsilon$. Next, fix $\delta > 0$ and use a union bound (i.e. that $\mathbb{P}(\cup_j A_j) \leqslant \sum_j \mathbb{P}(A_j)$) and $m_0 = g(e_2)$ to write

$$\begin{aligned}\mathbb{P}\big(G_{0,kne_2+\varepsilon kne_1} \geqslant nkg(e_2) + nk\delta\big) &\leqslant \sum_{\substack{x_{0,nk(1+\varepsilon)}\\ \text{up-right}}} \mathbb{P}\Big(\sum_{i=1}^{nk(1+\varepsilon)} \omega_{x_i} \geqslant nkg(e_2) + nk\delta\Big)\\ &= \sum_{\substack{x_{0,nk(1+\varepsilon)}\\ \text{up-right}}} \mathbb{P}\Big(\sum_{i=1}^{nk(1+\varepsilon)} (\omega_{x_i} - m_0) \geqslant -nk\varepsilon m_0 + nk\delta\Big).\end{aligned}$$

Now observe that once the up-right path $x_{0,nk(1+\varepsilon)}$ is fixed, weights ω_{x_i} are i.i.d. with mean m_0. Thus, all the probabilities in the last sum have the same value and we can continue by writing

$$\begin{aligned}&\mathbb{P}\big(G_{0,kne_2+\varepsilon kne_1} \geqslant nkg(e_2) + nk\delta\big)\\ &\qquad\leqslant \sqrt{\frac{1+\varepsilon}{2\pi nk\varepsilon}}e^{nkh(\varepsilon)} P\Big(\sum_{i=1}^{nk(1+\varepsilon)} Z_i \geqslant nk(\delta - \varepsilon m_0)\Big),\end{aligned} \tag{5.5}$$

where Z_i are i.i.d. centered (and bounded) random variables with the same distribution as $\omega_0 - m_0$.

If $m_0 > 0$, then pick $\varepsilon \in (0, \delta/m_0)$. Otherwise, just pick $\varepsilon > 0$. Then the event in the last probability contradicts the law of large numbers which says that the sample mean $\frac{1}{nk(1+\varepsilon)}\sum_{i=1}^{nk(1+\varepsilon)} Z_i$ should be close to 0. In fact, large deviation theory tells us that the probability of such an event decays exponentially fast. Indeed, fix a positive number $\lambda > 0$ and use Chebyshev's exponential inequality to write

$$\begin{aligned}P\Big(\sum_{i=1}^{nk(1+\varepsilon)} Z_i \geqslant nk(\delta - \varepsilon m_0)\Big) &\leqslant E\big[e^{\sum_{i=1}^{nk(1+\varepsilon)} \lambda Z_i - \lambda nk(\delta - \varepsilon m_0)}\big]\\ &= e^{-\lambda nk(\delta-\varepsilon m_0)} E\big[e^{\lambda Z_0}\big]^{nk(1+\varepsilon)}\\ &= \exp\Big\{-nk\Big(\lambda(\delta - \varepsilon m_0) - (1+\varepsilon)\log E\big[e^{\lambda Z_0}\big]\Big)\Big\}.\end{aligned}$$

A little calculus exercise shows that $e^x \leqslant 1 + x + x^2$ for x close to 0. Use this and $\log(1+x) \leqslant x$ (valid for all x) to continue the above computation, remembering that $E[Z_0] = 0$,

$$\begin{aligned}P\Big(\sum_{i=1}^{nk(1+\varepsilon)} Z_i \geqslant nk(\delta - \varepsilon m_0)\Big) \\ &\leqslant \exp\Big\{-nk\Big(\lambda(\delta - \varepsilon m_0) - (1+\varepsilon)\log\big(1 + \lambda^2 E[Z_0^2]\big)\Big)\Big\} \\ &\leqslant \exp\Big\{-nk\Big(\lambda(\delta - \varepsilon m_0) - (1+\varepsilon)\lambda^2 E[Z_0^2]\Big)\Big\}.\end{aligned}$$

Take $\lambda = (\delta - \varepsilon m_0)/(2 + 2\varepsilon)$ and the above becomes

$$P\Big(\sum_{i=1}^{nk(1+\varepsilon)} Z_i \geqslant nk(\delta - \varepsilon m_0)\Big) \leqslant e^{-(\delta-\varepsilon m_0)^2 nk/(4+4\varepsilon)}.$$

Going back to (5.5) we have

$$\begin{aligned}&\mathbb{P}\big(G_{0,kne_2+\varepsilon kne_1} \geqslant nkg(e_2) + nk\delta\big) \\ &\qquad\leqslant \sqrt{\frac{1+\varepsilon}{2\pi nk\varepsilon}}\exp\Big\{-nk\Big(\frac{(\delta - \varepsilon m_0)^2}{4(1+\varepsilon)} - (1+\varepsilon)\log(1+\varepsilon) + \varepsilon\log\varepsilon\Big)\Big\}.\end{aligned}$$

Taking ε small enough makes the right-hand side converge exponentially fast to 0 as $n \to \infty$. Combining this with the fact that $G_{0,kne_2+\varepsilon kne_2}/(nk)$ converges almost surely to $g(e_2 + \varepsilon e_1)$ gives

$$g(e_2 + \varepsilon e_1) - g(e_2) \leqslant \delta.$$

Together with (5.4) this proves continuity of $a \mapsto g(e_2 + ae_1)$ at $a = 0$. Symmetry gives the same result with e_1 and e_2 switched around. This and the homogeneity of g imply its continuity at the boundary of $\mathbb{R}_+^2$. Continuity in the interior is a known fact about concave functions. □

When weights ω_x are exponentially distributed (with rate $\theta > 0$) one can get an explicit formula for the shape:

$$g(\xi_1, \xi_2) = m_0(\xi_1 + \xi_2) + 2\sigma_0\sqrt{\xi_1\,\xi_2}\,, \quad \xi = (\xi_1, \xi_2) \in \mathbb{R}_+^2. \tag{5.6}$$

Here, $m_0 = \theta^{-1}$ is the mean of ω_x and $\sigma_0 = \theta^{-1}$ is its standard deviation. (The two quantities are equal for exponential random variables, but this is not true in general.)

A similar formula also holds when weights ω_x are geometrically distributed, i.e. when μ is supported on $\mathbb{N}$ and for some $p \in (0,1)$ and all $j \in \mathbb{N}$, $\mathbb{P}(\omega_x = j) = \mu(\{j\}) = p^{j-1}(1-p)$. In the exponential case this formula was first derived by Rost [**52**] (who presented the model in its coupling with TASEP without the last-passage formulation) while early derivations of the geometric case appeared in [**15, 40, 53**]. We will prove the formula at the end of Section 7 using equation (6.4) and the explicit knowledge of the distribution of Busemann functions. See also Theorem 3.4 in [**55**].

Other than the above two cases, no explicit formula is known for g. However, it is known that the above formula does hold in general near the boundary.

THEOREM 5.2. [**45**] *Assume*

$$\int_0^\infty \sqrt{\mathbb{P}(\omega_0 > s)}\,ds < \infty \quad \textit{and} \quad \int_{-\infty}^0 \sqrt{\mathbb{P}(\omega_0 \leqslant s)}\,ds < \infty.$$

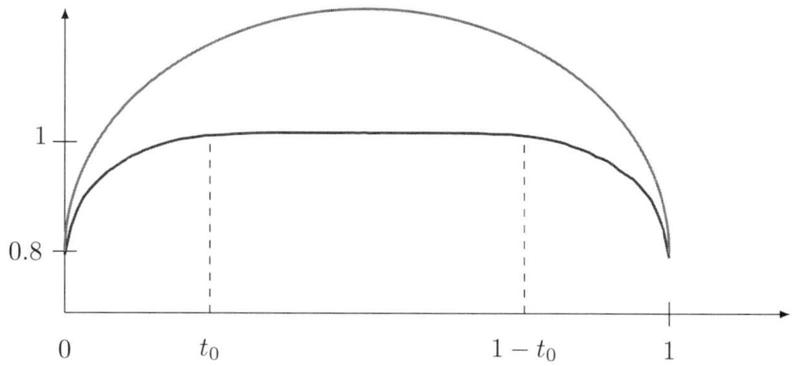

FIGURE 5.2. LPP with Bernoulli distributed vertex weights: $\mathbb{P}(\omega_0 = 1) = 1 - \mathbb{P}(\omega_0 = 0) = 0.8$. The bottom blue line is a simulation of the curve $t \mapsto g(te_1 + (1-t)e_2)$, $t \in [0,1]$. The top red curve is $t \mapsto 0.8 + \sqrt{0.8 \times 0.2t(1-t)}$, according to formula (5.6). The two curves are different: the top red one is strictly concave while the bottom blue one is flat on $[t_0, 1-t_0]$. But their asymptotics match near $t = 0$ and $t = 1$.

Let $m_0 = \mathbb{E}[\omega_0]$ *and* $\sigma_0^2 = \mathbb{E}[\omega_0^2] - m_0^2$. *Then*

$$g(1,\alpha) = m_0 + 2\sigma_0\sqrt{\alpha} + o(\sqrt{\alpha}) \quad \text{as } \alpha \searrow 0. \tag{5.7}$$

The fact that $g(1,\alpha) - m_0$ is of order $\sqrt{\alpha}$ for small α comes by a more careful look at the continuity proof above. An even more careful look involving comparison to the case with exponential weights gives the more precise formula (5.7).

The above has a nice consequence regarding the limiting shape.

COROLLARY 5.3. *The limiting shape* $\{x \in \mathbb{R}_2^+ : g(x) \leqslant 1\}$ *is not a polygon (with finitely many sides).*

It is tempting to think that perhaps formula (5.6) holds in general. The following situation shows that this is not the case.

Assume that the LPP weights satisfy $\omega_x \leqslant 1$ and $p = \mathbb{P}\{\omega_0 = 1\} > 0$. The classical Durrett-Liggett flat edge result implies that if p is large enough, g is linear on a whole cone. See Figure 5.2.

THEOREM 5.4. **[21]** *Suppose* $\mathbb{E}[|\omega_0|^{2+\varepsilon}] < \infty$ *for some* $\varepsilon > 0$ *and* $\mathbb{P}\{\omega_x \leqslant 1\} = 1$. *There exists a critical value* $p_c \in (0,1)$ *such that if* $p = \mathbb{P}\{\omega_0 = 1\} > p_c$, *then there exists* $t_0 \in (0, 1/2)$ *such that* $g(\xi) = |\xi|_1 = \xi \cdot (e_1 + e_2)$ *for all* $\xi \in \mathbb{R}_+^2$ *with* $\xi \cdot e_1/|\xi|_1 \in [t_0, 1-t_0]$.

Here is a heuristic argument to help understand why the above result holds. When probability p is large one can show that there is an infinite up-right path $x_{0,\infty}$ starting from the origin $x_0 = 0$ such that $\exists n_0$ with $\omega_{x_n} = 1$ for all $n \geqslant n_0$. Furthermore, one can guarantee that $x_n \cdot e_1/n \to t_0$ for some $t_0 \in (0, 1/2)$ as $n \to \infty$. The shape theorem then implies that $g(t_0, 1-t_0) = 1$. By concavity of g we have that $g(t, 1-t) = 1$ for all $t \in [t_0, 1-t_0]$ and the claim of the theorem follows from the homogeneity of g.

Before we close the section, it may be noteworthy that even though no closed formulas are known for g in general, some variational characterizations of g do exist in the general weights setting. See Section 2 of **[55]**.

6. Busemann functions

How does one prove (5.6)? Can one get any information on g, more than what is given by Theorem 5.1?

One way to approach such questions is by taking a closer look at the shape function. In precise terms, abbreviate

$$\mathcal{U} = \{(t, 1-t) : t \in [0,1]\} \quad \text{and} \quad \mathcal{U}^\circ = \{(t, 1-t) : 0 < t < 1\}$$

and fix a $\xi \in \mathcal{U}^\circ$. Consider the collection of random variables

$$\{G_{0,\lfloor n\xi \rfloor} - G_{0,\lfloor n\xi \rfloor + z} : |z|_1 \leqslant M\},$$

for some $M > 0$. In other words, we want to examine the passage times to points in the vicinity of $n\xi$, relative to the passage time to $n\xi$ itself. Presumably, as $n \to \infty$, (the distribution of) this vector of random variables converges weakly to some limit and this limit carries some useful information about the large scale behavior of the system "in direction ξ".

Since the point of reference is moving, there is really no hope of the convergence being almost sure. However, we can change our frame of reference and view things from $n\xi$ by considering

$$\{G_{-\lfloor n\xi \rfloor, 0} - G_{-\lfloor n\xi \rfloor, z} : |z|_1 \leqslant M\},$$

or equivalently

$$\{G_{-\lfloor n\xi \rfloor, 0} - G_{-\lfloor n\xi \rfloor, -z} : |z|_1 \leqslant M\}.$$

Since $\{\omega_{-x} : x \in \mathbb{Z}^2\}$ has the same distribution as $\{\omega_x : x \in \mathbb{Z}^2\}$, the above has the same law as

$$\{G_{0,\lfloor n\xi \rfloor} - G_{z,\lfloor n\xi \rfloor} : |z|_1 \leqslant M\}. \tag{6.1}$$

The advantage now is that there is a chance this random vector does converge to an almost sure limit. This is indeed the case under some mild regularity assumption on the shape g.

One little technicality: since when we defined $G_{x,y}$ in (3.1) we left out the weight ω_x, going through the above reflection argument leads us to slightly change our definition to become

$$G_{x,y} = \max\Big\{ \sum_{i=0}^{n-1} \omega_{x_i} : x_{0,n} \text{ up-right, } x_0 = x,\, x_n = y,\, n = |y - x|_1 \Big\}, \tag{6.2}$$

i.e. we now leave out the last weight ω_y instead. We will use this definition in the rest of the chapter. The reader should note, though, that since we are interested in the large scale behavior of the system, it is really immaterial whether we leave out or include the first or last weights.

Recall that g is a concave function. Then the set

$$\mathcal{C} = \{x \in \mathbb{R}^2_+ : g(x) \geqslant 1\}$$

is convex. It can then be decomposed into a union of closed faces (see Section 17 of [**51**]). For a given $\xi \in \mathcal{U}^\circ$ the point $\xi/g(\xi)$ is on the relative boundary of $\mathcal{C}$. Let

$$\mathcal{U}_\xi = \Big\{ \frac{\zeta}{|\zeta|_1} : \zeta \text{ belongs to the closed face containing } \xi/g(\xi) \Big\}.$$

If g is differentiable, then $\mathcal{U}_\xi$ is simply the largest connected subset of $\mathcal{U}^\circ$ containing ξ on which g is affine. Set $\mathcal{U}_\xi$ cannot contain e_1 or e_2 because we chose ξ in the relative interior of $\mathcal{U}$ and Theorem 5.2 prevents g from being linear on a neighborhood of e_1 or e_2.

Let $\mathcal{U}_{e_1} = \{e_1\}$ and $\mathcal{U}_{e_2} = \{e_2\}$. Clearly $\mathcal{U} = \cup_{\xi \in \mathcal{U}} \mathcal{U}_\xi$. When g is differentiable we have that $\forall \xi, \zeta \in \mathcal{U}$, either $\mathcal{U}_\xi = \mathcal{U}_\zeta$ or $\mathcal{U}_\xi \cap \mathcal{U}_\zeta = \varnothing$.

We will say that a sequence x_n is asymptotically directed into a subset $A \subset \mathcal{U}$ if all the limit points of x_n/n are inside A.

We are ready to state the theorem about the limits of the gradients in (6.1). The proof requires a detour into queuing theory and is thus deferred to Section 7.

THEOREM 6.1. [**29**] *Assume* $\mathbb{P}\{\omega_0 \geqslant c\} = 1$ *for some* $c \in \mathbb{R}$ *and* $\mathbb{E}[|\omega_0|^{2+\varepsilon}] < \infty$ *for some* $\varepsilon > 0$. *Assume g is differentiable on* $(0,\infty)^2$. *Then, for each* $\xi \in \mathcal{U}^\circ$ *the (random) limit*

$$B^\xi(\omega, x, y) = \lim_{n\to\infty} (G_{x,x_n} - G_{y,x_n}) \tag{6.3}$$

exists almost surely and in L^1 *for all* $x, y \in \mathbb{Z}^2$ *and sequence* $x_n \in \mathbb{Z}^2$ *that is directed into* $\mathcal{U}_\xi$. *Furthermore,*

$$\begin{aligned} &\text{for } i \in \{1,2\} \quad \mathbb{E}[B^\xi(0,e_i)] = e_i \cdot \nabla g(\xi) \\ &\text{and} \quad g(\xi) = \mathbb{E}[B^\xi(0,e_1)]\, \xi \cdot e_1 + \mathbb{E}[B^\xi(0,e_2)]\, \xi \cdot e_2. \end{aligned} \tag{6.4}$$

If $\xi, \zeta \in \mathcal{U}^\circ$ *are such that* $\mathcal{U}_\xi = \mathcal{U}_\zeta$, *then* $B^\xi \equiv B^\zeta$ *almost surely.*

(It is customary in probability theory to drop the dependence on ω from the notation of random variables, e.g. to write $B^\xi(x,y)$ instead of $B^\xi(\omega,x,y)$.)

The distribution of $B^\xi(0,e_i)$, $i \in \{1,2\}$, is known in the solvable cases. Formula (5.6) then follows from the second equation in (6.4). See the end of Section 7 for more details.

The condition that weights are bounded below by a deterministic constant is not really necessary. We assumed it in [**29**] because we used queuing theory for the proof, as we will see in Section 7, and then weights ω_x are service times and have to be nonnegative. The extension to weights that are bounded below is immediate. However, the math in the proof works just as well for general weights, even though then interpreting them as service times would not make sense.

The differentiability assumption is more serious. Although still an open question, differentiability is believed to be generally true. It can be directly verified from the explicit formula, when the weights are either exponentially or geometrically distributed. In the case when a flat segment occurs (see Theorem 5.4), it is known that g is differentiable at the two edges of the segment. See [**4**] for the standard first-passage percolation and [**29**] for the directed LPP. It is worthy to note that when $\{\omega_x\}$ are only ergodic (a generalization of being i.i.d.), the limiting shape can have corners and linear segments, and can even be a polygon with finitely many edges. See [**32**].

Limits B^ξ are called *Busemann functions.* This name is borrowed from metric geometry due to a connection between Busemann functions and geodesics, which is revealed in Section 8.

The first equation in (6.4) can be understood as follows: B^ξ is a (microscopic) gradient of passage times to "far away points in direction ξ". The shape function at ξ is the large scale (macroscopic) limit of passage times to these far away points.

(6.4) says that the mean of the microscopic gradient is exactly the macroscopic gradient.

The second equation in (6.4) is simply a consequence of the first one since differentiating $g(t\xi) = tg(\xi)$ in t gives

$$g(\xi) = \xi \cdot \nabla g(\xi). \tag{6.5}$$

Let us record right away a few important properties of processes B^ξ.

1. Cocycle:

$$B^\xi(x, y) + B^\xi(y, z) = B^\xi(x, y), \quad \text{almost surely and for all } x, y, z \in \mathbb{Z}^2. \tag{6.6}$$

It follows immediately from (6.3).

2. Recovery:

$$\omega_x = \min(B^\xi(x, x + e_1), B^\xi(x, x + e_2)) \quad \text{almost surely and for all } x \in \mathbb{Z}^2. \tag{6.7}$$

To see this holds start with an induction equation similar to (4.1) (but recall that passage times are now defined by (6.2)):

$$G_{x,y} = \omega_x + \max(G_{x+e_1,y}, G_{x+e_2,y}) \quad \text{for all } x \text{ and } y \geqslant x + e_i,\ i \in \{1, 2\}.$$

Rewrite this as

$$\omega_x = \min(G_{x,y} - G_{x+e_1,y}, G_{x,y} - G_{x+e_2,y}). \tag{6.8}$$

Now set $y = \lfloor n\xi \rfloor$ and take $n \to \infty$.

3. Stationarity: To express this, define the group action of $\mathbb{Z}^2$ on $\Omega = \mathbb{R}^{\mathbb{Z}^2}$ that consists of shifting the weights: for $z \in \mathbb{Z}^2$ and $\omega \in \Omega$, $T_z\omega \in \Omega$ is such that $(T_z\omega)_x = \omega_{x+z}$. In words, $T_z\omega$ is simply the weight configuration obtained from ω by placing the origin at z. Then we have

$$B^\xi(T_z\omega, x, y) = B^\xi(\omega, x + z, y + z) \quad \text{almost surely and for all } x, y, z \in \mathbb{Z}^2. \tag{6.9}$$

This follows directly from (6.3) and the fact that $G_{x,y}(T_z\omega) = G_{x+z,y+z}(\omega)$. (All these equations are just saying is that fixing the lattice and shifting the weight configuration is the same thing as fixing the weight configuration and shifting the lattice.)

4. Monotonicity: We record the last property as a lemma.

LEMMA 6.2. [**29**] *Make the same assumptions as in Theorem* 6.1. *Fix $\xi, \zeta \in \mathcal{U}^\circ$ with $\xi_1 < \zeta_1$. Then we have almost surely and for all $x \in \mathbb{Z}^2$*

$$B^\xi(x, x + e_1) \geqslant B^\zeta(x, x + e_1) \quad \textit{and} \quad B^\xi(x, x + e_2) \leqslant B^\zeta(x, x + e_2). \tag{6.10}$$

PROOF. The claim follows from a monotonicity of the passage times $G_{x,y}$ themselves: for all $x \in \mathbb{Z}^2$ and $y, z \in \mathbb{Z}_+^2$ such that $x + e_1 + e_2 \leqslant y$, $x + e_1 + e_2 \leqslant z$, $|y|_1 = |z|_1$, and $y \cdot e_1 < z \cdot e_1$ we have

(6.11)
$$G_{x,y} - G_{x+e_1,y} \geqslant G_{x,z} - G_{x+e_1,z} \quad \text{and} \quad G_{x,y} - G_{x+e_2,y} \leqslant G_{x,z} - G_{x+e_2,z}.$$

Indeed, once this is proved set $y = y_n$ and $z = z_n$ with $y_n, z_n \in \mathbb{Z}_+^2$ any two sequences with $|y_n|_1 = |z_n|_1 = n$, $y_n/n \to \xi$, and $z_n/n \to \zeta$, then send $n \to \infty$.

Inequalities (6.11) are due to paths crossing. Indeed, a geodesic from x to z must cross a geodesic from $x + e_1$ to y. Let u be the first point where the two paths cross, i.e. $u \cdot (e_1 + e_2)$ is the smallest possible. See Figure 6.1.

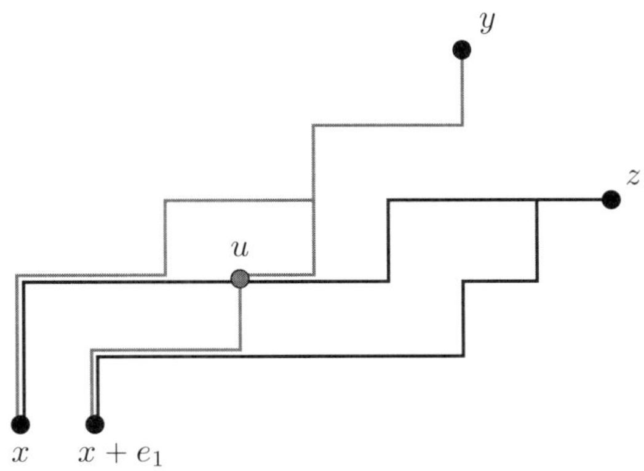

FIGURE 6.1. The paths crossing trick: a path from x to z must cross a path from $x+e_1$ to y.

Then superadditivity implies that

$$G_{x,u}+G_{u,y}\leqslant G_{x,y} \quad\text{and}\quad G_{x+e_1,u}+G_{u,z}\leqslant G_{x+e_1,z}.$$

Add the two inequalities and rearrange to get

$$G_{x,y}-G_{x+e_1,u}-G_{u,y}\geqslant G_{x,u}+G_{u,z}-G_{x+e_1,z}.$$

Use the fact that u is on geodesics to add the passage times and get the desired inequality:

$$G_{x,y}-G_{x+e_1,y}=G_{x,y}-G_{x+e_1,u}-G_{u,y}\geqslant G_{x,u}+G_{u,z}-G_{x+e_1,z}=G_{x,z}-G_{x+e_1,z}.$$

A similar proof works for the e_2-gradients and as we showed above, the claim of the theorem follows. This "crossing trick" has been used profitably in planar percolation, and goes back at least to [**2**, **3**]. □

7. Queuing fixed points

We now sketch how Theorem 6.1 is proved. By adding a constant to the weights, if necessary, we can assume they are nonnegative. Then we can use the queuing terminology. Let us note one small modification, though. Due to our new definition (6.2), which replaced (3.1), now $G_{0,(k,\ell)}$ is the time when customer k enters service at station ℓ and $G_{0,(k,\ell)}+\omega_{k,\ell}$, is the time when customer k departs station ℓ and joins the end of the queue at station $\ell+1$.

We are looking for stationary versions of the queuing process described in Section 4.2. For this, customers need to have been arriving for a long time. Thus, indexing them by $\mathbb{N}$ will not do and we need instead to have customers indexed by $\mathbb{Z}$. Furthermore, one cannot track the customer's arrival times at the queues (since the process started in the far past) and the right thing to do is to consider *inter-arrival* times.

Thus, the development begins with a processes $\{A_{n,0}: n\in\mathbb{Z}\}$ that records the time between the arrival of customers number n and $n+1$ to queue 0. This process is supposed to be stationary, i.e. the distribution of $\{A_{n+m,0}: n\in\mathbb{Z}\}$ does not depend on $m\in\mathbb{Z}$. We also assume it is *ergodic*. (Stationary measures form a convex set whose extreme points are said to be ergodic. A process is ergodic if its distribution is ergodic.) One special case is a constant inter-arrival process: $A_{n,0} = \alpha$ for all $n \in \mathbb{Z}$ and some $\alpha \in \mathbb{R}$.

We are also given service times $\{S_{n,k} : n \in \mathbb{Z}, k \in \mathbb{Z}_+\}$. They represent the time it takes to serve customer n at station k, once the customer is first in line at that station. These service times have the same joint distribution $\mathbb{P}$ as the weights in our directed LPP model. In particular, they are i.i.d. Service times are also independent of the inter-arrival process. In order for the system to be stable, we need to have customers served faster than they arrive. Hence, we require that

$$m_0 = \mathbb{E}[S_{0,0}] < \mathbb{E}[A_{0,0}] < \infty. \tag{7.1}$$

Given the inter-arrival and service times define waiting times at station 0 by

$$W_{n,0} = \Big(\sup_{j \leqslant n-1} \sum_{i=j}^{n-1} (S_{i,0} - A_{i,0}) \Big)^+. \tag{7.2}$$

$W_{n,0}$ is the time customer n waits at station 0 before their service starts. See further down for an explanation.

Process $\{S_{i,0} - A_{i,0} : i \in \mathbb{Z}\}$ is ergodic. This is because one can show that the product measure of an ergodic process with an i.i.d. one is ergodic. By the ergodic theorem the sample mean $(n-j)^{-1} \sum_{i=j}^{n-1} (S_{i,0} - A_{i,0})$ converges to the population mean $\mathbb{E}[S_{0,0} - A_{0,0}]$, which by (7.1) is negative. Therefore, almost surely, as $j \to -\infty$ the sum $\sum_{i=j}^{n-1} (S_{i,0} - A_{i,0})$ goes to $-\infty$ and $0 \leqslant W_{n,0} < \infty$ for all $n \in \mathbb{Z}$. It is immediate to check that these times satisfy Lindley's equation

$$W_{n+1,0} = (W_{n,0} + S_{n,0} - A_{n,0})^+. \tag{7.3}$$

This now explains why $W_{n,0}$ is the time customer n waits at station 0 before their service starts. Indeed, if $W_{n,0} + S_{n,0} < A_{n,0}$ then customer n will leave station 0 before the next customer $n+1$ arrives. As a result, customer $n+1$ does not wait and $W_{n+1,0} = 0$. If, on the other hand, $W_{n,0} + S_{n,0} \geqslant A_{n,0}$, then customer $n+1$ waits time $W_{n+1,0} = W_{n,0} + S_{n,0} - A_{n,0}$ before service begins.

Inter-departure times from queue 0 or, equivalently, inter-arrival times at queue 1 are given by

$$A_{n,1} = (W_{n,0} + S_{n,0} - A_{n,0})^- + S_{n+1,0}. \tag{7.4}$$

Again, if $W_{n+1,0} > 0$ then customer $n+1$ is already waiting and will start being serviced as soon as customer n departs. The time between the departure of customer n and that of customer $n+1$, from station 0, is then equal to $S_{n+1,0}$. In the other case, when $W_{n+1,0} = 0$, station 0 is empty before customer $n+1$ gets there, for exactly time $A_{n,0} - S_{n,0} - W_{n,0}$. The time between the departures of customers n and $n+1$ from queue 0 equals this idle time, plus the service time $S_{n+1,0}$.

From (7.3) and (7.4) we have

$$W_{n+1,0} + S_{n+1,0} + A_{n,0} = W_{n,0} + S_{n,0} + A_{n,1}$$

for all $n \in \mathbb{Z}$. This in turn gives

$$\sum_{m=0}^{n-1} A_{m,0} + S_{n,0} - S_{0,0} + W_{n,0} - W_{0,0} = \sum_{m=0}^{n-1} A_{m,1}.$$

Process $\{A_{n,1} : n \in \mathbb{Z}\}$ is again ergodic. Hence, dividing by n and taking it to ∞ the ergodic theorem tells us that $n^{-1} \sum_{m=0}^{n-1} A_{m,0}$ and $n^{-1} \sum_{m=0}^{n-1} A_{m,1}$ converge to $\mathbb{E}[A_{0,0}]$ and $\mathbb{E}[A_{0,1}]$, respectively. Since $S_{n,0}$ is a stationary process, $S_{n,0}/n$

converges to 0. The next lemma shows that also $W_{n,0}/n \to 0$ almost surely. Consequently, we have that $\mathbb{E}[A_{0,1}] = \mathbb{E}[A_{0,0}]$.

LEMMA 7.1. *We have $n^{-1}W_{n,0} \to 0$ almost surely, as $n \to \infty$.*

PROOF. Let $U_n = S_{n,0} - A_{n,0}$. Fix $\varepsilon > 0$, $a \geqslant 0$, and let $u_0 = \mathbb{E}[U_n] < 0$. ($u_0$ does not depend on n because U_n is stationary.) Set $W_0^\varepsilon(a) = a$ and define inductively $W_{n+1}^\varepsilon(a) = (W_n^\varepsilon(a) + U_n - u_0 + \varepsilon)^+$, $n \geqslant 0$. By induction

$$W_n^\varepsilon(0) = \Big(\max_{0 \leqslant m < n} \sum_{k=m}^{n-1} (U_k - u_0 + \varepsilon)\Big)^+.$$

Also, since the induction preserves monotonicity we have that

$$W_n^\varepsilon(a) \geqslant W_n^\varepsilon(0) \geqslant \sum_{k=0}^{n-1} (U_k - u_0 + \varepsilon).$$

Since $\mathbb{E}[U_k - u_0 + \varepsilon] = \varepsilon > 0$, the ergodic theorem tells us the last sum grows to infinity (linearly in n). Consequently, there exists an n_0 such that $W_n^\varepsilon(a) > 0$ for $n \geqslant n_0$. But then if $n \geqslant n_0$ we have $W_n^\varepsilon(a) = W_{n-1}^\varepsilon(a) + U_{n-1} - u_0 + \varepsilon$ and thus

$$n^{-1}W_n^\varepsilon(a) = n^{-1}W_{n_0}^\varepsilon(a) + n^{-1}\sum_{k=n_0}^{n-1} (U_k - u_0 + \varepsilon).$$

The ergodic theorem again tells us that the last term converges to ε as $n \to \infty$. As a result, we have shown that with probability one for any $a \geqslant 0$ we have $n^{-1}W_n^\varepsilon(a) \to 0$.

Recall now that $W_{0,0} < \infty$ and observe that $W_0^\omega(W_{0,0}) = W_{0,0}$. Induction then shows that $W_{n,0} \leqslant W_n^\varepsilon(W_{0,0})$ for all $n \geqslant 0$. Indeed,

$$\begin{aligned} W_{n+1,0} &= (W_{n,0} + U_n)^+ \leqslant (W_n^\varepsilon(W_{0,0}) + U_n)^+ \\ &\leqslant (W_n^\varepsilon(W_{0,0}) + U_n - u_0 + \varepsilon)^+ = W_{n+1}^\varepsilon(W_{0,0}). \end{aligned}$$

But then

$$0 \leqslant \varliminf_{n\to\infty} n^{-1}W_{n,0} \leqslant \varlimsup_{n\to\infty} n^{-1}W_{n,0} \leqslant \lim_{n\to\infty} n^{-1}W_n^\varepsilon(W_{0,0}) = \varepsilon.$$

Taking $\varepsilon \to 0$ completes the proof. □

Note that $A_{n,1}$ only used values $\{S_{m,0} : m \in \mathbb{Z}\}$ and is therefore independent of $\{S_{m,k} : m \in \mathbb{Z}, k \geqslant 1\}$. We can now repeat the above steps inductively: Say we already computed the (ergodic) inter-arrival process $\{A_{n,k} : n \in \mathbb{Z}\}$ at queue $k \geqslant 0$ and that it is independent of the service times $\{S_{n,\ell} : n \in \mathbb{Z}, \ell \geqslant k\}$. Say also that $\mathbb{E}[A_{0,k}] = \mathbb{E}[A_{0,0}]$. Then we define the waiting times

$$W_{n,k} = \Big(\sup_{j \leqslant n-1} \sum_{i=j}^{n-1} (S_{i,k} - A_{i,k})\Big)^+, \tag{7.5}$$

which satisfy

$$W_{n+1,k} = (W_{n,k} + S_{n,k} - A_{n,k})^+. \tag{7.6}$$

The inter-arrival process at queue $k+1$ is given by

$$A_{n,k+1} = (W_{n,k} + S_{n,k} - A_{n,k})^- + S_{n+1,k}. \tag{7.7}$$

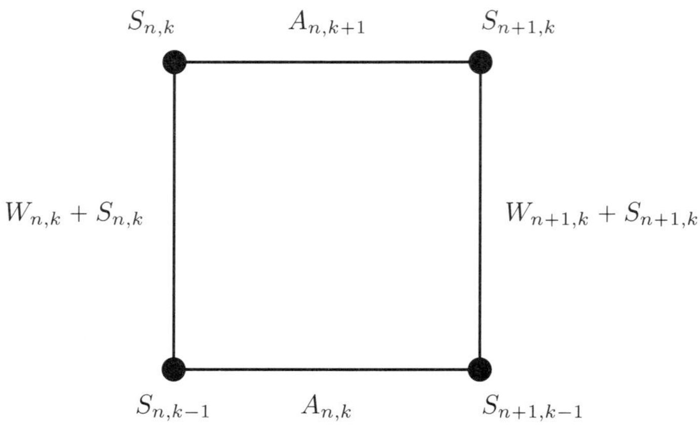

FIGURE 7.1. Assignment of work loads, inter-arrival times, and service times, to edges and vertices.

It is ergodic and has the same mean as $A_{n,k}$ (and thus as $A_{n,0}$). It is also independent of $\{S_{n,\ell} : n \in \mathbb{Z},\ \ell \geqslant k+1\}$ and we can continue the inductive process.

Using (7.6) and (7.7) we can check the conservation law

$$W_{n+1,k} + S_{n+1,k} + A_{n,k} = W_{n,k} + S_{n,k} + A_{n,k+1} \tag{7.8}$$

and the equation

$$S_{n+1,k} = \min(W_{n+1,k} + S_{n+1,k}, A_{n,k+1}). \tag{7.9}$$

Times $W_{n,k} + S_{n,k}$ are called *work load* of station k by customer n. Let us assign values to the edges of $\mathbb{Z} \times \mathbb{Z}_+$: on the horizontal edge $(n,k) - (n+1,k)$ put weight $A_{n,k}$ and on vertical edge $(n,k) - (n,k+1)$ put weight $W_{n,k} + S_{n,k}$. Also, put weights $S_{n,k}$ on vertices $(n,k+1)$. Recall now (6.6) and (6.7). Then (7.8) can be seen as a cocycle property and (7.9) is a recovery property (but in the south-west direction instead of the north-east direction). See Figure 7.1. This is where the connection to Busemann functions lies.

Observe that system $\{A_{n,k}, W_{n,k}, S_{n,k} : n \in \mathbb{Z},\ k \in \mathbb{Z}_+\}$ is invariant with respect to shifts in the first coordinate, i.e. the distribution of $\{A_{n+m,k}, W_{n+m,k}, S_{n+m,k} : n \in \mathbb{Z},\ k \in \mathbb{Z}_+\}$ is the same for all $m \in \mathbb{Z}$. It is in fact also ergodic under shifts in the first coordinate. However, it is not obvious at all (and in fact not true in general) that the system is invariant (let alone ergodic) under shifts in the second coordinate. That is, we do not know a priori that the distribution of $\{A_{n,k+\ell}, W_{n,k+\ell}, S_{n,k+\ell} : n \in \mathbb{Z},\ k \in \mathbb{Z}_+\}$ is independent of $\ell \in \mathbb{Z}_+$. For this to happen, clearly $\{A_{n,0} : n \in \mathbb{Z}\}$ needs to have some special distribution.

If we denote the distribution of $\{A_{n,0} : n \in \mathbb{Z}\}$ by ν then write $\Phi(\nu)$ for the distribution of $\{A_{n,1} : n \in \mathbb{Z}\}$. This is the so-called *queuing operator*. It takes a ergodic probability measure on $\mathbb{R}^{\mathbb{Z}}$ and transforms it to another ergodic probability measure on $\mathbb{R}^{\mathbb{Z}}$, while preserving the value of the mean (recall that $\mathbb{E}[A_{n,1}] = \mathbb{E}[A_{n,0}]$). Let Φ^k be the k-th iterate of Φ, i.e. $\Phi^1 = \Phi$ and $\Phi^{k+1}(\nu) = \Phi(\Phi^k(\nu))$.

For the invariance under shifts in the second coordinate to hold, what we need is $\Phi(\nu) = \nu$, i.e. that ν be an ergodic *fixed point* of the queuing operator.

Thus, the problem at hand is: Given $\alpha > m_0$ find ergodic measures ν_α on $\mathbb{R}^{\mathbb{Z}}$ such that $\Phi(\nu_\alpha) = \nu_\alpha$. And it would be good if along the way we can also answer the question of uniqueness of such measures.

One way to produce fixed points with a prescribed mean $\alpha > m_0$ is to start for example with the measure $\delta_\alpha^{\mathbb{Z}}$, the distribution of the constant process $\{A_{n,0} = \alpha : n \in \mathbb{Z}\}$, and hope that $\Phi^k(\delta_\alpha^{\mathbb{Z}})$ converges (weakly) to an ergodic fixed point that has mean α, as $k \to \infty$.

Mairesse and Prabhakar [**44**] proved that for each $\alpha > m_0$ there exists a unique stationary fixed point ν_α such that if one starts with any ergodic process ν with mean α, then the Césaro mean $k^{-1}\sum_{\ell=1}^k \Phi^k(\nu)$ converges weakly to ν_α. One consequence of this is that ergodic fixed points with a prescribed mean are unique. It is an open question whether or not for each $\alpha > m_0$ probability measure ν_α is ergodic. What is known, though, is that this is true if we have differentiability of the shape function g for the LPP problem with weights distribution $\mathbb{P}$ (the distribution of the service times) [**29**, Lemma 7.6(b)].

For a fixed $\alpha > m_0$ let $\{A_{n,0} : n \in \mathbb{Z}\}$ have distribution ν_α and let $\{S_{n,k} : n \in \mathbb{Z}, k \in \mathbb{Z}_+\}$ be i.i.d. with distribution $\mathbb{P}$, independent of $\{A_{n,0}\}$. Construct the processes $\{W_{n,k}, A_{n,k} : n \in \mathbb{Z}, k \in \mathbb{Z}_+\}$ by inductions (7.6) and (7.7). Then, because ν_α is a fixed point, process $\{W_{0,k} : k \in \mathbb{Z}_+\}$ is stationary and can thus be extended to a stationary process $\{W_{0,k} : k \in \mathbb{Z}\}$. (If this is not clear to the reader, it is an excellent exercise on a standard application of Kolmogorov's extension theorem.) Let

$$f(\alpha) = \mathbb{E}[W_{0,0}] + \mathbb{E}[S_{0,0}]. \tag{7.10}$$

Using the symmetry of the construction one can prove that the distribution of $\{W_{0,k} + S_{0,k} : k \in \mathbb{Z}\}$ is again a fixed point of the queuing operator and that it is ergodic if ν_α is ergodic [**29**, Lemma 7.7]. In this case, this distribution is no other than $\nu_{f(\alpha)}$. This tells us that not only the system $\{A_{n,k}, W_{n,k}, S_{n,k} : n \in \mathbb{Z}, k \in \mathbb{Z}_+\}$ is ergodic under shifting the n coordinate, but it is also ergodic under shifts of the k coordinate.

THEOREM 7.2. *Function f takes values in (m_0, ∞) and is a convex, continuous, and strictly decreasing involution (i.e. $f(f(\alpha)) = \alpha$). Furthermore, $f(\alpha)$ converges to ∞ as $\alpha \searrow m_0$ at to m_0 as $\alpha \to \infty$.*

Let g be the shape function for the LPP problem with weight distribution $\mathbb{P}$ (the distribution of the service times). Assume it is differentiable on $\mathcal{U}^\circ$. Then for all $\xi \in \mathcal{U}^\circ$ we have

$$g(\xi) = \inf_{\alpha > m_0} \big(\alpha\xi_1 + f(\alpha)\xi_2\big). \tag{7.11}$$

The infimum is minimized at $\alpha = e_1 \cdot \nabla g(\xi)$ and then $f(\alpha) = e_2 \cdot \nabla g(\xi)$.

PROOF. Let G_{me_1,ne_2} denote the last passage time from me_1 to ne_2 using weights $\{S_{m,k} : m \in \mathbb{Z}, k \in \mathbb{Z}_+\}$. Equation (7.2) for $n = 0$ can be rewritten as

$$S_{0,0} + W_{0,0} = \sup_{j \leqslant 0}\Big(G_{je_1,e_2} - \sum_{i=j}^{-1} A_{i,0}\Big),$$

with the convention that an empty sum is zero. We can extend this by induction to

$$\sum_{k=0}^{n-1}(S_{0,k}+W_{0,k})=\sup_{j\leqslant 0}\Big(G_{je_1,ne_2}-\sum_{i=j}^{-1}A_{i,0}\Big).$$

Divide by n, take it to infinity, and use the ergodic theorem and the shape theorem (plus a bit of work similar to what we did in the proof of continuity of g) we get

$$f(\alpha)=\sup_{s\geqslant 0}\big(g(s,1)-s\alpha\big)$$

with a maximizer at s such that $\alpha=e_1\cdot\nabla g(s,1)$.

The above formula shows that f is nonincreasing and convex function. Martin's asymptotic formula (5.7) (and homogeneity of g) implies that $m_0<f(\alpha)<\infty$ for $\alpha>m_0$. In particular, the convex function f is continuous. The symmetry of g implies that f is an involution and is thus strictly decreasing. The limits claimed in the theorem follow.

Inverting the above convex duality we get

$$g(s,1)=\inf_{\alpha>m_0}\big(\alpha s+f(\alpha)\big) \tag{7.12}$$

with a minimizer at $\alpha=e_1\cdot\nabla g(s,1)$. Variational formula (7.11) now comes by using homogeneity of g to write $g(\xi)=\xi_2 g(\xi_1/\xi_2,1)$. The minimizing α is given by $e_1\cdot g(\xi)$, as claimed, and then the fact that $f(\alpha)=e_2\cdot\nabla g(\xi)$ comes from (6.5). □

The above construction of fixed points can be carried out simultaneously for any given countable set of parameters $\alpha>m_0$, thus coupling the fixed points ν_α. A bit more precisely, fix a countable set $\mathcal{A}_0\subset(m_0,\infty)$ and start with inter-arrival times $A^{(\alpha)}_{n,0}=\alpha$, $\alpha\in\mathcal{A}_0$, $n\in\mathbb{Z}$, and service times $\{S_{n,k},n\in\mathbb{Z},\,k\in\mathbb{Z}_+\}$ that are independent and have the same distribution as the weights of the LPP model. (Note that the service times do not depend on α.) Use (7.5) and (7.7) to define inductively $W^{(\alpha)}_{n,k}$ and $A^{(\alpha)}_{n,k}$ on all of $\mathbb{Z}\times\mathbb{Z}_+$. Then, as $k\to\infty$ the Césaro mean of the distributions of $\{A^{(\alpha)}_{n,k}:n\in\mathbb{Z},\,\alpha\in\mathcal{A}_0\}$ converges weakly to a probability measure on $(\mathbb{R}^{\mathbb{Z}})^{\mathcal{A}_0}$ whose marginal for a fixed $\alpha\in\mathcal{A}_0$ is exactly ν_α.

We can now go back to proving existence of the limit (6.3).

Proof of Theorem 6.1. The first step is to extract from the above queuing objects candidates for the limits B^ξ. Then we prove that the Busemann limits indeed exist and equal these candidates.

Recall that we assume g is differentiable. For $\zeta\in\mathcal{U}^\circ$ define $\alpha(\zeta)=e_1\cdot\nabla g(\zeta)$. Then α is nonincreasing and continuous in $\zeta_1=\zeta\cdot e_1$. Also, it follows from (5.7) that $\alpha(\zeta)\to m_0$ as $\zeta\to e_1$ and $\alpha(\zeta)\to\infty$ as $\zeta\to e_2$. In particular, $\alpha(\zeta)$ is always strictly bigger than $m_0=\mathbb{E}[\omega_0]$. Fix $\xi\in\mathcal{U}^\circ$ and let $\mathcal{U}_0=\{\xi\}\cup(\mathcal{U}^\circ\cap\mathbb{Q}^2)$. Let $\mathcal{A}_0=\{\alpha(\zeta):\zeta\in\mathcal{U}_0\}$. This is a dense countable subset of (m_0,∞).

Let $\{A^{(\alpha)}_{n,0}:n\in\mathbb{Z},\,\alpha\in\mathcal{A}_0\}$ be distributed according to the above coupling of fixed points and let $\{S_{n,k}:n\in\mathbb{Z},\,k\in\mathbb{Z}_+\}$ be independent, with the same joint distribution $\mathbb{P}$ as the weights in the directed LPP model, and independent of the inter-arrival times. Construct times $\{A^{(\alpha)}_{n,k},W^{(\alpha)}_{n,k}:n\in\mathbb{Z},\,k\in\mathbb{N},\,\alpha\in\mathcal{A}_0\}$ using inductions (7.5) and (7.7). Since it comes from fixed points, system $\{A^{(\alpha)}_{n,k},W^{(\alpha)}_{n,k},S_{n,k}:n\in\mathbb{Z},\,k\in\mathbb{Z}_+,\alpha\in\mathcal{A}_0\}$ is stationary under shifts in both coordinates. We can then extend it to a system $\{A^{(\alpha)}_{n,k},W^{(\alpha)}_{n,k},S_{n,k}:n\in\mathbb{Z},\,k\in\mathbb{Z},\,\alpha\in\mathcal{A}_0\}$ on the whole lattice.

For $\zeta \in \mathcal{U}_0$ define

$$\omega_{ne_1+ke_2} = S_{-n,-k-1}, \quad B^\zeta(ne_1+ke_2,(n+1)e_1+ke_2) = A^{(\alpha(\zeta))}_{-n-1,-k}, \quad \text{and}$$
$$B^\zeta(ne_1+ke_2, ne_1+(k+1)e_2) = W^{(\alpha(\zeta))}_{-n,-k-1} + S_{-n,-k-1}\,.$$

Then $\{\omega_x : x \in \mathbb{Z}^2\}$ has distribution $\mathbb{P}$ and (7.9) says that B^ζ satisfies the recovery property (6.7). Equation (7.8) says that B^ζ satisfies

$$\begin{aligned} B^\zeta(x,x+e_1) + B^\zeta(x+e_1,x+e_1+e_2) \\ = B^\zeta(x,x+e_2) + B^\zeta(x+e_2,x+e_1+e_2) \end{aligned} \tag{7.13}$$

for all $x \in \mathbb{Z}^2$. Set $B^\zeta(x+e_i,x) = -B^\zeta(x,x+e_i)$, $i \in \{1,2\}$ and for $x,y \in \mathbb{Z}^2$ define

$$B^\zeta(x,y) = \sum_{i=0}^{n-1} B^\zeta(x_i,x_{i+1}),$$

where $x_{0,n}$ is any nearest-neighbor path from $x_0 = x$ to $x_n = y$. Thanks to (7.13) this definition does not depend on the choice of the path $x_{0,n}$. Now, B^ζ is an L^1 cocycle and $\mathbb{E}[B^\zeta(0,e_1)] = \mathbb{E}[A^{(\alpha(\zeta))}_{0,0}] = \alpha(\zeta)$. This equality says that B^ζ satisfies the first equation in (6.4), for the e_1 direction. The version for the e_2 direction comes from Theorem 7.2:

$$\mathbb{E}[B^\zeta(0,e_2)] = \mathbb{E}[W^{\alpha(\zeta)}_{0,-1} + S_{0,-1}] = f(\alpha(\zeta)) = e_2 \cdot \nabla g(\zeta).$$

As was mentioned earlier, the second equation in (6.4) is simply a consequence of the first one and (6.5).

Observe next that if we start with two sequences of inter-arrival times $A_{n,0} \leqslant A'_{n,0}$ for all $n \in \mathbb{Z}$, then (7.2) gives $W_{n,0} \geqslant W'_{n,0}$ for all $n \in \mathbb{Z}$. Then from (7.7) we get that $A_{n,1} \leqslant A'_{n,1}$. This monotonicity of the queuing operator leads to a monotonicity in the coupling of the fixed points. Combining this with the fact that if $\zeta,\eta \in \mathcal{U}_0$ are such that $\zeta_1 < \eta_1$, then $\alpha(\zeta) = e_1 \cdot \nabla g(\zeta) \geqslant e_1 \cdot \nabla g(\eta) = \alpha(\eta)$, we get that the B^ζ cocycles we constructed satisfy monotonicity (6.10).

The first equation in (6.4) and differentiability of g imply that $\zeta \mapsto \mathbb{E}[B^\zeta(0,e_i)]$ is continuous, for $i \in \{1,2\}$. Combined with the above monotonicity we get that with probability one

$$\lim_{\mathcal{U}_0 \ni \zeta \to \xi} B^\zeta(x,x+e_i) = B^\xi(x,x+e_i), \quad i \in \{1,2\}. \tag{7.14}$$

Note that the set of full $\mathbb{P}$-measure on which the above event holds depends on ξ. In fact, we will see in Corollary 8.10 that this continuity does not hold on all $\mathcal{U}^\circ$ simultaneously (i.e. with one null set thrown away).

Lastly, by another monotonicity argument, not too different from the one we used in the proof of Lemma 6.2, we can show that for $x \in \mathbb{Z}^2$, a sequence x_n directed in $\mathcal{U}_\xi$, directions $\zeta,\eta \in \mathcal{U}_0 \backslash \mathcal{U}_\xi$ with $\zeta_1 < \xi_1 < \eta_1$, and a large integer n, we have the stochastic inequalities

$$\begin{aligned} B^\eta(x,x+e_1) \leqslant G_{x,x_n} - G_{x+e_1,x_n} \leqslant B^\zeta(x,x+e_1) \quad \text{and} \\ B^\eta(x,x+e_2) \geqslant G_{x,x_n} - G_{x+e_2,x_n} \geqslant B^\eta(x,x+e_2). \end{aligned} \tag{7.15}$$

(A random variable Y is said to be stochastically smaller than a random variable Z if for all $a \in \mathbb{R}$ we have $P(Z \leqslant a) \leqslant P(Y \leqslant a)$. If moreover one shows that

$E[Z] = E[Y]$, then Y and Z have the same distribution.) Taking $n \to \infty$ then η and ζ to ξ and applying (7.14) we get

$$B^\xi(x, x+e_i) = \lim_{n\to\infty}(G_{x,x_n} - G_{x+e_i,x_n}).$$

Then the cocycle property (7.13) gives (6.3).

If $\mathcal{U}_\xi = \mathcal{U}_\zeta$ for some $\xi, \zeta \in \mathcal{U}^\circ$, then $\alpha(\xi) = e_1 \cdot \nabla g(\xi) = e_1 \cdot \nabla g(\zeta) = \alpha(\zeta)$ and thus $B^\xi = B^\zeta$. The proof of Theorem 6.1 is complete. □

The next lemma develops the very first inequality in (7.15), the others being similar. We focus on the case $x = 0$, the general case coming from shift-invariance of $\mathbb{P}$.

LEMMA 7.3. *Fix $\xi \in \mathcal{U}^\circ$ and a (possibly random) sequence x_n directed in $\mathcal{U}_\xi$. Fix $\eta \in \mathcal{U}^\circ \backslash \mathcal{U}_\xi$ with $\xi_1 < \eta_1$. Then almost surely for large n*

$$B^\eta(0, e_1) \leqslant G_{0,x_n} - G_{e_1,x_n}.$$

PROOF. Consider the two rectangle with common south-west corner at 0 and with north-east corners at x_n and $x_n + e_1 + e_2$. Put weights ω_x at all vertices $x \leqslant x_n$. Let $\bar\omega_x = \omega_x$ for such vertices. At sights $x = x_n + e_1 + e_2 - ke_1$, $k \in \mathbb{N}$, put weights $\bar\omega_x = B^\eta(x, x+e_1)$. Similarly, at sights $x = x_n + e_1 + e_2 - ke_2$, $k \in \mathbb{N}$, put weights $\bar\omega_x = B^\eta(x, x+e_2)$. We will write $\overline{G}_{y,x_n}$ for the passage time from y to x_n using weights ω_x, as defined in (6.2). We also use the passage times from y to $x_n + e_1 + e_2$ that use weights the combination of weights ω_x, $x \leqslant x_n$, and □

To close this section let us describe the situation for solvable models. Here, fixed points ν_α can be described explicitly. For example, in the case of exponentially distributed service times with mean $m_0 > 0$ (rate $1/m_0$) one can check directly that for any $\alpha > m_0$ inter-arrival times $\{A_{n,0} : n \in \mathbb{Z}\}$ that are i.i.d. exponentially distributed with mean α (rate $1/\alpha$) furnish an ergodic fixed point of the queuing operator. Because of the uniqueness of ergodic fixed points, this identifies ν_α. Another direct computation verifies that $f(\alpha) = \mathbb{E}[W_{0,0} + S_{0,0}] = m_0\alpha/(\alpha - m_0)$ and the symmetry observed below (7.10) says that $\{W_{0,k} + S_{0,k} : k \in \mathbb{Z}\}$ are i.i.d. exponentially distributed with mean $f(\alpha)$. One more miracle occurs: It turns out that $\{A_{n,0} : n \in \mathbb{Z}_+\}$ and $\{W_{0,k} + S_{0,k} : k \in \mathbb{Z}_+\}$ are independent of each other. See Theorem 3.1 in [**55**] for the proofs of all these distributional claims. Once the explicit formula for f is known solving the variational formula (7.12) leads to Rost's formula (5.6).

A consequence of the above is that for $\xi \in \mathcal{U}^\circ$, $\{B^\xi(ne_1, (n+1)e_1) : n \in \mathbb{Z}_+\}$ are independent exponentially distributed with rate $\frac{\sqrt{\xi_1}}{m_0(\sqrt{\xi_1}+\sqrt{\xi_2})}$, $\{B^\xi(ne_2, (n+1)e_2) : n \in \mathbb{Z}_+\}$ are independent exponentially distributed with rate $\frac{\sqrt{\xi_2}}{m_0(\sqrt{\xi_1}+\sqrt{\xi_2})}$, and the two sets of random variables are independent of each other.

Information about the distribution of the Busemann functions is powerful. For example, it allows to get bounds on the coalescence time of geodesics [**11**, **49**]. We we will see in Section 9 that it enables calculation of the distribution of the asymptotic direction of the competition interface. It is also used in proving bounds on the fluctuations of passage times and geodesics, as is done in Section 5 of [**55**].

8. Geodesics

In this section, let us assume the conditions of Theorem 6.1 to be satisfied. In particular, the shape g is differentiable on $(0,\infty)^2$.

One of the important questions in LPP concerns infinite geodesics: an infinite path is a geodesic if every finite segment of it is a geodesic between its endpoints. The following existence result comes quite easily.

LEMMA 8.1. *With probability one, for every $x \in \mathbb{Z}^2$ there is at least one infinite geodesics starting at that point.*

PROOF. Fix $x \in \mathbb{Z}^2$. Take any sequence $x_n \in \mathbb{Z}_+^2$ with $|x_n|_1 \to \infty$ and consider for each n a geodesic from x to x_n. Denote it by $x_{0,n}^{(n)}$. Fix $m \geqslant 1$. We have only finitely many possible up-right paths of length m. Hence, there exists a subsequence along which $x_{0,n}^{(n)}$ all share the same initial m steps. Using the diagonal trick, we can find a subsequence n_j such that for all $m \geqslant 1$ there exists a j_m such that , $\{x_{0,n_j}^{(n_j)} : j \geqslant j_m\}$ share the first m steps. This constructs an infinite path $x_{0,\infty}$ such that for all $m \geqslant 1$, $x_{0,m}$ is the path shared by $\{x_{0,n_j}^{(n_j)} : j \geqslant j_m\}$. In particular, $x_{0,m}$ is a geodesic between $x_0 = x$ and x_m, for all $m \geqslant 1$, and thus $x_{0,\infty}$ is an infinite geodesic starting at $x_0 = x$. □

Now we know that infinite geodesics exist. But how many infinite geodesics starting at a given point are there? And what are their properties? Does every infinite geodesic $x_{0,\infty}$ have to have an asymptotic direction, i.e. is it necessary that $x_n/n \to \xi$ for some $\xi \in \mathcal{U}$? Can there be multiple geodesics that go in the same asymptotic direction $\xi \in \mathcal{U}$? Do geodesics starting at different points and going in a given direction ξ cross? Does there exist a bi-infinite geodesic, i.e. an up-right path $x_{-\infty,\infty}$ whose finite segments are all geodesics?

Busemann functions can help answer some (if not all) of the above questions. Take a look, for example, at Theorems 8.3, 8.5, 8.7, 8.9, 8.11, 8.13 and Corollary 8.8 below.

To see the connection between Busemann functions and geodesics start with formula (6.8). Say, for simplicity, weights ω_x have a continuous distribution. Then $G_{x+e_1,y} = G_{x+e_2,y}$ happens with zero probability and thus there is a unique $i \in \{1,2\}$ for which

$$\omega_x = G_{x,y} - G_{x+e_i,y}.$$

The geodesic path from x to y will follow this increment and go from x to $x+e_i$. Then, from there the procedure can be repeated, until the path reaches the north-east boundary with corner y, i.e. until one gets to an $x \in \{y-ke_1 : k \in \mathbb{N}\} \cup \{y-ke_2 : k \in \mathbb{N}\}$. From there, the geodesic marches straight to y using only e_1 or only e_2 steps. This description of geodesics motivates the following.

LEMMA 8.2. **[29]** *Let $B : \mathbb{Z}^2 \times \mathbb{Z}^2 \to \mathbb{R}$ satisfy the cocycle and recovery properties* (6.6) *and* (6.7). *Let $x_{0,\infty}$ be a path such that for every $i \geqslant 0$ we have*

$$\omega_{x_i} = B(x_i, x_{i+1}).$$

In other words, the path goes along the "minimal gradient" of B. Then, $x_{0,\infty}$ is a geodesic.

PROOF. Fix $n \geqslant 1$. Consider an arbitrary up-right path $y_{0,n}$ with $y_0 = x_0$ and $y_n = x_n$. Write

$$\sum_{i=0}^{n-1} \omega_{x_i} = \sum_{i=0}^{n-1} B(x_i, x_{i+1}) = B(x_0, x_n) = \sum_{i=0}^{n-1} B(y_i, y_{i+1}) \geqslant \sum_{i=0}^{n-1} \omega_{y_i}.$$

(The first equality is from recovery, the second and third use the cocycle property, and the fourth uses recovery again.) Take a maximum over all up-right paths $y_{0,n}$ between x_0 and x_n to get

$$\sum_{i=0}^{n-1} \omega_{x_i} \geqslant G_{x_0,x_n},$$

which says that $x_{0,n}$ is a geodesic. Since n was arbitrary, the lemma is proved. □

As a bonus, we get in the above proof that when $x_{0,\infty}$ follows the smallest gradient of a cocycle B that recovers, we have for $0 \leqslant m \leqslant n$

$$G_{x_m,x_n} = \sum_{i=m}^{n-1} \omega_{x_i} = B(x_m, x_n). \tag{8.1}$$

The above lemma says in particular that Busemann functions B^ξ from (6.3) provide us with a "machine" to produce infinite geodesics starting from any given point. Given a starting point u, a direction $\xi \in \mathcal{U}^\circ$, and an integer $j \in \{1, 2\}$, let $x_{0,\infty}^{u,\xi,j}$ be the path produced by the following inductive mechanism: $x_0 = u$ and for $k \geqslant 0$, if $B^\xi(x_k, x_k + e_1) \neq B^\xi(x_k, x_k + e_2)$, then let $x_{k+1} = x_k + e_i$ for the unique $i \in \{1, 2\}$ such that $\omega_{x_k} = B^\xi(x_k, x_k + e_i)$. If, on the other hand, $B^\xi(x_k, x_k + e_1) = B^\xi(x_k, x_k + e_2)$, then break the tie by letting $x_{k+1} = x_k + e_j$.

Now that we know how to produce geodesics we ask about whether or not these geodesics have an asymptotic direction. We have the following theorem.

THEOREM 8.3. [**29**] *Make the same assumptions as in Theorem* 6.1. *For each* $\xi \in \mathcal{U}^\circ$ *we have with probability one that for all* $u \in \mathbb{Z}^2$ *and* $j \in \{1, 2\}$

$$\mathbb{P}\{\textit{geodesic } x_{0,\infty}^{u,\xi,j} \textit{ is asymptotically directed into } \mathcal{U}_\xi\} = 1.$$

In particular, if the boundary of $\mathcal{C}$ *is strictly convex at* ξ, *then* $x_{0,\infty}^{u,\xi,j}$ *has asymptotic direction* ξ: $n^{-1}x_n^{u,\xi,j} \to \xi$ *as* $n \to \infty$.

The proof of the above theorem needs a fact about stationary L^1 cocycles, i.e. measurable functions $B : \Omega \times \mathbb{Z}^2 \times \mathbb{Z}^2 \to \mathbb{R}$ that satisfy (6.6) and (6.9) and are such that for each $x, y \in \mathbb{Z}^2$ we have $\mathbb{E}[|B(\omega, x, y)|] < \infty$.

THEOREM 8.4. [**29, 30, 50**] *Let* B *be a stationary* L^1 *cocycle. Define* $\overline{B} = \mathbb{E}[B(0, e_1)]e_1 + \mathbb{E}[B(0, e_2)]e_2$. *Then for any* $\xi \in \mathbb{R}_+^2$ *we have* $\mathbb{P}$*-almost surely*

$$\lim_{n\to\infty} \frac{B(0, \lfloor n\xi \rfloor)}{n} = \overline{B} \cdot \xi. \tag{8.2}$$

If furthermore B *recovers (i.e. satisfies* (6.7)*), then we also have*

$$\lim_{n\to\infty} \max_{x\in\mathbb{Z}_+^2 : |x|_1 = n} \frac{|B(0, x) - \overline{B} \cdot x|}{n} = 0 \quad \mathbb{P}\textit{-almost surely}. \tag{8.3}$$

SKETCH OF PROOF OF THEOREM 8.4. As it was the case in Theorem 5.1, we will show how the proof of (8.2) goes when $\xi \in \mathbb{Z}_+^2$. The case $\xi \in \mathbb{R}_+^2$ comes with some more work and (8.3) comes with considerably more work. Assume thus that $\xi \in \mathbb{Z}_+^2$. Then we can use the cocycle and stationarity properties to write

$$B(\omega, 0, n\xi) = \sum_{i=0}^{n-1} B(\omega, i\xi, (i+1)\xi) = \sum_{i=0}^{n-1} B(T_{i\xi}\omega, 0, \xi).$$

Terms $B(T_{i\xi}\omega, 0, \xi)$ are just shifted copies of the first term $B(\omega, 0, \xi)$. As such, $n^{-1}B(\omega, 0, n\xi)$ can be thought of as a sample mean of, albeit dependent, samples of $B(\omega, 0, \xi)$. A generalization of the law of large numbers, called the ergodic theorem, tells us then that this sample mean converges to the population mean $\mathbb{E}[B(\omega, 0, \xi)]$. In other words, for $\mathbb{P}$-almost every ω

$$\lim_{n\to\infty} \frac{B(\omega, n\xi)}{n} = \mathbb{E}[B(0, \xi)]. \tag{8.4}$$

Now use the cocycle property again to write

$$B(0, \xi) = \sum_{i=0}^{\xi\cdot e_1 - 1} B(ie_1, (i+1)e_1) + \sum_{j=0}^{\xi\cdot e_2 - 1} B((\xi\cdot e_1)e_1 + je_2, (\xi\cdot e_1)e_1 + (j+1)e_2).$$

The summands in the first sum are shifted copies of $B(0, e_1)$ and the summands in the second sum are shifted copies of $B(0, e_2)$. Hence, taking expectation we get $\mathbb{E}[B(0, \xi)] = \mathbb{E}[B(0, e_1)]\xi \cdot e_1 + \mathbb{E}[B(0, e_2)]\xi \cdot e_2 = \overline{B} \cdot \xi$, which combined with (8.4) proves the claim of the theorem. □

PROOF OF THEOREM 8.3. Let us abbreviate and write x_n for $x_n^{u,\xi,j}$. Let $\zeta \in \mathcal{U}$ be a (possibly random) limit point of x_n/n, i.e. there exists a (possibly random) subsequence n_j such that $x_{n_j}/n_j \to \zeta$.

By Lemma 8.2 we know $x_{0,\infty}$ is a geodesic and by its definition it moves along the smallest gradient of B^ξ. By (8.1)

$$G_{0,x_n} = B^\xi(x_0, x_n).$$

Divide by n then apply this to $n = n_j$ and use (5.3) and (8.3) to deduce that

$$g(\zeta) = \mathbb{E}[B^\xi(0, e_1)]\zeta \cdot e_1 + \mathbb{E}[B^\xi(0, e_2)]\zeta \cdot e_2.$$

Apply the first equality in (6.4) to get

$$g(\zeta) = \zeta \cdot \nabla g(\xi).$$

Combine with (6.5) to get

$$g(\zeta) - g(\xi) = (\zeta - \xi) \cdot \nabla g(\xi).$$

Then, function

$$f(t) = g(t\xi + (1-t)\zeta) - g(\xi) - (t\xi + (1-t)\zeta - \xi) \cdot \nabla g(\xi), \quad t \in [0, 1],$$

satisfies $f(0) = f(1) = 0$ and

$$\lim_{\varepsilon \searrow 0} \frac{f(1) - f(1-\varepsilon)}{\varepsilon} = -\lim_{\varepsilon \searrow 0} \frac{g(\xi + \varepsilon(\zeta - \xi)) - g(\xi)}{\varepsilon} + (\zeta - \xi) \cdot \nabla g(\xi) = 0.$$

Since f is also concave it is identically 0 and thus for all $t \in [0, 1]$

$$g(t\xi + (1-t)\zeta) - g(\xi) = (t\xi + (1-t)\zeta - \xi) \cdot \nabla g(\xi) = (1-t)(\zeta - \xi) \cdot \nabla g(\xi).$$

This says g is affine on $\{t\xi + (1-t)\zeta : 0 \leqslant t \leqslant 1\}$ and thus $\zeta \in \mathcal{U}_\xi$. The theorem is proved. □

Now that we know that geodesics generated using Busemann functions B^ξ have an asymptotic direction we can prove the same thing about all geodesics.

THEOREM 8.5. [**29**] *Make the same assumptions as in Theorem* 6.1. *With probability one, any geodesic is asymptotically directed into $\mathcal{U}_\xi$ for some $\xi \in \mathcal{U}$.*

The proof will need one more fact about geodesics $x_{0,\infty}^{u,\xi,j}$.

LEMMA 8.6. [**29**] *For all $n \geqslant m$, $x_{m,n}^{u,\xi,1}$ is the right-most geodesic between its two endpoints: if $y_{m,n}$ is a geodesic between $x_m^{u,\xi,1}$ and $x_n^{u,\xi,1}$, then we have $y_k \cdot e_1 \leqslant x_k^{u,\xi,1} \cdot e_1$ for $m \leqslant k \leqslant n$. Similarly, $x_{m,n}^{u,\xi,2}$ is the left-most geodesic between its two endpoints.*

PROOF. We will prove the claim about $x_{m,n}^{u,\xi,1}$, the other one being symmetric. Abbreviate this path by writing $x_{m,n}$. Take $y_{m,n}$ as in the claim. In particular, $y_m = x_m$. For starters we want to prove that $y_{m+1} \cdot e_1 \leqslant x_{m+1} \cdot e_1$. For this, we only need to consider the case when $x_{m+1} = x_m + e_2$, for the inequality clearly holds in the other case. Since we are using a superscript $j = 1$ it cannot be that $B^\xi(x_m, x_m + e_1) = B^\xi(x_m, x_m + e_2)$, for otherwise the path would have taken an e_1-step out of x_m. Since the path always takes a step along the smaller B^ξ gradient and since B^ξ recovers, we conclude that in the case at hand we have $\omega_{x_m} = B^\xi(x_m, x_m + e_2) < B^\xi(x_m, x_m + e_1)$.

Now, recovery and the cocycle property imply that $G_{x,y} \leqslant B^\xi(x, y)$ for any $x \leqslant y$. Combine this with (8.1) and the cocycle property again to get

$$\begin{aligned}\omega_{x_m} + G_{x_m+e_1,x_n} &\leqslant B^\xi(x_m, x_m + e_2) + B^\xi(x_m + e_1, x_n)\\ &< B^\xi(x_m, x_m + e_1) + B^\xi(x_m + e_1, x_n) = B^\xi(x_m, x_n) = G_{x_m,x_n}.\end{aligned}$$

Therefore, no geodesic from x_m to x_n can go through $x_m + e_1$ and we have $y_{m+1} = x_m + e_2 = x_{m+1}$.

Now repeat this argument every time $x_{m,n}$ and $y_{m,n}$ intersect to see that the latter never goes to the "right" of the former. The lemma is proved. □

One can squeeze the proof of the above lemma a little bit more to get the following interesting result.

THEOREM 8.7. [**29**] *Make the same assumptions as in Theorem* 6.1.

(i) *Fix $\xi \in \mathcal{U}^\circ$. With probability one and for all $u \in \mathbb{Z}^2$, $x_{0,\infty}^{u,\xi,1}$ is the right-most geodesic directed into $\mathcal{U}_\xi$ and $x_{0,\infty}^{u,\xi,2}$ is the left-most geodesic directed into $\mathcal{U}_\xi$.*
(ii) *With probability one and for any $u \in \mathbb{Z}^2$, every infinite geodesic out of u stays between $x_{0,\infty}^{u,\xi,1}$ and $x_{0,\infty}^{u,\xi,2}$ for some $\xi \in \mathcal{U}^\circ$.*

PROOF OF THEOREM 8.5. First, observe that although we proved Theorem 8.3 and Lemma 8.6 for a fixed $\xi \in \mathcal{U}^\circ$, they both hold simultaneously (i.e. with one null set thrown away) for all $\xi \in \mathcal{U}^\circ \cap \mathbb{Q}^2$, which is countable and dense in $\mathcal{U}^\circ$.

Assume that for some geodesic $x_{0,\infty}$, x_n/n has limit points in both $\mathcal{U}_\zeta$ and $\mathcal{U}_\eta$ with $\zeta, \eta \in \mathcal{U}$ and $\mathcal{U}_\zeta \neq \mathcal{U}_\eta$. We can assume $\zeta \cdot e_1 < \eta \cdot e_1$. Since we have assumed

g to be differentiable, there must exist at least one (and in fact infinitely many) point(s) $\xi \in \mathcal{U}^\circ \cap \mathbb{Q}^2$ such that

$$\zeta \cdot e_1 < \xi \cdot e_1 < \eta \cdot e_1, \quad \mathcal{U}_\xi \neq \mathcal{U}_\zeta, \quad \text{and} \quad \mathcal{U}_\xi \neq \mathcal{U}_\eta. \tag{8.5}$$

Let $x_0 = u$ (the starting point of the geodesic under study). Since we have shown that geodesic $x_{0,\infty}^{u,\xi,1}$ has asymptotic direction ξ, the ordering in (8.5) implies that $x_{0,\infty}$ passes infinitely often to the left of $x_{0,\infty}^{u,\xi,1}$. But then Lemma 8.6 implies that once the former goes strictly to the left of the latter, it has to remain (weakly) on that side forever. A similar argument shows that $x_{0,\infty}$ must also eventually stay to the right of $x_{0,\infty}^{u,\xi,2}$. In other words, $x_{0,\infty}$ eventually stays between $x_{0,\infty}^{u,\xi,1}$ and $x_{0,\infty}^{u,\xi,2}$. But both these geodesics are directed into $\mathcal{U}_\xi$. Hence, so is $x_{0,\infty}$, which contradicts the assumption that x_n/n has limit points in $\mathcal{U}_\zeta$ and $\mathcal{U}_\eta$. □

Theorems 8.3 and 8.5 have a very nice consequence when we know more about the regularity of g.

COROLLARY 8.8. [**29**] *Make the same assumptions as in Theorem 6.1. Assume also that g is strictly concave. Then*

(i) *For any given direction, with probability one, out of any given point, there exists an infinite geodesic going in this direction:*

$$\forall \xi \in \mathcal{U}: \quad \mathbb{P}\{\forall u \in \mathbb{Z}^2\ \exists x_{0,\infty} \text{ geodesic } : x_n/n \to \xi\} = 1;$$

(ii) *With probability one, every infinite geodesic has an asymptotic direction:*

$$\mathbb{P}\{\forall x_{0,\infty} \text{ geodesic } \exists \xi \in \mathcal{U} : x_n/n \to \xi\} = 1.$$

This simply follows from the fact that if g is strictly concave, then $\mathcal{U}_\xi = \{\xi\}$ for all $\xi \in \mathcal{U}$. Strict concavity is still an open question, but it is believed to hold in general, either when the maximum of ω_0 does not percolate or outside the flat segment that occurs when the maximum does percolate (see Theorem 5.4).

The claims in the above corollary appeared before in Proposition 7 of [**25**] for the solvable model where weights ω_x are exponentially distributed. Note that in this case formula (5.6) gives an explicit expression for g and we can check directly that g is indeed strictly concave.

The approach used by [**25**] follows the ideas of Licea and Newman [**42**] for nearest-neighbor first-passage percolation (FPP). In [**42**] the authors assume a certain global curvature assumption on g and use it to control how much infinite geodesics can wander, proving existence and directedness of infinite geodesics. They also use a lack-of-space argument to prove coalescence (i.e. merger) of geodesics with a given asymptotic direction. This is the only method known to date for proving coalescence of directional geodesics. The same idea was adapted by [**25**] to the directed LPP model with exponential weights and then by [**29**] to the general weights setting.

THEOREM 8.9. [**29**] *Make the same assumptions as in Theorem 6.1. Fix $\xi \in \mathcal{U}^\circ$. With probability one and for all $u, v \in \mathbb{Z}^2$ the right-most geodesics directed into $\mathcal{U}_\xi$ and starting at u and at v coalesce: there exist $m, n \geqslant 0$ such that $x_{m,\infty}^{u,\xi,1} = x_{n,\infty}^{v,\xi,1}$. The same claim holds for the left-most geodesics.*

Here is a very rough and high level sketch of how such a coalescence result is proved. Details can be found in Appendix A of [**27**] (which is an extended version of [**28**]). See also the proof of Theorem 2.3 in [**34**]. First observe that if $x_{0,\infty}^{u,\xi,1}$ and $x_{0,\infty}^{v,\xi,1}$ ever intersect, then from there on they follow the same evolution (smallest B^ξ increment and increment e_1 in case of a tie). Therefore, the task is really to prove that they eventually intersect. By stationarity the assumption of two nonintersecting geodesics implies we can find at least three nonintersecting ones. A local modification of the weights turns the middle geodesic of the triple into a geodesic that stays disjoint from all geodesics that emanate from sufficiently far away. By stationarity again at least δL^2 such disjoint geodesics emanate from an $L \times L$ square. This gives a contradiction because there are only $2L$ boundary points for these geodesics to exit through.

COROLLARY 8.10. *Make the same assumptions as in Theorem* 6.1. *Then with probability one*

$$\{B^\xi(0,e_1) : \xi \in \mathcal{U}^\circ\} \subset \{G_{0,z} - G_{e_1,z} : z \in e_1 + \mathbb{Z}_+^2\}.$$

In particular the map $\xi \mapsto B^\xi(0,e_1)$ *cannot be continuous on all of* $\mathcal{U}^\circ$.

PROOF. For each $\xi \in \mathcal{U}^\circ \cap \mathbb{Q}^2$ consider geodesics $x_{0,\infty}^{0,\xi,1}$ and $x_{0,\infty}^{e_1,\xi,1}$. Theorem 8.3 says that with probability one these geodesics are asymptotically directed into $\mathcal{U}_\xi$. Then Theorem 6.1 implies that almost surely and for all $\xi \in \mathcal{U}^\circ \cap \mathbb{Q}^2$

$$B^\xi(0,e_1) = \lim_{n\to\infty}(G_{0,x_n^{0,\xi,1}} - G_{e_1,x_n^{0,\xi,1}}).$$

By Theorem 8.9 the two geodesics $x_{0,\infty}^{0,\xi,1}$ and $x_{0,\infty}^{e_1,\xi,1}$ coalesce at say a point we denote by z^ξ. This leads to

$$B^\xi(0,e_1) = G_{0,z^\xi} - G_{e_1,z^\xi}.$$

Therefore, we have almost surely

$$\{B^\xi(0,e_1) : \xi \in \mathcal{U}^\circ \cap \mathbb{Q}^2\} \subset \{G_{0,z} - G_{e_1,z} : z \in e_1 + \mathbb{Z}_+^2\}.$$

The claim now follows from monotonicity (6.2). □

Note that the above result does not contradict (7.14). Together, the two results say that there is zero probability that a given (fixed) $\xi \in \mathcal{U}^\circ$ happens to be a discontinuity point of $\xi \mapsto B^\xi(0,e_1)$.

When the weights have a continuous distribution one has a *unique* geodesic between any two given points. What about infinite directional geodesics? The answer is also in the positive.

THEOREM 8.11. [**29**] *Make the same assumptions as in Theorem* 6.1. *Assume also that* ω_0 *has a continuous distribution. Fix* $\xi \in \mathcal{U}^\circ$. *Then with probability one, out of any* $u \in \mathbb{Z}^2$, *there exists a unique infinite geodesic directed into* $\mathcal{U}_\xi$.

PROOF. In view of Theorem 8.7(ii), it is enough to show that $x_{0,\infty}^{u,\xi,1} = x_{0,\infty}^{u,\xi,2}$. This in turn follows from the coalescence result. Indeed, assume the two geodesics do not match. We can assume that they separate right away, otherwise just consider the paths starting at the separation point. Under this assumption, it must be the case that $\omega_u = B^\xi(u,u+e_1) = B^\xi(u,u+e_2)$, for otherwise both geodesics would have followed the smaller B^ξ-gradient and thus stayed together. Now, the above coalescence result implies that $x_{0,\infty}^{u,\xi,1}$ and $x_{0,\infty}^{u+e_2,\xi,1}$ will eventually coalesce, say at point $v = x_n^{u,\xi,1} = x_{n-1}^{u+e_2,\xi,1}$. But then applying (8.1) we would have

$$\sum_{i=0}^{n-1} \omega(x_i^{u,\xi,1}) = B^\xi(u,v) = B^\xi(u,u+e_2) + B^\xi(u+e_2,v) = \omega_u + \sum_{i=0}^{n-2} \omega(x_i^{u+e_2,\xi,1}),$$

which says that the weights add up to the same amount along two different paths. This happens with zero probability if weights have a continuous distribution. Hence the two geodesics can never separate and the theorem is proved. (We used $\omega(x)$ to denote ω_x, for aesthetic reasons.) □

One of the things Theorem 8.11 is really saying is that one should not need to worry about breaking ties among B^ξ gradients. Let us spell this out as a separate result.

THEOREM 8.12. **[29]** *Make the same assumptions as in Theorem* 6.1. *Assume also that* ω_0 *has a continuous distribution. Fix* $\xi \in \mathcal{U}^\circ$. *Then* $\mathbb{P}\{\exists u : B^\xi(u,u+e_1) = B^\xi(u,u+e_2)\} = 0$.

PROOF. The proof is quite simple: when a tie happens at u geodesics $x_{0,\infty}^{u,\xi,1}$ and $x_{0,\infty}^{u,\xi,2}$ separate right away. Since we just showed this cannot happen when weights have a continuous distribution, the theorem follows. □

As the last result of this section we address existence doubly-infinite geodesics (or rather lack thereof).

THEOREM 8.13. **[29]** *Make the same assumptions as in Theorem* 6.1. *Assume also that* ω_0 *has a continuous distribution. Fix* $\xi \in \mathcal{U}^\circ$. *Then*

$$\mathbb{P}\{\exists x_{-\infty,\infty} \textit{ geodesic} : x_{0,\infty} \textit{ is directed into } \mathcal{U}_\xi\} = 0.$$

SKETCH OF THE PROOF. If such a doubly-infinite geodesic existed, with positive probability, then by stationarity we would have another (different) one $y_{-\infty,\infty}$, also directed into $\mathcal{U}_\xi$. By the coalescence result we would have that $x_{0,\infty}$ and $y_{0,\infty}$ coalesce. Modulo renumbering the indices, we can assume that $x_{0,\infty} = y_{0,\infty}$ but $x_{-1} \neq y_{-1}$. Because weights have a continuous distribution, it cannot be that $x_{-n} = y_{-n}$ for any $n > 1$ (otherwise the weights would add up to the same amount $G_{x_{-n},0}$ along two different paths $x_{-n,0}$ and $y_{-n,0}$). We thus have a bi-infinite three-armed "fork" embedded in $\mathbb{Z}^2$. But by stationarity this picture will repeat infinitely often, allowing us to embed a binary tree into $\mathbb{Z}^2$, with its vertices having a positive density. This embedding is not possible and thus we have a contradiction. (Such a tree grows exponentially fast, while the boundary of a box in $\mathbb{Z}^2$ grows only linearly in the diameter of the box.) See Figure 8.1 for an illustration and **[29]** for the details. □

The proof of the above theorem used the coalescence result, which requires us to fix the direction of the geodesics. But we will see in the next section that there are random directions in which there are multiple geodesics out of say the origin. So are there then doubly-infinite geodesics in these random directions? The answer is still expected to be in the negative, but a proof remains elusive.

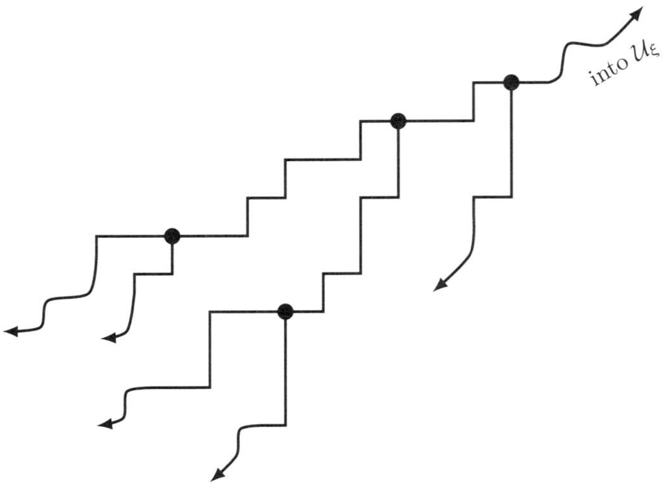

FIGURE 8.1. The tree of bi-infinite geodesics. Bullets mark the triple split points. They have a positive density.

9. The competition interface

We continue to assume the conditions of Theorem 6.1 to be satisfied. In particular, the shape g is still assumed differentiable on $(0,\infty)^2$. Let us also assume in this section that weights ω_x have a continuous distribution.

Recall our earlier definition of the competition interface separating the two geodesic trees rooted at e_1 and e_2 (see Figure 4.2). Denote this up-right path of sites in $\mathbb{Z}^2 + (e_1+e_2)/2$ by φ_n. In particular, $\varphi_0 = (e_1+e_2)/2$.

Does φ_n have an asymptotic direction? What can we say about this direction? Can we describe φ_n using Busemann functions, as we did for geodesics in the previous section?

By monotonicity (6.10) of the Busemann functions we have that

$$B^\xi(0,e_1) - B^\xi(0,e_2)$$

is monotone in $\xi \in \mathcal{U}^\circ \cap \mathbb{Q}^2$. Namely, the above is nonincreasing as $\xi \cdot e_1$ increases in $(0,1) \cap \mathbb{Q}$. (The reason for only considering rational directions is that the limit in (6.3) holds for configurations ω outside a set of measure zero, but this null set depends on the direction ξ. Thus, the limit can be claimed to hold almost surely for only countably many directions at once.)

By Theorem 8.12 we have $B^\xi(0,e_1) \neq B^\xi(0,e_2)$ almost surely for all $\xi \in \mathcal{U}^\circ \cap \mathbb{Q}^2$. If

$$\mathbb{P}\{B^\xi(0,e_1) > B^\xi(0,e_2)\ \forall \xi \in \mathcal{U}^\circ \cap \mathbb{Q}^2\} > 0,$$

then, due to Theorem 8.7(ii), for the configurations in the above event all infinite geodesics out 0 must start with an e_2-step. This can be contradicted by following an argument similar to the proof of Lemma 8.1 to show that, with probability one, there exists at least one geodesic out of 0 that takes a first step e_1. A similar reasoning applies for the case where with positive probability $B^\xi(0,e_1) < B^\xi(0,e_2)$ for all $\xi \in \mathcal{U}^\circ$.

Then, with probability one there exists a unique $\xi^* \in \mathcal{U}^\circ$ such that for all $\xi \in \mathcal{U}^\circ \setminus \{\xi^*\}$

$$B^\xi(0,e_1) > B^\xi(0,e_2) \text{ if } \xi \cdot e_1 < \xi^* \cdot e_1 \quad \textit{and} \quad B^\xi(0,e_1) < B^\xi(0,e_2) \text{ if } \xi \cdot e_1 > \xi^* \cdot e_1.$$

One thing this says is that geodesics originating at 0 that are directed into $\mathcal{U}_\xi$ with $\xi \cdot e_1 < \xi^* \cdot e_1$ (i.e. ξ to the left of ξ^*) must start with an e_2 step. Similarly, geodesics originating at 0 that are directed into $\mathcal{U}_\xi$ with $\xi \cdot e_1 > \xi^* \cdot e_1$ (i.e. ξ to the right of ξ^*) must start with an e_1 step. A slightly sharper version of this argument leads to the following.

THEOREM 9.1. [**29**] *Make the same assumptions as in Theorem* 6.1. *Assume also that ω_0 has a continuous distribution.*

(i) *With probability one, competition interface φ_n has asymptotic direction ξ^*:*

$$\mathbb{P}\{\varphi_n/n \to \xi^*\} = 1.$$

(ii) *With probability one, there exist two infinite geodesics out of* 0 *with asymptotic direction ξ^*, one going through e_1 and the other through e_2:*

$$\mathbb{P}\{\exists x^1_{0,\infty}, x^2_{0,\infty} \textit{ geodesics } : x^1_1 = e_1,\ x^2_1 = e_2,\ x^1_n/n \to \xi^*,\ x^2_n/n \to \xi^*\} = 1.$$

(iii) *ξ^* is a genuine random variable that has a continuous distribution and is supported outside the linear segments of g (if any):*

$$\forall \xi \in \mathcal{U}: \quad \mathbb{P}\{\xi^* = \xi\} \leqslant \mathbb{P}\{\xi^* \in \mathcal{U}_\xi\} = 0.$$

(iv) *ξ^* is supported on all of $\mathcal{U}$, take away the linear segments of g (if any): for any open interval (i.e. connected subset) $\mathcal{V} \subset \mathcal{U}$ such that $\mathcal{U}_\xi = \{\xi\}$ $\forall \xi \in \mathcal{V}$, we have $\mathbb{P}\{\xi^* \in \mathcal{V}\} > 0$.*

It is conjectured that g is strictly concave when weights ω_x have a continuous distribution. Hence, ξ^* is expected to be supported on all of $\mathcal{U}$.

Since weights are assumed to be continuous, the two geodesics in Theorem 9.1(ii) cannot coalesce. This does not contradict Theorem 8.9 because direction ξ^* is random while the coalescence result was about fixed deterministic directions. In fact, this is one way to see why claim (iii) is true.

In the solvable case of exponentially distributed weights, one can compute the distribution of ξ^* explicitly.

LEMMA 9.2. *Assume ω_0 is exponentially distributed with rate $\theta > 0$. Then for $a \in (0,1)$*

$$\mathbb{P}\{\xi^* \cdot e_1 > a\} = \frac{\sqrt{1-a}}{\sqrt{a}+\sqrt{1-a}}. \tag{9.1}$$

If we define the angle ξ^* makes with e_1 by $\theta^* = \tan^{-1}(\xi^* \cdot e_2/\xi^* \cdot e_1)$, then

$$\mathbb{P}\{\theta^* \leqslant t\} = \frac{\sqrt{\sin t}}{\sqrt{\sin t}+\sqrt{\cos t}}. \tag{9.2}$$

PROOF. Recall that $B^\xi(0,e_i)$ has an exponential distribution with rate

$$\lambda_i = \frac{\theta\sqrt{\xi \cdot e_i}}{\sqrt{\xi \cdot e_1}+\sqrt{\xi \cdot e_2}}.$$

Furthermore, $B^\xi(0,e_1)$ and $B^\xi(0,e_2)$ are independent. See Section 7 below. Now compute

$$\begin{aligned}\mathbb{P}\{\xi^*\cdot e_1 > a\} &= \mathbb{P}\{B^{ae_1+(1-a)e_2}(0,e_1) > B^{ae_1+(1-a)e_2}(0,e_2)\}\\ &= \int_0^\infty \lambda_2 e^{-\lambda_2 s}\mathbb{P}\{B^{ae_1+(1-a)e_2}(0,e_1) > s\}\,ds\\ &= \int_0^\infty \lambda_2 e^{-\lambda_2 s}e^{-\lambda_1 s}\,ds = \frac{\lambda_2}{\lambda_1+\lambda_2}\,,\end{aligned}$$

from which (9.1) follows. For the distribution of θ^* we have for $t\in(0,\pi/2)$

$$\mathbb{P}\{\theta^*\leqslant t\} = \mathbb{P}\{\xi^*\cdot e_2 \leqslant \xi^*\cdot e_1\tan t\} = \mathbb{P}\{\xi^*\cdot e_1 \geqslant 1/(1+\tan t)\} = \frac{\sqrt{\tan t}}{1+\sqrt{\tan t}}\,,$$

which is (9.2). □

The competition interface of the exponential corner growth model maps to a certain object called the *second-class particle* in TASEP, so this object has been studied from both perspectives. In this case, a weak-limit version of Theorem 9.1(i) follows from translating a result of Ferrari and Kipnis [**23**] on the limit of the scaled location of the second-class particle in TASEP to LPP language. Almost sure convergence was shown by Mountford and Guiol [**46**] using concentration inequalities and the TASEP variational formula of Seppäläinen [**54**]. Concurrently, [**25**] gave a different proof of almost sure convergence of ϕ_n/n by applying the techniques of directed geodesics and then obtained the distribution (9.2) of the angle of the asymptotic direction ξ^* from the TASEP results of [**23**].

Later, these results on the direction of the competition interface were extended from the quadrant to larger classes of initial profiles in two rounds: first by [**24**] still with TASEP and geodesic techniques, and then by [**14**] using their earlier results on Busemann functions [**13**].

Coupier [**17**] also relied on the TASEP connection to sharpen the geodesics results of [**25**]. He showed that with probability one there are no triple geodesics (out of the origin) in any direction.

10. History

We now give a quick overview of the last twenty or so years of research on Busemann functions and geodesics in percolation.

As we mentioned above, Licea and Newman [**42**] were the first to introduce a technique for proving existence, uniqueness, and coalescence of directional geodesics under a global curvature assumption on the limit shape to control how much geodesics deviate from a straight line, and then as a consequence deducing the existence of Busemann functions. See also the summary in Newman's ICM paper [**48**]. Although verifying the curvature assumption for percolation models with general weights remains an open problem, it can be done in a number of special cases. Thus, Licea and Newman's approach was applied to directed LPP with exponential weights [**25**] and to several other (non-lattice) models built on homogeneous Poisson processes [**8, 12, 38, 57**]. In the case of the exponential corner growth model, another set of tools comes from its connection with TASEP, as explained at the end of Section 9.

The idea of deducing existence and uniqueness of stationary processes by studying geodesic-like objects has also been used in random dynamical systems. For example, this is how [**22**] and its extensions [**5, 6, 9, 35, 39**] show existence of invariant measures for the Burgers equation with random forcing. These works treated cases where space is compact or essentially compact. To make progress in a non-compact case, the approach of Newman et al. was adopted again in [**7, 8, 10**].

The approach we presented in this chapter is the one we took in [**28, 29**] and is the very opposite of the above. Using the connection to queues in tandem the Busemann limits are constructed a priori in the form of stationary cocycles that come from certain *invariant* measures of the queuing system. Using a certain monotonicity the cocycles are then compared to the gradients of passage times. The monotonicity, ergodicity, and differentiability (rather than curvature) of the limit shape give the control that proves the Busemann limits. After establishing existence of Busemann functions, we use them to prove existence, directedness, coalescence, and uniqueness results about the geodesics.

A similar approach was carried out by Damron and Hanson [**19, 20**] for the standard first-passage percolation model. They first construct (generalized) Busemann functions from weak subsequential limits of first-passage time differences. These weak Busemann limits can be regarded as a counterpart of our stationary cocycles. This then gives access to properties of geodesics, while weakening the need for the global curvature assumption.

An independent line of work is that of Hoffman [**36, 37**] on the standard first-passage percolation, with general weights and without any regularity assumptions on the limit shape. Assuming all semi-infinite geodesics coalesce, [**36**] constructed a Busemann function and used it to get a contradiction, concluding that there are at least two semi-infinite geodesics. ([**26**] gave an independent proof with a different method.) [**37**] extended this to at least four geodesics.

11. Next: fluctuations

The results we presented can be thought of as analogues of the law of large numbers. A natural follow-up is to study the analogue of the central limit theorem, i.e. questions concerning the size of deviations of the passage times $G_{0,\lfloor n\xi\rfloor}$ from their asymptotic limit $ng(\xi)$ and of the geodesics, both finite (i.e. from 0 to $\lfloor n\xi\rfloor$) and infinite (i.e. going in direction ξ), from a straight line.

Due to the maximum in the definition of the passage times, $G_{0,\lfloor n\xi\rfloor} - ng(\xi)$ should be tighter than in the case of just a sum of identically distributed independent random variables. That is, its fluctuations should be smaller than order $n^{1/2}$. On the other hand, one would expect the geodesic paths to wander a great deal in search of favorable weights. For example, if $x_{0,\infty}$ is a path that follows the smallest B^ξ gradient, then it should be the case that $x_n - n\xi$ has fluctuations of order larger than $n^{1/2}$.

We will see in articles [**55**] and [**16**] how the above can be answered quite precisely for the solvable models: $G_{0,\lfloor n\xi\rfloor} - ng(\xi)$ fluctuates on order $n^{1/3}$ (and we can even determine the limiting *Tracy-Widom distribution* of $n^{-1/3}(G_{0,\lfloor n\xi\rfloor} - ng(\xi))$ and $x_n - n\xi$ fluctuates on the order $n^{2/3}$ (but here, a limiting distribution of $(x_n - n\xi)/n^{2/3}$ is only conjectured).

Just like it is the case for the central limit theorem, this behavior is believed to be universal, going beyond solvable models, but this is far from being proved. See, however, [**1**] and [**41**] for results towards this universality conjecture.

References

[1] Tom Alberts, Konstantin Khanin, and Jeremy Quastel, *The intermediate disorder regime for directed polymers in dimension* $1+1$, Ann. Probab. **42** (2014), no. 3, 1212–1256, DOI 10.1214/13-AOP858. MR3189070

[2] Sven Erick Alm, *A note on a problem by Welsh in first-passage percolation*, Combin. Probab. Comput. **7** (1998), no. 1, 11–15, DOI 10.1017/S0963548397003301. MR1611098

[3] Sven Erick Alm and John C. Wierman, *Inequalities for means of restricted first-passage times in percolation theory*, Combin. Probab. Comput. **8** (1999), no. 4, 307–315, DOI 10.1017/S0963548399003843. Random graphs and combinatorial structures (Oberwolfach, 1997). MR1723645

[4] Antonio Auffinger and Michael Damron, *Differentiability at the edge of the percolation cone and related results in first-passage percolation*, Probab. Theory Related Fields **156** (2013), no. 1-2, 193–227, DOI 10.1007/s00440-012-0425-4. MR3055257

[5] Yuri Bakhtin, *Burgers equation with random boundary conditions*, Proc. Amer. Math. Soc. **135** (2007), no. 7, 2257–2262, DOI 10.1090/S0002-9939-07-08736-9. MR2299503

[6] Yuri Bakhtin, *The Burgers equation with Poisson random forcing*, Ann. Probab. **41** (2013), no. 4, 2961–2989, DOI 10.1214/12-AOP747. MR3112935

[7] ———, *Inviscid Burgers equation with random kick forcing in noncompact setting*, Electron. J. Probab. **21** (2016), Paper No. 37, 50.

[8] Yuri Bakhtin, Eric Cator, and Konstantin Khanin, *Space-time stationary solutions for the Burgers equation*, J. Amer. Math. Soc. **27** (2014), no. 1, 193–238, DOI 10.1090/S0894-0347-2013-00773-0. MR3110798

[9] Yuri Bakhtin and Konstantin Khanin, *Localization and Perron-Frobenius theory for directed polymers* (English, with English and Russian summaries), Mosc. Math. J. **10** (2010), no. 4, 667–686, 838. MR2791052

[10] Yuri Bakhtin, *Ergodic theory of the Burgers equation*, Probability and statistical physics in St. Petersburg, Proc. Sympos. Pure Math., vol. 91, Amer. Math. Soc., Providence, RI, 2016, pp. 1–49. MR3526825

[11] Riddhipratim Basu, Sourav Sarkar, and Allan Sly, *Coalescence of geodesics in exactly solvable models of last passage percolation.*, (2017), (`arXiv 1704.05219`).

[12] Eric Cator and Leandro P. R. Pimentel, *A shape theorem and semi-infinite geodesics for the Hammersley model with random weights*, ALEA Lat. Am. J. Probab. Math. Stat. **8** (2011), 163–175. MR2783936

[13] Eric Cator and Leandro P. R. Pimentel, *Busemann functions and equilibrium measures in last passage percolation models*, Probab. Theory Related Fields **154** (2012), no. 1-2, 89–125, DOI 10.1007/s00440-011-0363-6. MR2981418

[14] Eric Cator and Leandro P. R. Pimentel, *Busemann functions and the speed of a second class particle in the rarefaction fan*, Ann. Probab. **41** (2013), no. 4, 2401–2425, DOI 10.1214/11-AOP709. MR3112921

[15] Henry Cohn, Noam Elkies, and James Propp, *Local statistics for random domino tilings of the Aztec diamond*, Duke Math. J. **85** (1996), no. 1, 117–166, DOI 10.1215/S0012-7094-96-08506-3. MR1412441

[16] Ivan Corwin, *Exactly solving the KPZ equation*, Proc. Sympos. Appl. Math. **75** (2018).

[17] David Coupier, *Multiple geodesics with the same direction*, Electron. Commun. Probab. **16** (2011), 517–527, DOI 10.1214/ECP.v16-1656. MR2836758

[18] Michael Damron, *Random growth models: shape and convergence rate*, Proc. Sympos. Appl. Math. **75** (2018).

[19] Michael Damron and Jack Hanson, *Busemann functions and infinite geodesics in two-dimensional first-passage percolation*, Comm. Math. Phys. **325** (2014), no. 3, 917–963, DOI 10.1007/s00220-013-1875-y. MR3152744

[20] Michael Damron and Jack Hanson, *Bigeodesics in first-passage percolation*, Comm. Math. Phys. **349** (2017), no. 2, 753–776, DOI 10.1007/s00220-016-2743-3. MR3595369

[21] Richard Durrett and Thomas M. Liggett, *The shape of the limit set in Richardson's growth model*, Ann. Probab. **9** (1981), no. 2, 186–193. MR606981

[22] Weinan E, K. Khanin, A. Mazel, and Ya. Sinai, *Invariant measures for Burgers equation with stochastic forcing*, Ann. of Math. (2) **151** (2000), no. 3, 877–960, DOI 10.2307/121126. MR1779561

[23] P. A. Ferrari and C. Kipnis, *Second class particles in the rarefaction fan* (English, with English and French summaries), Ann. Inst. H. Poincaré Probab. Statist. **31** (1995), no. 1, 143–154. MR1340034

[24] Pablo A. Ferrari, James B. Martin, and Leandro P. R. Pimentel, *A phase transition for competition interfaces*, Ann. Appl. Probab. **19** (2009), no. 1, 281–317, DOI 10.1214/08-AAP542. MR2498679

[25] Pablo A. Ferrari and Leandro P. R. Pimentel, *Competition interfaces and second class particles*, Ann. Probab. **33** (2005), no. 4, 1235–1254, DOI 10.1214/009117905000000080. MR2150188

[26] Olivier Garet and Régine Marchand, *Coexistence in two-type first-passage percolation models*, Ann. Appl. Probab. **15** (2005), no. 1A, 298–330, DOI 10.1214/105051604000000503. MR2115045

[27] Nicos Georgiou, Firas Rassoul-Agha, and Timo Seppäläinen, *Geodesics and the competition interface for the corner growth model*, Probab. Theory Related Fields **169** (2017), no. 1-2, 223–255, DOI 10.1007/s00440-016-0734-0. MR3704769

[28] Nicos Georgiou, Firas Rassoul-Agha, and Timo Seppäläinen, *Geodesics and the competition interface for the corner growth model*, Probab. Theory Related Fields **169** (2017), no. 1-2, 223–255, DOI 10.1007/s00440-016-0734-0. MR3704769

[29] Nicos Georgiou, Firas Rassoul-Agha, and Timo Seppäläinen, *Stationary cocycles and Busemann functions for the corner growth model*, Probab. Theory Related Fields **169** (2017), no. 1-2, 177–222, DOI 10.1007/s00440-016-0729-x. MR3704768

[30] Nicos Georgiou, Firas Rassoul-Agha, Timo Seppäläinen, and Atilla Yilmaz, *Ratios of partition functions for the log-gamma polymer*, Ann. Probab. **43** (2015), no. 5, 2282–2331, DOI 10.1214/14-AOP933. MR3395462

[31] Peter W. Glynn and Ward Whitt, *Departures from many queues in series*, Ann. Appl. Probab. **1** (1991), no. 4, 546–572. MR1129774

[32] Olle Häggström and Ronald Meester, *Asymptotic shapes for stationary first passage percolation*, Ann. Probab. **23** (1995), no. 4, 1511–1522. MR1379157

[33] J. M. Hammersley and D. J. A. Welsh, *First-passage percolation, subadditive processes, stochastic networks, and generalized renewal theory*, Proc. Internat. Res. Semin., Statist. Lab., Univ. California, Berkeley, Calif., Springer-Verlag, New York, 1965, pp. 61–110. MR0198576

[34] Jack Hanson, *Infinite geodesics, asymptotic directions, and busemann functions in first-passage percolation*, Proc. Sympos. Appl. Math. **75** (2018).

[35] Viet Ha Hoang and Konstantin Khanin, *Random Burgers equation and Lagrangian systems in non-compact domains*, Nonlinearity **16** (2003), no. 3, 819–842, DOI 10.1088/0951-7715/16/3/303. MR1975784

[36] Christopher Hoffman, *Coexistence for Richardson type competing spatial growth models*, Ann. Appl. Probab. **15** (2005), no. 1B, 739–747, DOI 10.1214/105051604000000729. MR2114988

[37] Christopher Hoffman, *Geodesics in first passage percolation*, Ann. Appl. Probab. **18** (2008), no. 5, 1944–1969, DOI 10.1214/07-AAP510. MR2462555

[38] C. Douglas Howard and Charles M. Newman, *Geodesics and spanning trees for Euclidean first-passage percolation*, Ann. Probab. **29** (2001), no. 2, 577–623, DOI 10.1214/aop/1008956685. MR1849171

[39] R. Iturriaga and K. Khanin, *Burgers turbulence and random Lagrangian systems*, Comm. Math. Phys. **232** (2003), no. 3, 377–428, DOI 10.1007/s00220-002-0748-6. MR1952472

[40] William Jockusch, James Propp, and Peter Shor, *Random domino tilings and the arctic circle theorem*, (1998), `arXiv:math/9801068`.

[41] Arjun Krishnan and Jeremy Quastel, *Tracy-Widom fluctuations for perturbations of the log-gamma polymer in intermediate disorder*, (2016), to appear in Annals of Applied Probability (`arXiv 1610.06975`).

[42] Cristina Licea and Charles M. Newman, *Geodesics in two-dimensional first-passage percolation*, Ann. Probab. **24** (1996), no. 1, 399–410, DOI 10.1214/aop/1042644722. MR1387641

[43] Carolyn T MacDonald, Julian H Gibbs, and Allen C Pipkin, *Kinetics of biopolymerization on nucleic acid templates*, Biopolymers **6** (1968), no. 1, 1–25.

[44] Jean Mairesse and Balaji Prabhakar, *The existence of fixed points for the $\cdot/GI/1$ queue*, Ann. Probab. **31** (2003), no. 4, 2216–2236, DOI 10.1214/aop/1068646383. MR2016617

[45] James B. Martin, *Limiting shape for directed percolation models*, Ann. Probab. **32** (2004), no. 4, 2908–2937, DOI 10.1214/009117904000000838. MR2094434

[46] Thomas Mountford and Hervé Guiol, *The motion of a second class particle for the TASEP starting from a decreasing shock profile*, Ann. Appl. Probab. **15** (2005), no. 2, 1227–1259, DOI 10.1214/105051605000000151. MR2134103

[47] Eginhard J. Muth, *The reversibility property of production lines*, Management Sci. **25** (1979/80), no. 2, 152–158, DOI 10.1287/mnsc.25.2.152. MR537320

[48] Charles M. Newman, *A surface view of first-passage percolation*, Proceedings of the International Congress of Mathematicians, Vol. 1, 2 (Zürich, 1994), Birkhäuser, Basel, 1995, pp. 1017–1023. MR1404001

[49] Leandro P. R. Pimentel, *Duality between coalescence times and exit points in last-passage percolation models*, Ann. Probab. **44** (2016), no. 5, 3187–3206, DOI 10.1214/15-AOP1044. MR3551194

[50] Firas Rassoul-Agha, Timo Seppäläinen, and Atilla Yilmaz, *Quenched free energy and large deviations for random walks in random potentials*, Comm. Pure Appl. Math. **66** (2013), no. 2, 202–244, DOI 10.1002/cpa.21417. MR2999296

[51] R. Tyrrell Rockafellar, *Convex analysis*, Princeton Mathematical Series, No. 28, Princeton University Press, Princeton, N.J., 1970. MR0274683

[52] H. Rost, *Nonequilibrium behaviour of a many particle process: density profile and local equilibria*, Z. Wahrsch. Verw. Gebiete **58** (1981), no. 1, 41–53, DOI 10.1007/BF00536194. MR635270

[53] T. Seppäläinen, *Hydrodynamic scaling, convex duality and asymptotic shapes of growth models*, Markov Process. Related Fields **4** (1998), no. 1, 1–26. MR1625007

[54] Timo Seppäläinen, *Existence of hydrodynamics for the totally asymmetric simple K-exclusion process*, Ann. Probab. **27** (1999), no. 1, 361–415, DOI 10.1214/aop/1022677266. MR1681094

[55] Timo Seppäläinen, *The corner growth model with exponential weights*, Proc. Sympos. Appl. Math. **75** (2018).

[56] Frank Spitzer, *Interaction of Markov processes*, Advances in Math. **5** (1970), 246–290 (1970), DOI 10.1016/0001-8708(70)90034-4. MR0268959

[57] Mario V. Wüthrich, *Asymptotic behaviour of semi-infinite geodesics for maximal increasing subsequences in the plane*, In and out of equilibrium (Mambucaba, 2000), Progr. Probab., vol. 51, Birkhäuser Boston, Boston, MA, 2002, pp. 205–226. MR1901954

University of Utah, Mathematics Department, 155 S 1400 E, Salt Lake City, Utah 84109

Email address: `firas@math.utah.edu`

Proceedings of Symposia in Applied Mathematics
Volume **75**, 2018
https://doi.org/10.1090/psapm/075/00660

The corner growth model with exponential weights

Timo Seppäläinen

Abstract. We discuss several related features of the exactly solvable two-dimensional corner growth model with exponentially distributed weights. A key property of this model is the availability of a fairly explicit stationary version that possesses useful independence properties. With the help of couplings and estimates, we prove the existence of Busemann functions for this model, and the precise values of the longitudinal and transversal fluctuation exponents for the stationary corner growth model. The Busemann functions in turn furnish extremals for variational formulas that describe limiting shape functions.

Contents

1. Introduction

Overview. This article discusses a circle of ideas associated with random growth models. The results are developed in the context of one of the oldest exactly solvable models in the 1+1 dimensional Kardar-Parisi-Zhang (KPZ) class, namely the *corner growth model* (CGM) with exponential weights. The centerpiece of the development is an increment-stationary stochastic process that represents a stationary version of the random growth model. Section 2 describes how this stochastic process arises as the extremal in a variational formula for the limiting shape function. This part of the text treats a very general last-passage percolation (LPP) model on the d-dimensional integer lattice.

From Section 3 onwards we restrict ourselves to the exponential CGM in two dimensions. Section 3 constructs the stationary version of the exponential CGM and shows how it enables an explicit calculation of the limiting shape function of

2010 *Mathematics Subject Classification.* 60K35, 60K37, 82B43.

The author was partially supported by National Science Foundation grant DMS-1602486 and by the Wisconsin Alumni Research Foundation.

both the point-to-point and the point-to-line version of the CGM. Additionally we prove the shape theorem and show that macroscopically geodesics are straight lines.

Section 4 proves the existence of limiting Busemann functions. These Busemann functions are extremals for the variational formulas described in Section 2. As an application of the results on Busemann functions we prove a CGM version of the midpoint problem.

A feature of the KPZ class is that two fluctuation exponents take universal values for all models in the class. The longitudinal exponent χ describes the magnitude of fluctuations of particle current in conservative particle systems, last-passage and first-passage times in growth models, and free energy fluctuations in positive temperature polymers. The transversal exponent ζ describes the magnitude of spatial correlations and the fluctuations of random paths such as geodesics in percolation models and polymer paths.

In 1+1 dimensions the values of the KPZ exponents are $\chi = \frac{1}{3}$ and $\zeta = \frac{2}{3}$. So far these exact exponent values have been proved only for solvable models in the KPZ class, with the possible exception of the class of zero-range processes treated in [**10**]. In Section 5 we prove $\chi = \frac{1}{3}$ and $\zeta = \frac{2}{3}$ in the 1+1 dimensional stationary exponential CGM. The key to these coupling proofs is the explicit stationary version of the CGM.

Appendix A develops some useful couplings for the CGM.

This article does not have to be read in its entirety in linear order. Section 2 sets the stage but strictly speaking is not needed for the remainder. Section 3 is a prerequisite for Sections 4 and 5, but these last two sections are independent of each other.

Assuming standard background in measure theory and probability theory, this article is fairly self-contained, except for the following points in Section 2. We take for granted the existence of limiting shape functions (Theorem 2.4). This existence theorem is an application of subadditive ergodic theory, together with some estimation to produce limit functions defined on $\mathbb{R}^d$. Three further results are quoted from the literature: Lemma 2.5, Theorem 2.9 which is an ergodic theorem for cocycles, and the existence of a minimizer in Theorem 2.10.

Literature. Here are some brief notes on the history of the topics of this article, without any claim to completeness.

The first derivation of the limit shape of the corner growth model with exponential weights came in Rost's seminal 1981 paper [**47**] on the hydrodynamic limit of the totally asymmetric simple exclusion process (TASEP) with step initial condition. An even earlier instance of work on similar limit shapes involved the longest common subsequence of a random permutation [**32, 40, 54**]. The Poissonization of this problem can be cast as a last-passage growth model on a homogeneous planar Poisson point process. The study of last-passage percolation began to flourish in the 1990s. A number of papers appeared on limit shapes and large deviations, examples of which include [**2, 16, 23, 24, 34, 37, 48–51**].

The notion that the fluctuations of these models do not follow the standard central limit theorem, but have a different exponent 1/3, was a well-established conjecture from physics, but mathematical proof was lacking. The first rigorous results of KPZ fluctuations came at the turn of the millennium in the breakthrough papers [**7, 35**].

Variational formulas of the type presented in Section 2 arose first in homogenization and random walk in random environment (RWRE). See equation (4.19) in the review paper [**38**] and also Definition 4 on page 5 of [**46**]. Versions of these formulas for last-passage percolation came in [**29**]. Another related variational formula was introduced independently in undirected first-passage percolation in [**39**].

Busemann function is a notion from geometry. It was introduced into first-passage percolation by Hoffman [**33**] and Newman [**43**]. Newman and coauthors developed an argument for the existence of Busemann functions based on the coalescence of directed geodesics. This method relies on an assumption on the curvature of the limit shape. For the exponential corner growth model the existence of Busemann functions was first proved by Ferrari and Pimentel [**27**] with Newman's approach. (The limiting Busemann function appears in equation (38) of [**27**].) Newman's method was applicable because the limit shape of the exponential corner growth model is known explicitly. In Section 4 we apply an alternative method that uses a stationary version of the LPP process. This approach has been applied to the CGM with general weights in [**30**]. See [**8, 14, 26**] for a further selection of papers that use Busemann functions to study percolation, growth models, and interacting particle systems.

The seminal papers [**7, 35**] on KPZ fluctuations proved Tracy-Widom limits with refined combinatorics and asymptotic analysis. A more robust probabilistic proof of only the fluctuation exponents appeared a few years later in [**13**]. This approach was adapted to the corner growth model in [**9**] and to a positive-temperature polymer model in [**53**].

Review articles and lecture notes on KPZ universality include the following: [**12, 18, 36**]. Fluctuation exponents in both the KPZ class and the EW (Edwards-Wilkinson) class are developed in lectures [**52**]. Lecture notes [**17, 22**] give overviews of the state of the art in directed polymer models, which are the positive temperature counterpart of directed percolation. Articles [**19, 21**] are introductions to KPZ universality and growth models aimed at general mathematical audiences.

Some general notation and terminology. $\mathbb{Z}_{\geqslant 0} = \{0, 1, 2, 3, \dots\}$ and $\mathbb{Z}_{>0} = \{1, 2, 3, \dots\}$. For $n \in \mathbb{Z}_{>0}$ we abbreviate $[n] = \{1, 2, \dots, n\}$. A sequence of n points is denoted by $x_{0,n} = (x_k)_{k=0}^n = \{x_0, x_1, \dots, x_n\}$, and in case it is a path of length n also by $x_{\bullet}$. $a \vee b = \max\{a, b\}$. C is a constant whose value can change from line to line.

The standard basis vectors of $\mathbb{R}^2$ are $e_1 = (1, 0)$ and $e_2 = (0, 1)$. For a point $x = (x_1, x_2) \in \mathbb{R}^2$ the ℓ^1-norm is $|x|_1 = |x_1| + |x_2|$ and integer parts are taken coordinatewise: $\lfloor x \rfloor = (\lfloor x_1 \rfloor, \lfloor x_2 \rfloor)$. We call the x-axis occasionally the e_1-axis, and similarly the y-axis and the e_2-axis are the same thing. Inequalities on $\mathbb{R}^2$ are interpreted coordinatewise: for $x = (x_1, x_2) \in \mathbb{R}^2$ and $y = (y_1, y_2) \in \mathbb{R}^2$, $x \leqslant y$ means $x_1 \leqslant y_1$ and $x_2 \leqslant y_2$. Notation $[x, y]$ represents both the line segment $[x, y] = \{tx + (1-t)y : 0 \leqslant t \leqslant 1\}$ and the rectangle $[x, y] = \{(z_1, z_2) \in \mathbb{R}^2 : x_i \leqslant z_i \leqslant y_i \text{ for } i = 1, 2\}$. The context will make clear which case is used. 0 denotes the origin of both $\mathbb{R}$ and $\mathbb{R}^2$.

$X \sim \text{Exp}(\lambda)$ for $0 < \lambda < \infty$ means that random variable X has exponential distribution with rate λ. This X is a positive random variable whose probability distribution satisfies $P(X > t) = e^{-\lambda t}$ for $t \geqslant 0$. It has mean $E(X) = \lambda^{-1}$ and variance $\text{Var}(X) = \lambda^{-2}$.

We write ω_x and $\omega(x)$ interchangeably for the weight attached to lattice point x. In general, $\overline{X} = X - EX$ denotes a random variable X centered at its mean.

2. Variational formulas for LPP shapes

2.1. Directed last-passage percolation on $\mathbb{Z}^d$. We consider here a general setting before specializing to the two-dimensional corner growth model. Let $(\Omega, \mathfrak{S}, \mathbb{P})$ be a Polish product probability space $\Omega = \Gamma^{\mathbb{Z}^d}$ of random environments $\omega = (\omega_x)_{x\in\mathbb{Z}^d} \in \Omega$, with Borel σ-algebra $\mathfrak{S}$, and a product probability measure $\mathbb{P}$ under which the coordinates are i.i.d. random variables: for any distinct lattice points $x_1, \dots, x_n \in \mathbb{Z}^d$ and any Borel sets $B_1, \dots, B_n \subset \Gamma$,

$$\mathbb{P}\{\omega : \omega_{x_i} \in B_i \text{ for } i = 1, \dots, n\} = \prod_{i=1}^{n} \mathbb{P}\{\omega : \omega_0 \in B_i\}. \tag{2.1}$$

The group of translations or shifts $\{\theta_x\}_{x\in\mathbb{Z}^d}$ act on Ω by $(\theta_x\omega)_y = \omega_{x+y}$.

Let $\mathcal{R}$ be a finite subset of $\mathbb{Z}^d$. A lattice path $x_{0,n} = (x_k)_{k=0}^n \subset \mathbb{Z}^d$ is admissible if its steps satisfy $z_k = x_k - x_{k-1} \in \mathcal{R}$. Let $\mathcal{U} = \text{co}\,\mathcal{R}$ be the convex hull of $\mathcal{R}$ in $\mathbb{R}^d$, and $\text{ri}\,\mathcal{U}$ the relative interior of $\mathcal{U}$. We put ourselves in the directed setting by assuming that

$$0 \notin \mathcal{U}. \tag{2.2}$$

This implies the existence of a vector $\hat{u} \in \mathbb{R}^d$ and $\delta > 0$ such that $z \cdot \hat{u} \geqslant \delta$ for all $z \in \mathcal{R}$.

For convenience we also assume that $\mathbb{Z}^d$ is the smallest additive group that contains $\mathcal{R}$. Without this assumption we would carry along the group generated by $\mathcal{R}$ in the development.

The weights of admissible steps z are determined by a measurable function $V : \Omega \times \mathcal{R} \to \mathbb{R}$ about which we assume the following:

$$\begin{aligned} &\forall z \in \mathcal{R},\ V(\omega, z) \text{ is a local function of } \omega \\ &\text{and for some } p > d,\ V(\cdot\,, z) \in L^p(\mathbb{P}). \end{aligned} \tag{2.3}$$

By definition, a local function of ω is one that depends on only finitely many coordinates of ω.

Example 2.1. The basic example to think about is the two-dimensional corner growth model, with real weights on the vertices: $\omega = (\omega_x)_{x\in\mathbb{Z}^2} \in \Omega = \mathbb{R}^{\mathbb{Z}^2}$. The set of admissible steps is $\mathcal{R} = \{e_1, e_2\}$, and potential given by the weight at the origin: $V(\omega, z) = \omega_0$. The set of possible limiting velocities of paths is the closed line segment $\mathcal{U} = [e_1, e_2]$, and its relative interior is the open line segment $\text{ri}\,\mathcal{U} = (e_1, e_2)$. △

Example 2.2. The formulation covers also weights on directed edges. Let $\mathcal{R} = \{e_1, e_2, \dots, e_d\}$ and let $\vec{\mathcal{E}}_d = \{(x, y) \in \mathbb{Z}^d \times \mathbb{Z}^d : y - x \in \mathcal{R}\}$ be the set of directed nearest-neighbor edges on $\mathbb{Z}^d$. Let $\omega = (\omega(e))_{e\in\vec{\mathcal{E}}_d}$ be a configuration of weights on directed nearest-neighbor edges. The potential picks out the edge weight: $V(\theta_x\omega, z) = \omega(x, x + z)$ for $x \in \mathbb{Z}^d$ and $z \in \mathcal{R}$. △

The point-to-level last-passage percolation with *external field* or *tilt* $h \in \mathbb{R}^d$ is defined by

$$G_n(h) = \max_{x_{0,n}:\, x_0=0} \Big\{ \sum_{k=0}^{n-1} V(\theta_{x_k}\omega, z_{k+1}) + h \cdot x_n \Big\}, \qquad h \in \mathbb{R}^d. \tag{2.4}$$

The maximum is over admissible n-step paths $x_{0,n} = (x_k)_{k=0}^n$ that start at the origin $x_0 = 0$ and whose steps are denoted by $z_k = x_k - x_{k-1}$.

The point-to-point last-passage percolation with restricted path length is defined by

$$G_{x,(n),y} = \max_{x_{0,n}:\, x_0=x,\, x_n=y} \sum_{k=0}^{n-1} V(\theta_{x_k}\omega,\, z_{k+1}), \qquad x \in \mathbb{Z}^d. \tag{2.5}$$

The maximum is over admissible n-step paths $x_{0,n} = (x_k)_{k=0}^n$ that start at x and end at y. If y cannot be reached from x with an admissible n-step path then set $G_{x,(n),y} = -\infty$. Our convention is $G_{x,(0),x} = 0$.

Any path $x_{0,n} = (x_k)_{k=0}^n$ with steps $z_{k+1} = x_{k+1} - x_k$ is a *geodesic* or a *maximizing path* if

$$\sum_{k=0}^{n-1} V(\theta_{x_k}\omega,\, z_{k+1}) = G_{x_0,(n),\,x_n}. \tag{2.6}$$

REMARK 2.3. The number of steps in an admissible path from x to y is determined uniquely by x and y for all pairs x, y iff 0 does not lie in the affine hull of $\mathcal{R}$. This is true for natural directed examples such as $\mathcal{R} = \{e_1, e_2, \dots, e_d\}$. Then we can write $G_{x,y} = G_{x,(n),y}$ where n is the unique number of admissible steps from x to y. △

We take the existence of the limiting shape functions for granted, as stated in the next theorem.

THEOREM 2.4. *Let $\mathbb{P}$ be an i.i.d. product probability measure and assume* (2.2) *and* (2.3).

(i) *There exists a nonrandom finite, convex, Lipschitz function $g_{\text{pl}} : \mathbb{R}^d \to \mathbb{R}$ such that*

$$g_{\text{pl}}(h) = \lim_{n\to\infty} n^{-1} G_n(h) \quad \mathbb{P}\text{-}a.s. \tag{2.7}$$

(ii) *There exists a nonrandom finite, concave, continuous function $g_{\text{pp}} : \mathcal{U} \to \mathbb{R}$ such that*

$$g_{\text{pp}}(\xi) = \lim_{n\to\infty} n^{-1} G_{0,(n),[n\xi]}, \qquad \xi \in \mathcal{U} \tag{2.8}$$

where $[n\xi]$ is a point reachable in n steps and approximately $n\xi$. The limits satisfy

$$g_{\text{pl}}(h) = \sup_{\xi\in\mathcal{U}} \{g_{\text{pp}}(\xi) + h\cdot\xi\}. \tag{2.9}$$

The theorem above is a part of Theorem 2.4 in [**29**].

Here is a sketch of the argument for the duality (2.9) between the point-to-point and point-to-line shape functions.

$$\begin{aligned} g_{\text{pl}}(h) &= \lim_{n\to\infty} \frac{1}{n} G_n(h) = \lim_{n\to\infty} \max_{x_{0,n}:\, x_0=0} \frac{1}{n}\Big\{ \sum_{k=0}^{n-1} V(\theta_{x_k}\omega, z_{k+1}) + h\cdot x_n \Big\} \\ &= \lim_{n\to\infty} \max_x \frac{1}{n}\{G_{0,(n),x} + h\cdot x\} = \lim_{n\to\infty} \sup_{\xi\in\mathcal{U}} \Big\{ \frac{1}{n} G_{0,(n),[n\xi]} + h\cdot \frac{[n\xi]}{n} \Big\} \\ &= \sup_{\xi\in\mathcal{U}} \lim_{n\to\infty} \Big\{ \frac{1}{n} G_{0,(n),[n\xi]} + h\cdot\frac{[n\xi]}{n} \Big\} = \sup_{\xi\in\mathcal{U}} \{\, g_{\text{pp}}(\xi) + h\cdot\xi \,\}. \end{aligned}$$

The interchange of the limit and the supremum is justified by a discretization argument and estimates of the kind used for positive temperature polymer models in Lemma 2.9 of [**44**].

We can extend equation (2.9) to a convex duality by setting $g_{\text{pp}} \equiv -\infty$ on $\mathcal{U}^c$ so that then the supremum extends over all $\xi \in \mathbb{R}^d$. Then by convex duality,

$$g_{\text{pp}}(\xi) = \inf_{h \in \mathbb{R}^d} \{g_{\text{pl}}(h) - h \cdot \xi\}, \qquad \xi \in \mathcal{U}. \tag{2.10}$$

Let us say that $\xi \in \mathcal{U}$ and $h \in \mathbb{R}^d$ are *dual* if

$$g_{\text{pl}}(h) = g_{\text{pp}}(\xi) + h \cdot \xi.$$

LEMMA 2.5. *Every $\xi \in \text{ri}\,\mathcal{U}$ has a dual $h \in \mathbb{R}^d$.*

The lemma is proved by arguing that the infimum in (2.10) can be restricted to a compact set. See Lemma 4.3 in [**29**].

In order to develop variational formulas for the limits g_{pp} and g_{pl}, we introduce a class of stationary processes we call cocycles and state an ergodic theorem for them.

2.2. Stationary cocycles.

DEFINITION 2.6 (Cocycles). A measurable function $B : \Omega \times \mathbb{Z}^d \times \mathbb{Z}^d \to \mathbb{R}$ is a stationary cocycle if it satisfies these two conditions for $\mathbb{P}$-a.e. ω and all $x, y, z \in \mathbb{Z}^d$:

$$\begin{aligned} B(\omega, x+z, y+z) &= B(\theta_z\omega, x, y) \qquad &\text{(stationarity)}\\ B(\omega, x, y) + B(\omega, y, z) &= B(\omega, x, z) \qquad &\text{(additivity)}. \end{aligned}$$

$\mathcal{K}$ denotes the space of stationary cocycles B such that $\mathbb{E}|B(x,y)| < \infty$ $\forall x, y \in \mathbb{Z}^d$. $\mathcal{K}_0$ denotes the subspace of $F \in \mathcal{K}$ such that $\mathbb{E}[F(x,y)] = 0$ $\forall x, y \in \mathbb{Z}^d$.

An equivalent way to define a stationary cocycle is as a shift-invariant gradient of a function on $\Omega \times \mathbb{Z}^d$. For such a function f, define $\nabla_u f(\omega, x) = f(\omega, x+u) - f(\omega, x)$. Say that f has a *shift-invariant gradient* if $\nabla_u f(\omega, x) = \nabla_u f(\theta_x\omega, 0)$ for all $u, x \in \mathbb{Z}^d$.

LEMMA 2.7. *Let B be a real function on $\Omega \times \mathbb{Z}^d \times \mathbb{Z}^d$. Then B is a stationary cocycle if and only if there exists a real function f on $\Omega \times \mathbb{Z}^d$ with a shift-invariant gradient and such that $B(\omega, x, y) = \nabla_{y-x} f(\omega, x) = f(\omega, y) - f(\omega, x)$ for all $x, y \in \mathbb{Z}^d$.*

PROOF. Suppose f exists. Additivity of B is immediate. For stationarity,

$$B(\omega, u+x, u+y) = \nabla_{y-x} f(\omega, u+x) = \nabla_{y-x} f(\theta_u\omega, x) = B(\theta_u\omega, x, y).$$

Conversely, suppose B is a stationary cocycle. Set $f(\omega, x) = B(\omega, 0, x)$. Then $f(\omega, 0) = B(\omega, 0, 0) = 0$.

$$\begin{aligned} \nabla_u f(\omega, x) &= f(\omega, x+u) - f(\omega, x) = B(\omega, 0, x+u) - B(\omega, 0, x) = B(\omega, x, x+u)\\ &= B(\theta_x\omega, 0, u) = f(\theta_x\omega, u) = f(\theta_x\omega, u) - f(\theta_x\omega, 0) = \nabla_u f(\theta_x\omega, 0). \end{aligned}$$

Thus f has a shift-invariant gradient. $\square$

The second part of the proof above illustrates that the first lattice variable in our definition of a cocycle is superfluous. If we put $\widetilde{B}(\omega, y) = B(\omega, 0, y)$ then $B(\omega, x, y) = B(\theta_x\omega, 0, y-x) = \widetilde{B}(\theta_x\omega, y-x)$.

A special class of cocycles is given by gradients $B(\omega, x, y) = \varphi(\theta_y\omega) - \varphi(\theta_x\omega)$ of functions φ on Ω. The next lemma shows that $\mathcal{K}_0$ is the $L^1(\mathbb{P})$-closure of such gradients.

LEMMA 2.8. *Given $F \in \mathcal{K}_0$, there exist $\varphi_n \in L^1(\mathbb{P})$ such that for each fixed $z \in \mathbb{Z}^d$, $\varphi_n(\theta_z\omega) - \varphi_n(\omega) \to F(\omega, 0, z)$ as $n \to \infty$, both almost surely and in $L^1(\mathbb{P})$. If $1 < p < \infty$ and $F(x, y) \in L^p(\mathbb{P})$ for all $x, y \in \mathbb{Z}^d$, then the convergence holds also in $L^p(\mathbb{P})$.*

PROOF. Let $\Lambda_n = [-n, n]^d \cap \mathbb{Z}^d$ and

$$\varphi_n(\omega) = -\frac{1}{|\Lambda_n|}\sum_{x\in\Lambda_n} F(\omega, 0, x).$$

Then

$$\begin{aligned}\varphi_n(\theta_z\omega) - \varphi_n(\omega) &= \frac{1}{|\Lambda_n|}\sum_{x\in\Lambda_n}\big[-F(\omega, z, x+z) + F(\omega, 0, x)\big] \\ &= \frac{1}{|\Lambda_n|}\sum_{x\in\Lambda_n}\big[F(\omega, 0, z) - F(\omega, x, x+z)\big] = F(\omega, 0, z) - \frac{1}{|\Lambda_n|}\sum_{x\in\Lambda_n} F(\theta_x\omega, 0, z).\end{aligned}$$

The second equality is true term by term by additivity. As $n \to \infty$, the last term converges to $\mathbb{E}F(0, z) = 0$ by the multidimensional ergodic theorem (for example, almost sure convergence by the multidimensional pointwise ergodic theorem given in Theorem 14.A8 in **[28]** and then L^p convergence by truncation). □

Our convention for centering non-mean-zero cocycles is the following. For $B \in \mathcal{K}$ there exists a vector $h(B) \in \mathbb{R}^d$ such that

$$\mathbb{E}[B(0, x)] = -h(B)\cdot x \qquad \forall x \in \mathbb{Z}^d. \tag{2.11}$$

Existence of $h(B)$ follows because $c(x) = \mathbb{E}[B(0, x)]$ is an additive function on the group $\mathbb{Z}^d$. Then

$$F(\omega, x, y) = -h(B)\cdot(y - x) - B(\omega, x, y), \qquad x, y \in \mathbb{Z}^d \tag{2.12}$$

is a centered stationary $L^1(\mathbb{P})$ cocycle.

Consider this assumption on a given $F \in \mathcal{K}_0$.

$$\begin{aligned}&\exists\, \overline{F} : \Omega \times \mathcal{R} \to \mathbb{R} \text{ such that the following properties hold} \\ &\text{for } z \in \mathcal{R}\backslash\{0\} \text{ and } \mathbb{P}\text{-a.s.:}\quad F(\omega, 0, z) \leqslant \overline{F}(\omega, z) \quad \text{and} \\ &\overline{\lim_{\delta\searrow 0}}\ \overline{\lim_{n\to\infty}}\ \max_{|x|_1\leqslant n}\frac{1}{n}\sum_{0\leqslant i\leqslant n\delta}|\overline{F}(\theta_{x+iz}\omega, z)| = 0.\end{aligned} \tag{2.13}$$

A sufficient condition for the limit above is that $\mathbb{E}|\overline{F}(\omega, z)|^{d+\varepsilon} < \infty$ for some $\varepsilon > 0$ and the shifts of $\overline{F}$ have finite range of dependence. The latter condition means that there exists $r_0 < \infty$ such that the random vectors $\{(\overline{F}(\theta_{x_i}\omega, z))_{z\in\mathcal{R}} : 1 \leqslant i \leqslant m\}$ are independent whenever $|x_i - x_j| \geqslant r_0$ for each pair $i \neq j$.

THEOREM 2.9. *Let $F \in \mathcal{K}_0$. Under assumption* (2.13) *we have the following uniform ergodic theorem:*

$$\lim_{n\to\infty}\max_{|x|_1\leqslant n}\frac{|F(\omega, 0, x)|}{n} = 0 \qquad \mathbb{P}\text{-a.s.}$$

For a proof see Appendix A.3 of **[31]**.

2.3. Variational formulas. In this section we derive variational formulas for the restricted path length point-to-level and point-to-point last-passage values.

THEOREM 2.10.

$$g_{\mathrm{pl}}(h) = \inf_{F \in \mathcal{K}_0} \mathbb{P}\text{-}\operatorname{ess\,sup}_{\omega} \max_{z \in \mathcal{R}} \{V(\omega, z) + h \cdot z + F(\omega, 0, z)\}. \tag{2.14}$$

A minimizing $F \in \mathcal{K}_0$ exists for each $h \in \mathbb{R}^d$.

Abbreviate

$$K(F) = \mathbb{P}\text{-}\operatorname{ess\,sup}_{\omega} \max_{z \in \mathcal{R}} \{V(\omega, z) + h \cdot z + F(\omega, 0, z)\}. \tag{2.15}$$

PROOF. **Upper bound.** Let $F \in \mathcal{K}_0$. Assume $K(F) < \infty$. Then

$$F(\omega, 0, z) \leqslant -V(\omega, z) - h \cdot z + K(F)$$

together with assumption (2.3) on V imply that F satisfies assumption (2.13) and therefore the uniform ergodic theorem (Theorem 2.9) applies.

$$\begin{aligned}
g_{\mathrm{pl}}(h) &= \lim_{n \to \infty} \max_{x_{0,n}} \frac{1}{n} \Big\{ \sum_{k=0}^{n-1} V(\theta_{x_k}\omega, z_{k+1}) + h \cdot x_n \Big\} \\
&= \lim_{n \to \infty} \max_{x_{0,n}} \frac{1}{n} \Big\{ \sum_{k=0}^{n-1} V(\theta_{x_k}\omega, z_{k+1}) + h \cdot x_n + F(\omega, 0, x_n) \Big\} \\
&= \lim_{n \to \infty} \max_{x_{0,n}} \frac{1}{n} \sum_{k=0}^{n-1} \big[V(\theta_{x_k}\omega, z_{k+1}) + h \cdot z_{k+1} + F(\theta_{x_k}\omega, 0, z_{k+1}) \big] \\
&\leqslant K(F)
\end{aligned}$$

because the last upper bound is valid $\mathbb{P}$-a.s. for each term. We have shown that

$$g_{\mathrm{pl}}(h) \leqslant \inf_{F \in \mathcal{K}_0} K(F).$$

Lower bound. Let $\lambda > g_{\mathrm{pl}}(h)$. Set $u_n(\omega) = e^{G_n(h) - n\lambda}$ with the interpretation that $u_0 = 1$. Since $n^{-1}G_n(h) \to g_{\mathrm{pl}}(h)$ almost surely, $u_n(\omega) < e^{-n\varepsilon}$ for large n and a fixed $\varepsilon > 0$. Hence f below is a well-defined finite function:

$$\begin{aligned}
f(\omega) \;&=\; \sum_{n=0}^{\infty} u_n(\omega) \;=\; 1 + \sum_{n=1}^{\infty} \exp\Big\{ \max_{x_{0,n}} \Big[V(\omega, x_1) + h \cdot x_1 - \lambda \\
&\qquad\qquad + \sum_{k=1}^{n-1} V(\theta_{x_k}\omega, z_{k+1}) + h \cdot (x_n - x_1) - (n-1)\lambda \Big] \Big\} \\
&\geqslant\; 1 + \max_z e^{V(\omega,z) + h \cdot z - \lambda} \sum_{n=1}^{\infty} u_{n-1}(\theta_z\omega) \;\geqslant\; \max_z e^{V(\omega,z) + h \cdot z - \lambda} f(\theta_z\omega) \\
&=\; e^{-\lambda} \cdot e^{\max_z [V(\omega,z) + h \cdot z + \log f(\theta_z\omega)]}.
\end{aligned}$$

In the first inequality above we take the sum inside the maximum over the first step z. Rearrange this to

$$\lambda \geqslant \max_z \{V(\omega, z) + h \cdot z + \log f(\theta_z\omega) - \log f(\omega)\} \qquad \text{a.s.}$$

Utilizing the notation $\nabla \log f(\omega, x, y) = \log f(\theta_y\omega) - \log f(\theta_x\omega)$, we can now write

$$\lambda \geqslant K(\nabla \log f) \geqslant \inf_{F \in \mathcal{K}_0} K(F)$$

where the second inequality holds provided $\nabla \log f \in \mathcal{K}_0$. This point is implied by the next lemma. Hence we can let $\lambda \searrow g_{\text{pl}}(h)$ to get

$$g_{\text{pl}}(h) \geqslant \inf_{F\in\mathcal{K}_0} K(F).$$

The existence of a minimizer is proved by a weak convergence argument that we skip. It is given for positive temperature polymer models in Theorem 2.3 of [**45**]. □

LEMMA 2.11. *For a measurable function* $\varphi : \Omega \to \mathbb{R}$ *define* $\nabla\varphi(\omega, x, y) = \varphi(\theta_y\omega) - \varphi(\theta_x\omega)$. *Then* $K(\nabla\varphi) < \infty$ *implies* $\nabla\varphi(\cdot\,, z) \in L^1(\mathbb{P})$ *and* $\mathbb{E}[\nabla\varphi(\cdot\,, z)] = 0$, *for all* $z \in \mathcal{R}$.

PROOF. For each $z \in \mathcal{R}$,

$$V(\omega, z) + h\cdot z + \nabla\varphi(\omega, 0, z) \leqslant K(\nabla\varphi) < \infty \qquad \text{a.s.}$$

and so $(\nabla\varphi(\cdot\,, z))^+ \in L^1(\mathbb{P})$. Suppose $(\nabla\varphi(\cdot\,, z))^- \notin L^1$. Then a contradiction arises as follows, where the first equality comes by the pointwise ergodic theorem:

$$\begin{aligned} -\infty &= \lim_{n\to\infty} \frac{1}{n}\sum_{k=0}^{n-1} \nabla\varphi(\theta_{kz}\omega, z) = \lim_{n\to\infty} \frac{1}{n}\sum_{k=0}^{n-1} [\varphi(\theta_{(k+1)z}\omega) - \varphi(\theta_{kz}\omega)] \\ &= \lim_{n\to\infty} \frac{1}{n}[\varphi(\theta_{nz}\omega) - \varphi(\omega)] = 0 \qquad \text{in probability.} \end{aligned}$$ □

Once we know $\nabla\varphi(\cdot\,, z) \in L^1(\mathbb{P})$, the ergodic theorem and the calculation above give $\mathbb{E}[\nabla\varphi(\cdot\,, z)] = 0$.

THEOREM 2.12. *For each* $\xi \in \operatorname{ri}\mathcal{U}$ *we have this variational formula.*

$$g_{\text{pp}}(\xi) \;=\; \inf_{B\in\mathcal{K}} \mathbb{P}\text{-}\operatorname*{ess\,sup}_{\omega} \max_{z\in\mathcal{R}} \{V(\omega, z) - B(\omega, 0, z) - h(B)\cdot\xi\}. \tag{2.16}$$

For each $\xi \in \operatorname{ri}\mathcal{U}$ *there exists a minimizing* $B \in \mathcal{K}$ *such that* $h(B)$ *is dual to* ξ.

PROOF. From duality (2.10),

$$\begin{aligned} g_{\text{pp}}(\xi) &= \inf_{h\in\mathbb{R}^d} \{g_{\text{pl}}(h) - h\cdot\xi\} \\ &= \inf_{h\in\mathbb{R}^d} \inf_{F\in\mathcal{K}_0} \mathbb{P}\text{-}\operatorname*{ess\,sup}_{\omega} \max_{z\in\mathcal{R}} \{V(\omega, z) + h\cdot z + F(\omega, 0, z) - h\cdot\xi\} \\ &= \inf_{B\in\mathcal{K}} \mathbb{P}\text{-}\operatorname*{ess\,sup}_{\omega} \max_{z\in\mathcal{R}} \{V(\omega, z) - B(\omega, 0, z) - h(B)\cdot\xi\} \end{aligned}$$

where we let

$$B(\omega, x, y) = -h\cdot(y - x) - F(\omega, x, y)$$

define an element of $\mathcal{K}$ with

$$h(B)\cdot z = -\mathbb{E}[B(0, z)]\cdot z = h\cdot z \quad \Longrightarrow \quad h(B) = h.$$

The first infimum above is achieved at some h dual to ξ whose existence is given in Lemma 2.5, and for this h a minimizing F exists for $g_{\text{pl}}(h)$. Thus the B defined above has $h(B)$ dual to ξ. □

2.4. Cocycles adapted to the potential. The cocycles and the potential V that defines the percolation both live on a general product space with a product probability measure $\mathbb{P}$. It is not evident how these two structures are connected. Next we identify a local condition that characterizes those cocycles that are relevant to the percolation problem in various ways.

DEFINITION 2.13. A cocycle $B \in \mathcal{K}$ is adapted to the potential V if

$$\max_{z\in\mathcal{R}} [V(\omega, z) - B(\omega, 0, z)] = 0 \quad \mathbb{P}\text{-a.s.} \tag{2.17}$$

This condition is linked to (i) minimizing cocycles, (ii) geodesics, (iii) Busemann functions and (iv) stationary percolation. We discuss briefly these four issues before moving to an exactly solvable model.

Minimizing cocycles. Suppose $B \in \mathcal{K}$ satisfies (2.17). Define a mean-zero cocycle $F \in \mathcal{K}_0$ by

$$F(\omega, x, y) = -h(B) \cdot (y - x) - B(\omega, x, y).$$

Then (2.17) becomes

$$0 = \max_{z\in\mathcal{R}}[V(\omega, z) + h(B) \cdot z + F(\omega, 0, z)] \quad \mathbb{P}\text{-a.s.}$$

The one-sided bound $B(\omega, 0, z) \geqslant V(\omega, z)$ is enough for the uniform ergodic theorem to work for F. Thus we can iterate the identity above and take a limit.

$$\begin{aligned} 0 &= \max_{x_{0,n}} \Big\{ \frac{1}{n} \sum_{k=0}^{n-1} V(\theta_{x_k}\omega, z_{k+1}) + \frac{1}{n} h(B) \cdot x_n + \frac{1}{n} F(\omega, 0, x_n) \Big\} \\ &= \frac{1}{n} G_n(h(B)) + o(1) \longrightarrow g_{\mathrm{pl}}(h(B)) \quad \text{as } n \to \infty. \end{aligned}$$

We can conclude that

$$g_{\mathrm{pl}}(h(B)) = 0 = \max_{z\in\mathcal{R}} \big[V(\omega, z) + h(B) \cdot z + F(\omega, 0, z)\big] \quad \mathbb{P}\text{-a.s.}$$

The equality above shows that F minimizes in the variational formula

$$g_{\mathrm{pl}}(h) = \inf_{F\in\mathcal{K}_0} \mathbb{P}\text{-}\operatorname*{ess\,sup}_{\omega} \max_{z\in\mathcal{R}} \{V(\omega, z) + h \cdot z + F(\omega, 0, z)\}$$

for $h = h(B)$, even without the essential supremum over ω.

Furthermore: suppose $h(B)$ and ξ are dual. Then from above, almost surely,

$$g_{\mathrm{pp}}(\xi) = g_{\mathrm{pl}}(h(B)) - h(B) \cdot \xi = -h(B) \cdot \xi = \max_{z\in\mathcal{R}}[V(\omega, z) - B(\omega, 0, z)] - h(B) \cdot \xi.$$

Thus B is a minimizer for $g_{\mathrm{pp}}(\xi)$.

In summary, we see that (2.17) is a criterion that finds cocycles that serve as minimizers in the variational formulas. For future use we record the outcome of the calculation above in the next lemma.

LEMMA 2.14. *Let $B \in \mathcal{K}$ be an integrable stationary cocycle adapted to the potential V as required by* (2.17) *and let $h(B)$ be the negative of the mean vector of B as defined in* (2.11)*. Then $g_{\mathrm{pl}}(h(B)) = 0$.*

Geodesics. Turns out that a cocycle B satisfying (2.17) is involved not only in optimization on the macroscopic level but also on the pathwise level. Suppose the path $(x_k)_{k=0}^n$ with steps $z_{k+1} = x_{k+1} - x_k$ follows the maximal increments specified in (2.17), in other words, satisfies

$$V(\theta_{x_k}\omega, z_{k+1}) - B(\omega, x_k, x_{k+1}) = 0 \qquad \forall k = 0, 1, \ldots, n-1. \tag{2.18}$$

Then this path is a geodesic from x_0 to x_n. Here is the simple argument. Consider any path $y_\bullet$ from $y_0 = x_0$ to $y_n = x_n$. Then, by (2.17), the stationarity and additivity of B, and (2.18),

$$\begin{aligned}\sum_{k=0}^{n-1} V(\theta_{y_k}\omega, y_{k+1} - y_k) &\leqslant \sum_{k=0}^{n-1} B(\theta_{y_k}\omega, 0, y_{k+1} - y_k) = \sum_{k=0}^{n-1} B(\omega, y_k, y_{k+1}) \\ &= B(\omega, x_0, x_n) = \sum_{k=0}^{n-1} B(\omega, x_k, x_{k+1}) = \sum_{k=0}^{n-1} V(\theta_{x_k}\omega, z_{k+1}).\end{aligned}$$

Busemann functions. Having seen the usefulness of condition (2.17), we must ask how cocycles that satisfy (2.17) arise. One way of obtaining such cocycles is through limits of local gradients of passage times, called *Busemann functions.*

Suppose that we are in a setting where admissible paths that connect two given points have a uniquely determined number of steps. Let $G_{x,y} = G_{x,(m),y}$ denote the point-to-point last-passage value where m is the unique number of steps from x to y. Fix a direction $\xi \in \mathrm{ri}\,\mathcal{U}$. Assume that we have a process B that satisfies the following: for all sequences $\{v_n\} \subset \mathbb{Z}^d$ such that $v_n/|v_n| \to \xi$ this almost sure limit holds:

$$B(\omega, x, y) = \lim_{n\to\infty} [G_{x,\,v_n} - G_{y,\,v_n}] \quad \forall x, y \in \mathbb{Z}^d \quad \mathbb{P}\text{-a.s.} \tag{2.19}$$

B is called a Busemann function in direction ξ. It is a stationary cocycle. Additivity is immediate from the limit (2.19). Stationarity comes from the fact that shifting v_n by a fixed amount does not alter its limiting direction ξ. Under additional assumptions (see for example Theorem 5.1 in [**29**]) this cocycle is integrable.

At this time we simply wish to observe that B satisfies (2.17):

$$\max_{z\in\mathcal{R}}[V(\omega, z) - B(\omega, 0, z)] = \lim_{n\to\infty} \max_{z\in\mathcal{R}}[V(\omega, z) - G_{0,\,v_n} + G_{z,\,v_n}] = 0$$

where the last equality follows from

$$G_{0,\,v_n} = \max_{z\in\mathcal{R}}[V(\omega, z) + G_{z,\,v_n}]. \tag{2.20}$$

Proof of the limit in (2.19) is highly nontrivial. The route that we will take to finding cocycles that satisfy (2.17) will involve the next item below.

Stationary percolation. Once again we assume cocycle B satisfies (2.17) and develop this identity in a different direction. Let $v \in \mathbb{Z}^d$ be fixed.

$$\begin{aligned}0 &= \max_{z\in\mathcal{R}}[V(\theta_x\omega, z) - B(\theta_x\omega, 0, z)] = \max_{z\in\mathcal{R}}[V(\theta_x\omega, z) - B(\omega, x, x+z)] \\ &= \max_{z\in\mathcal{R}}[V(\theta_x\omega, z) - B(\omega, x, v) + B(\omega, x+z, v)]\end{aligned}$$

from which we write

$$B(\omega, x, v) = \max_{z\in\mathcal{R}} [V(\theta_x\omega, z) + B(\omega, x+z, v)] \qquad \forall x, v \in \mathbb{Z}^d. \tag{2.21}$$

Iterate this. Fix also $u \in \mathbb{Z}^d$ and write $x_n = u + z_1 + \ldots + z_n$ for an admissible path from u with steps z_i.

$$\begin{aligned} B(\omega, u, v) &= \max_{z_1 \in \mathcal{R}} [V(\theta_{x_0}\omega, z_1) + B(\omega, x_1, v)] \\ &= \max_{z_1, z_2 \in \mathcal{R}} [V(\theta_{x_0}\omega, z_1) + V(\theta_{x_1}\omega, z_2) + B(\omega, x_2, v)] \\ &= \cdots = \max_{z_1, \ldots, z_n \in \mathcal{R}} \Big[\sum_{k=0}^{n-1} V(\theta_{x_k}\omega, z_{k+1}) + B(\omega, x_n, v) \Big] \\ &= \max_x [\, G_{u,(n),x} + B(\omega, x, v)]. \end{aligned} \tag{2.22}$$

We can turn this into a boundary value problem. Assume again that the number of steps on an admissible path is determined uniquely by the endpoints so that we can write

$$G_{x,y} = \begin{cases} G_{x,(n),y} & \text{if } y \text{ is reachable from } x \text{ along an admissible path of } n \text{ steps} \\ -\infty & \text{if } y \text{ is not reachable from } x \text{ along any admissible path.} \end{cases}$$

Let $\mathcal{H}$ and $\partial\mathcal{H}$ be finite subsets of $\mathbb{Z}^d$ with the property that any infinite admissible path from $\mathcal{H}$ intersects $\partial\mathcal{H}$. So in a sense $\partial\mathcal{H}$ is a "boundary" of $\mathcal{H}$.

For example, suppose we are in the directed case $\mathcal{R} = \{e_1, \ldots, e_d\}$. If $\mathcal{H}$ is the rectangle $\mathcal{H} = \prod_{i=1}^d \{0, 1, \ldots, N_i\}$, then $\partial\mathcal{H}$ could be its "northeast" boundary $\partial\mathcal{H} = \bigcup_{i=1}^d \{x \in \mathcal{H} : x_i = N_i\}$.

LEMMA 2.15. *Assume that the two endpoints of any admissible path determine uniquely the number of steps in the path. Assume that the stationary cocycle B satisfies (2.17). Fix $v \in \mathbb{Z}^d$ and finite subsets $\mathcal{H}$ and $\partial\mathcal{H}$ of $\mathbb{Z}^d$ such that every long enough admissible path from $\mathcal{H}$ intersects $\partial\mathcal{H}$. Then*

$$B(\omega, u, v) = \max_{x \in \partial\mathcal{H}} [\, G_{u,x} + B(\omega, x, v)] \qquad \textit{for all } u \in \mathcal{H}. \tag{2.23}$$

PROOF. Fix $u \in \mathcal{H}$. Equation (2.22) gives

$$B(\omega, u, v) \geqslant G_{u,x} + B(\omega, x, v)$$

whenever there is an admissible path from u to x. This tells us that $\geqslant$ holds in (2.23). It also shows that if $u \in \mathcal{H} \cap \partial\mathcal{H}$ then the maximum in (2.23) is assumed at $x = u$ and the equality holds. (For this last statement we use the convention $G_{u,u} = 0$.)

We prove inequality $\leqslant$ in (2.23). By the directedness assumption (2.2) we can take n in (2.22) large enough so that every n-path from u intersects $\partial\mathcal{H}$. Fix a maximizing path $u = x_0, x_1, \ldots, x_n$ on the second last line of (2.22). Let $x_m \in \partial\mathcal{H}$. Then

$$\begin{aligned} B(\omega, u, v) &= \sum_{k=0}^{m-1} V(\theta_{x_k}\omega, z_{k+1}) + \sum_{k=m}^{n-1} V(\theta_{x_k}\omega, z_{k+1}) + B(\omega, x_n, v) \\ &\leqslant G_{u,x_m} + G_{x_m,x_n} + B(\omega, x_n, v) \leqslant G_{u,x_m} + B(\omega, x_m, v) \\ &\leqslant \text{right-hand side of (2.23)}. \end{aligned}$$

In the second-last inequality above we applied (2.22) with x_m in place of u and $n - m$ in place of n. □

Our interpretation is that equation (2.23) determines the values $\{B(u,v) : u \in \mathcal{H}\}$ from the last-passage percolation $G_{x,y}$ and given boundary values $\{B(x,v) : x \in \partial\mathcal{H}\}$. Then we can search for cases where the solution is tractable. In particular, we can look for distributional invariance. We can make this program work in exactly solvable cases.

The analogy the reader should have in mind is finding the invariant distribution of a Markov process such as an interacting particle system. The analogue of boundary values are the state variables at time zero. The analogue of the weights $V(\theta_x\omega, z)$ are the Poisson clocks or other random variables that govern the evolution of the particles.

3. Exponential corner growth model in two dimensions

For the remainder of this article we restrict the discussion to the two-dimensional corner growth model (CGM), with real weights on the vertices: $\omega = (\omega_x)_{x\in\mathbb{Z}^2} \in \Omega = \mathbb{R}^{\mathbb{Z}^2}$. The set of admissible steps is $\mathcal{R} = \{e_1, e_2\}$, and the potential is given by the weight at the origin: $V(\omega, z) = \omega_0$. The set of possible limiting velocities of paths is the closed line segment $\mathcal{U} = [e_1, e_2] = \{(s, 1-s) : 0 \leqslant s \leqslant 1\}$, and its relative interior is the open line segment $\operatorname{ri}\mathcal{U} = (e_1, e_2) = \{(s, 1-s) : 0 < s < 1\}$.

In this setting we alter slightly the earlier definition (2.5) of the point-to-point last-passage time to include both endpoints of the path. This makes no difference to large-scale properties. Given an environment ω and two points $x, y \in \mathbb{Z}^2$ with $x \leqslant y$ coordinatewise, define

$$G_{x,y} = \max_{x_\bullet \in \Pi_{x,y}} \sum_{k=0}^{|y-x|_1} \omega_{x_k}. \tag{3.1}$$

$\Pi_{x,y}$ is the set of paths $x_\bullet = (x_k)_{k=0}^n$ that start at $x_0 = x$, end at $x_n = y$ with $n = |y-x|_1$, and have increments $x_{k+1} - x_k \in \{e_1, e_2\}$. Call such paths *admissible* or *up-right*. See Figure 3.1 for an illustration. The zero-length path case is $G_{x,x} = \omega_x$. Our convention is that

$$G_{x,y} = -\infty \qquad \text{if } x \leqslant y \text{ fails.} \tag{3.2}$$

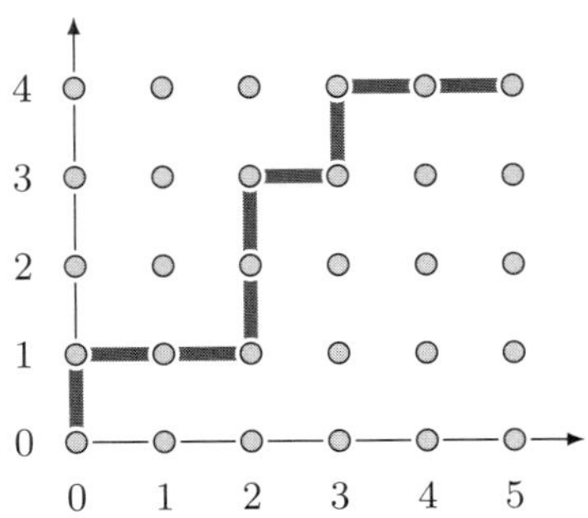

FIGURE 3.1. An up-right path from $(0,0)$ to $(5,4)$ on the lattice $\mathbb{Z}^2$.

We work with the exponentially distributed weights, and so make the following assumption:

$$\text{weights } \omega_x \text{ are independent rate 1 exponentially distributed random variables.} \tag{3.3}$$

This means that $\mathbb{P}\{\omega_x > t\} = e^{-t}$ for $t \geqslant 0$. This is abbreviated as $\omega_x \sim \text{Exp}(1)$.

In this section we construct a stationary version of the exponential CGM and then calculate a closed-form expression of the limiting shape function

$$g_{\text{pp}}(\xi) = \lim_{N\to\infty} N^{-1} G_{0,\lfloor N\xi \rfloor} \qquad \text{almost surely for } \xi \in \mathbb{R}^2_{\geqslant 0}. \tag{3.4}$$

Existence of the limit is a special case of the previous Theorem 2.4. Note though that in contrast with (2.8) the limit is now defined not only for $\xi \in \mathcal{U}$ but for all ξ in the closed first quadrant. (Theorem 2.3 of [**41**] gives (3.4) under a slightly weaker hypothesis than the $2+\varepsilon$ moment assumption (2.3) assumed for Theorem 2.4.) General soft arguments show that g_{pp} is concave, continuous and homogeneous (which means that $g_{\text{pp}}(c\xi) = c g_{\text{pp}}(\xi)$ for $c \geqslant 0$). In the exponential case the explicit formula for g_{pp} reveals that this function is strictly concave, except on rays from the origin.

In the remainder of this section we prove two complements to the limit (3.4). We prove a version of (3.4) with uniformity over the lattice. Based on this we obtain the *shape theorem*, which gives the limit of the randomly growing cluster $A_t = \{\zeta \in \mathbb{R}^2_{\geqslant 0} : G_{0,\lfloor \zeta \rfloor} \leqslant t\}$ as $t \to \infty$ and the cluster is scaled to $t^{-1}A_t$. As the last theorem of the section we show that the maximizing path in (3.1) stays within $o(|y-x|_1)$ of the straight line segment $[x,y]$, with high probability as $|y-x|_1 \to \infty$.

3.1. Stationary exponential corner growth model. Our first task is to construct a coupling of i.i.d. rate 1 exponential weights and a stationary integrable cocycle, for a given value of a parameter $0 < \rho < 1$, that together satisfy (2.17), essentially by solving the boundary value problem in (2.23). This construction will be performed on quadrants $u + \mathbb{Z}^2_{\geqslant 0}$ with a specified origin $u \in \mathbb{Z}^2$.

Fix a parameter $0 < \rho < 1$ and an origin $u \in \mathbb{Z}^2$. Assume given a collection of mutually independent random variables

$$\{\omega_x,\, I_{u+ie_1},\, J_{u+je_2} : x \in u + \mathbb{Z}^2_{>0},\, i,j \in \mathbb{Z}_{>0}\} \tag{3.5}$$

with these marginal distributions:

$$\omega_x \sim \text{Exp}(1), \quad I_{u+ie_1} \sim \text{Exp}(1-\rho), \quad \text{and} \quad J_{u+je_2} \sim \text{Exp}(\rho). \tag{3.6}$$

The interpretation is that I_x is a weight for the edge $(x-e_1, x)$, J_x is a weight for edge $(x-e_2, x)$, and ω_x is a vertex weight. The edge weights are on the boundary of the quadrant $u + \mathbb{Z}^2_{\geqslant 0}$ and the vertex weights in the bulk.

From the given variables (3.5) we define further variables as follows, proceeding inductively to the north and east from the origin u: for all $x \in u + \mathbb{Z}^2_{>0}$,

$$\check{\omega}_{x-e_1-e_2} = I_{x-e_2} \wedge J_{x-e_1} \tag{3.7}$$

$$I_x = \omega_x + (I_{x-e_2} - J_{x-e_1})^+ \tag{3.8}$$

$$J_x = \omega_x + (I_{x-e_2} - J_{x-e_1})^-. \tag{3.9}$$

The mapping above from $(\omega_x, I_{x-e_2}, J_{x-e_1})$ to $(\check{\omega}_{x-e_1-e_2}, I_x, J_x)$ is illustrated in

Figure 3.2. Note that (3.8)–(3.9) imply the symmetric counterpart of (3.7)

$$\omega_x = I_x \wedge J_x \tag{3.10}$$

and the additivity around the unit square:

$$I_{x-e_2} + J_x = J_{x-e_1} + I_x. \tag{3.11}$$

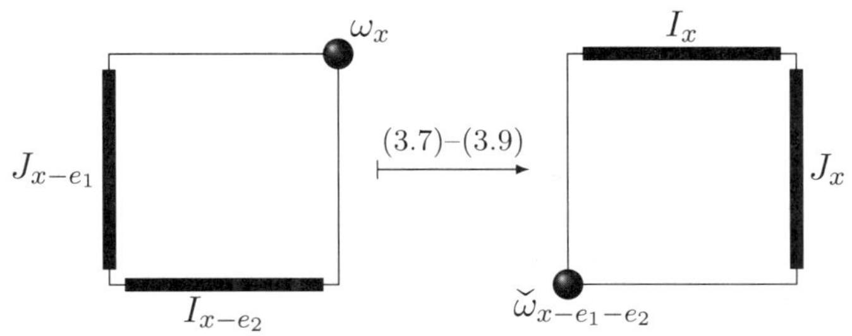

FIGURE 3.2. Mapping (3.7)–(3.9) on a single lattice square. The figure illustrates how southwest corners are flipped into northeast corners in the inductive construction of the increment variables.

Utilizing (3.7)–(3.9) we extend (3.5) to the larger collection

$$\{\omega_x,\, I_{x-e_2},\, J_{x-e_1},\, \check{\omega}_{x-e_1-e_2} : x \in u + \mathbb{Z}^2_{>0}\}. \tag{3.12}$$

This larger collection has an $\check{\omega}_x$ variable for each vertex in the quadrant $u + \mathbb{Z}^2_{\geqslant 0}$, an I_x variable for each horizontal nearest-neighbor edge in the quadrant $u + \mathbb{Z}^2_{\geqslant 0}$, a J_x variable for each vertical nearest-neighbor edge in the quadrant $u + \mathbb{Z}^2_{\geqslant 0}$, and the originally given ω_x variables for vertices in the bulk $u + \mathbb{Z}^2_{>0}$.

As the next theorem will show, process (3.12) is related to another last-passage process $G^{\rho}_{u,x}$ whose origin is fixed at u and that utilizes both edge weights on the boundary and bulk weights in the interior of the quadrant. First put $G^{\rho}_{u,u} = 0$ and on the boundaries

$$G^{\rho}_{u,\,u+me_1} = \sum_{i=1}^{m} I_{u+ie_1} \quad \text{and} \quad G^{\rho}_{u,\,u+ne_2} = \sum_{j=1}^{n} J_{u+je_2}. \tag{3.13}$$

Then in the bulk for $x = u + (m,n) \in u + \mathbb{Z}^2_{>0}$,

$$\begin{aligned} G^{\rho}_{u,\,x} = {} & \max_{1\leqslant k\leqslant m} \Big\{ \sum_{i=1}^{k} I_{u+ie_1} + G_{u+ke_1+e_2,\,x} \Big\} \\ & \bigvee \max_{1\leqslant \ell\leqslant n} \Big\{ \sum_{j=1}^{\ell} J_{u+je_2} + G_{u+\ell e_2+e_1,\,x} \Big\}. \end{aligned} \tag{3.14}$$

$G_{a,x}$ inside the braces is the last-passage value defined in (3.1). The superscript ρ in $G^{\rho}_{u,x}$ distinguishes this last-passage value from the one in (3.1) that uses only i.i.d. bulk weights. The first subscript u in $G^{\rho}_{u,x}$ specifies that the I and J edge weights are placed on the axes $u + \mathbb{Z}_{>0}e_k$, $k = 1, 2$.

An equivalent definition of $G^{\rho}_{u,x}$ would be to give the boundary conditions (3.13) and the inductive equation

$$G^{\rho}_{u,x} = \omega_x + G^{\rho}_{u,\,x-e_1} \vee G^{\rho}_{u,\,x-e_2}, \qquad x \in u + \mathbb{Z}^2_{>0}. \tag{3.15}$$

The next theorem summarizes the properties of process (3.12) and the connection with the process $G^{\rho}_{u,x}$. Point (i) of the theorem uses the following definitions. A bi-infinite sequence $\mathcal{Y} = (y_k)_{k\in\mathbb{Z}}$ in $u + \mathbb{Z}^2_{\geqslant 0}$ is a *down-right path* if $y_k - y_{k-1} \in \{e_1, -e_2\}$ for all $k \in \mathbb{Z}$. $\mathcal{Y}$ decomposes the vertices of the quadrant into a disjoint union $u + \mathbb{Z}^2_{\geqslant 0} = \mathcal{G}_- \cup \mathcal{Y} \cup \mathcal{G}_+$ where

$$\mathcal{G}_- = \{x \in u + \mathbb{Z}^2_{\geqslant 0} : \exists j \in \mathbb{Z}_{>0} \text{ such that } x + (j,j) \in \mathcal{Y}\}$$

is the region strictly to the south and west of $\mathcal{Y}$ and

$$\mathcal{G}_+ = \{x \in u + \mathbb{Z}^2_{\geqslant 0} : \exists j \in \mathbb{Z}_{>0} \text{ such that } x - (j,j) \in \mathcal{Y}\}$$

is the region strictly to the north and east of $\mathcal{Y}$. Note that $\mathcal{G}_+$ is necessarily unbounded but $\mathcal{G}_-$ is finite iff all but finitely many of the points y_k lie on the axes $u + \mathbb{Z}_{>0}e_k$, $k = 1, 2$. In the extreme case $\mathcal{Y} = \{u + ie_1, u + je_2 : 0 \leqslant i, j < \infty\}$ consists of the axes, $\mathcal{G}_- = \varnothing$ and $\mathcal{G}_+ = u + \mathbb{Z}^2_{>0}$.

For an undirected nearest-neighbor edge e on $u + \mathbb{Z}^2_{\geqslant 0}$, we denote the weight by

$$t(e) = \begin{cases} I_x & \text{if } e = \{x - e_1, x\} \\ J_x & \text{if } e = \{x - e_2, x\}. \end{cases} \tag{3.16}$$

THEOREM 3.1. *Fix $u \in \mathbb{Z}^2$ and $0 < \rho < 1$ and assume given independent variables* (3.5) *with marginal distributions* (3.6). *Then the variables in* (3.12) *have the following properties.*

(i) *For any down-right path $\mathcal{Y}$ in $u + \mathbb{Z}^2_{\geqslant 0}$, the random variables*

$$\{\check{\omega}_z,\ t(\{y_{k-1}, y_k\}),\ \omega_x : z \in \mathcal{G}_-,\ k \in \mathbb{Z},\ x \in \mathcal{G}_+\} \tag{3.17}$$

are mutually independent with marginal distributions

$$\omega_x, \check{\omega}_x \sim \text{Exp}(1), \quad I_x \sim \text{Exp}(1-\rho), \quad \textit{and} \quad J_x \sim \text{Exp}(\rho). \tag{3.18}$$

(ii) *The I_x and J_x variables are the increments of the $G^{\rho}_{u,x}$ last-passage process:*

$$\begin{aligned} I_x &= G^{\rho}_{u,x} - G^{\rho}_{u,x-e_1} \qquad \textit{for } x \in u + (\mathbb{Z}_{>0}) \times (\mathbb{Z}_{\geqslant 0}), \\ J_x &= G^{\rho}_{u,x} - G^{\rho}_{u,x-e_2} \qquad \textit{for } x \in u + (\mathbb{Z}_{\geqslant 0}) \times (\mathbb{Z}_{>0}). \end{aligned} \tag{3.19}$$

Theorem 3.1 rests on an inductive argument based on the next lemma, which describes the joint distribution preserved by the mapping in Figure 3.2.

LEMMA 3.2. *Let $0 < \rho < 1$. Assume given independent variables $W \sim \text{Exp}(1)$, $I \sim \text{Exp}(1-\rho)$, and $J \sim \text{Exp}(\rho)$. Define*

$$\begin{aligned} W' &= I \wedge J \\ I' &= W + (I - J)^+ \\ J' &= W + (I - J)^-. \end{aligned} \tag{3.20}$$

Then the triple (W', I', J') has the same distribution as (W, I, J).

This lemma is proved by calculating a joint transform such as the Laplace transform or characteristic function, or by transforming the joint density. The mapping $(W, I, J) \mapsto (W', I', J')$ is an involution, that is, its own inverse.

PROOF OF THEOREM 3.1. Part (i). This is proved inductively on $\mathcal{Y}$. The base case is $\mathcal{Y} = \{u+ie_1, u+je_2 : 0 \leqslant i, j < \infty\}$, in which case the claim simply amounts to the initial condition in (3.6).

Now assume given $\mathcal{Y} = \{y_k\}$ for which the claim in part (i) holds. We show that this claim continues to hold for any $\mathcal{Y}'$ obtained from $\mathcal{Y}$ by "flipping a southwest corner into a northeast corner". So pick any $x \in u + \mathbb{Z}^2_{>0}$ and $m \in \mathbb{Z}$ such that $(y_{m-1}, y_m, y_{m+1}) = (x-e_1, x-e_1-e_2, x-e_2)$ are points along $\mathcal{Y}$. Define $\mathcal{Y}' = \{y'_k\}$ by setting

$$y'_k = y_k \qquad \text{for } k \neq m, \qquad \text{and} \qquad y'_m = y_m + e_1 + e_2 = x.$$

In other words, $\mathcal{Y}$ has a southwest corner at $x - e_1 - e_2$, and $\mathcal{Y}'$ has a northeast corner at x.

Transforming $\mathcal{Y}$ into $\mathcal{Y}'$ changes $\mathcal{G}_-$ to $\mathcal{G}'_- = \mathcal{G}_- \cup \{x - e_1 - e_2\}$ and $\mathcal{G}_+$ to $\mathcal{G}'_+ = \mathcal{G}_+ \backslash \{x\}$. Thus constructing the variables (3.17) for $\mathcal{Y}'$ involves transforming the triple $(\omega_x, I_{x-e_2}, J_{x-e_1})$ into $(\breve{\omega}_{x-e_1-e_2}, I_x, J_x)$ according to equations (3.7)–(3.9), and copying the remaining variables from (3.17) for $\mathcal{Y}$. The claim now follows for $\mathcal{Y}'$ by the induction assumption and Lemma 3.2. By the induction assumption, variables $(\omega_x, I_{x-e_2}, J_{x-e_1})$ have the independent exponential distributions required for the hypothesis of Lemma 3.2, and so by the lemma the triple $(\breve{\omega}_{x-e_1-e_2}, I_x, J_x)$ also has the independent exponential distributions required for (3.18).

Part (ii). The claim is true by construction for variables I_{u+ie_1} and J_{u+je_2} on the axes. Here is the inductive argument for I_x, assuming that the claim holds for I_{x-e_2} and J_{x-e_1} and utilizing (3.15):

$$\begin{aligned} I_x &= \omega_x + (I_{x-e_2} - J_{x-e_1})^+ = \omega_x + (G^\rho_{u,\,x-e_2} - G^\rho_{u,\,x-e_1})^+ \\ &= \omega_x + G^\rho_{u,\,x-e_1} \vee G^\rho_{u,\,x-e_2} - G^\rho_{u,\,x-e_1} = G^\rho_{u,\,x} - G^\rho_{u,\,x-e_1}. \end{aligned}$$

A similar argument works for J_x under the same inductive assumption. □

Let us observe some immediate and valuable consequences of Theorem 3.1.

By taking $\mathcal{Y} = \{y_k\}$ as the axes at a new origin $v \in u+\mathbb{Z}^2_{\geqslant 0}$, given by $y_k = v+ke_1$ and $y_{-k} = v + ke_2$ for $k \geqslant 0$, part (i) of the theorem implies that the process $\{\omega_{v+x}, I_{v+x-e_2}, J_{v+x-e_1} : x \in \mathbb{Z}^2_{>0}\}$ has the same distribution for all $v \in u + \mathbb{Z}^2_{\geqslant 0}$. Thus $B(x, y) = G^\rho_{u,y} - G^\rho_{u,x}$ is a stationary cocycle, restricted to the quadrant $x, y \in u + \mathbb{Z}^2_{\geqslant 0}$.

The variables $\{\breve{\omega}_x : x \in u + \mathbb{Z}^2_{\geqslant 0}\}$ are i.i.d. Exp(1) distributed.

3.2. Explicit shape functions. We compute the limit shape functions for both last-passage percolation processes, the stationary one and the one with i.i.d. weights. For the stationary process define the function

$$g^\rho(s,t) = \frac{s}{1-\rho} + \frac{t}{\rho}. \tag{3.21}$$

PROPOSITION 3.3. *Fix $0 < \rho < 1$. The stationary corner growth model satisfies these properties: $\mathbb{E}[G^\rho_{0,(m,n)}] = g^\rho(m,n)$ for all $m, n \in \mathbb{Z}_{\geqslant 0}$ and the law of large numbers*

$$\lim_{N\to\infty} N^{-1} G^\rho_{0,(\lfloor Ns\rfloor, \lfloor Nt\rfloor)} = g^\rho(s,t) \qquad \textit{almost surely for each } (s,t) \in \mathbb{R}^2_{\geqslant 0}. \tag{3.22}$$

PROOF. Rewrite in terms of nearest-neighbor increments:

$$G^{\rho}_{0,(m,n)} = \sum_{i=1}^{m} I_{(i,0)} + \sum_{j=1}^{n} J_{(m,j)}. \tag{3.23}$$

Then use the translation invariance of the distributions which says that each nearest-neighbor increment has the exponential distribution imposed on the boundary variables in (3.6):

$$\mathbb{E}[G^{\rho}_{0,(m,n)}] = \sum_{i=1}^{m} \mathbb{E}I_{ie_1} + \sum_{j=1}^{n} \mathbb{E}J_{(m,j)} = \frac{m}{1-\rho} + \frac{n}{\rho}. \tag{3.24}$$

The limit of the stationary last-passage process is an application of the classical law of large numbers and some large deviation estimates, applied separately to the two sums: the limit below holds almost surely for any given $(s,t) \in \mathbb{R}^2_{\geqslant 0}$.

$$\begin{aligned} \lim_{N\to\infty} N^{-1} G^{\rho}_{0,(\lfloor Ns\rfloor,\lfloor Nt\rfloor)} &= \lim_{N\to\infty} \Big\{ N^{-1} \sum_{i=1}^{\lfloor Ns\rfloor} I_{(i,0)} + N^{-1} \sum_{j=1}^{\lfloor Nt\rfloor} J_{(\lfloor Ns\rfloor,\, j)} \Big\} \\ &= s\mathbb{E}(I_{e_1}) + t\mathbb{E}(J_{e_2}) = \frac{s}{1-\rho} + \frac{t}{\rho}. \end{aligned} \tag{3.25}$$

The proposition is proved. □

Next we take a limit in the coupling between the last-passage processes $G_{x,y}$ and $G^{\rho}_{0,x}$. Fix $s,t>0$ and use (3.14) for $x = (\lfloor Ns\rfloor, \lfloor Nt\rfloor)$ to write

$$\begin{aligned} G^{\rho}_{0,(\lfloor Ns\rfloor,\lfloor Nt\rfloor)} = &\sup_{0\leqslant a\leqslant s} \Big\{ \sum_{i=1}^{\lfloor Na\rfloor} I_{(i,0)} + G_{(\lfloor Na\rfloor,1),(\lfloor Ns\rfloor,\lfloor Nt\rfloor)} \Big\} \\ &\bigvee \sup_{0\leqslant b\leqslant t} \Big\{ \sum_{j=1}^{\lfloor Nb\rfloor} J_{(0,j)} + G_{(1,\lfloor Nb\rfloor),(\lfloor Ns\rfloor,\lfloor Nt\rfloor)} \Big\}. \end{aligned} \tag{3.26}$$

After letting $N\to\infty$, with some estimation on the right-hand side, utilizing limits (3.4) and (3.22), we have

$$\frac{s}{1-\rho} + \frac{t}{\rho} = \sup_{0\leqslant a\leqslant s} \Big\{ \frac{a}{1-\rho} + g_{\mathrm{pp}}(s-a,t) \Big\} \vee \sup_{0\leqslant b\leqslant t} \Big\{ \frac{b}{\rho} + g_{\mathrm{pp}}(s,t-b) \Big\}. \tag{3.27}$$

In the next theorem we take advantage of the connection above to find the shape function g_{pp} for the LPP process (3.1) with i.i.d. Exp(1) weights.

THEOREM 3.4. *Assume (3.3). Then we have the following law of large numbers. For each $\xi \in \mathbb{R}^2_{\geqslant 0}$ the limit below holds with probability 1, with the shape function g_{pp} as given.*

$$\lim_{N\to\infty} N^{-1} G_{0,\lfloor N\xi\rfloor} = g_{\mathrm{pp}}(\xi) \equiv \big(\sqrt{\xi_1} + \sqrt{\xi_2}\big)^2. \tag{3.28}$$

PROOF. By the general law of large numbers Theorem 2.4 for last-passage percolation, we know that the limit in (3.28) exists and that g_{pp} is finite, concave and continuous. Begin with (3.27) for $s=t$:

$$\frac{t}{1-\rho} + \frac{t}{\rho} = \sup_{0\leqslant a\leqslant t} \Big\{ \frac{a}{1-\rho} + g_{\mathrm{pp}}(t-a,t) \Big\} \vee \sup_{0\leqslant b\leqslant t} \Big\{ \frac{b}{\rho} + g_{\mathrm{pp}}(t,t-b) \Big\}.$$

Use the symmetry of g_{pp} and assume that $0 < \rho \leqslant 1/2$:

$$\frac{t}{1-\rho} + \frac{t}{\rho} = \sup_{0 \leqslant a \leqslant t} \left\{ \frac{a}{1-\rho} + g_{\text{pp}}(t-a,t) \right\} \vee \sup_{0 \leqslant b \leqslant t} \left\{ \frac{b}{\rho} + g_{\text{pp}}(t-b,t) \right\}$$
$$= \sup_{0 \leqslant b \leqslant t} \left\{ \frac{b}{\rho} + g_{\text{pp}}(t-b,t) \right\}.$$

Let

$$f(b) = \begin{cases} -g_{\text{pp}}(t-b,t), & 0 \leqslant b \leqslant t \\ \infty, & b < 0 \text{ or } b > t. \end{cases} \tag{3.29}$$

Then f is convex and lower semicontinuous. After a change of variable $x = 1/\rho \in [2, \infty)$, the equation above becomes

$$t\Big(x + 1 + \frac{1}{x-1}\Big) = \sup_{b \in \mathbb{R}} \{bx - f(b)\}, \qquad x \geqslant 2.$$

This is an instance of convex duality, so the convex conjugate f^* of f satisfies

$$f^*(x) = t\Big(x + 1 + \frac{1}{x-1}\Big) \qquad \text{for } x \geqslant 2.$$

The derivatives $(f^*)'(2+) = 0$ and $(f^*)'(\infty-) = t$ tell us that we can restrict the supremum in the double convex duality as in the second equality below, for $0 \leqslant b \leqslant t$. Then find the supremum by calculus:

$$f(b) = f^{**}(b) = \sup_{x \geqslant 2}\{xb - f^*(x)\} = -\frac{b\sqrt{t(t-b)} + bt - t(\sqrt{t} + \sqrt{t-b}\,)^2}{\sqrt{t(t-b)}}.$$

Taking $b = t - s$ for $s \in [0,t]$ in the definition (3.29) of f gives

$$g_{\text{pp}}(s,t) = -f(t-s) = (\sqrt{s} + \sqrt{t}\,)^2 \qquad \text{for } 0 \leqslant s \leqslant t.$$

Symmetry of g_{pp} completes the proof. □

Next we use (3.27) to identify the *characteristic direction* $\xi(\rho) \in \mathcal{U}$ associated with parameter value $\rho \in (0,1)$. The edge weights I and J are stochastically larger than the builk weights ω. Hence both boundaries attract the path. There is a unique direction in which the pulls of the boundaries balance out. By definition, $\xi(\rho)$ is the value of $\xi = (s, 1-s)$ for which the right-hand side of (3.27) with $(s,t) = (s, 1-s)$ is maximized at $a = b = 0$. This direction $\xi(\rho) \in \mathcal{U}$ is uniquely given by

$$\xi(\rho) = \left(\frac{(1-\rho)^2}{(1-\rho)^2 + \rho^2}, \frac{\rho^2}{(1-\rho)^2 + \rho^2} \right). \tag{3.30}$$

It then follows that in this direction $\xi(\rho)$ the optimal path for $G^{\rho}_{0,\lfloor N\xi(\rho)\rfloor}$ in (3.27) takes $o(N)$ steps on the coordinate axes, as $N \to \infty$.

An alternative characterization of the characteristic direction $\xi(\rho)$ is by comparison of the stationary and i.i.d. limit shapes. In general $g_{\text{pp}}(s,t) \leqslant g^{\rho}(s,t)$ for all $s, t \geqslant 0$, and

$$g_{\text{pp}}(s,t) = g^{\rho}(s,t) \text{ if and only if } (s,t) = c\big((1-\rho)^2, \rho^2\big) \text{ for some } c \geqslant 0. \tag{3.31}$$

We can also record the limit of the point-to-line LPP process, defined for $h \in \mathbb{R}^2$ by

$$G_n(h) = \max_{x_{0,n}:\, x_0=0} \Big\{ \sum_{k=0}^{n-1} \omega_{x_k} + h \cdot x_n \Big\}.$$

THEOREM 3.5. *Assume* (3.3). *Then for every $h \in \mathbb{R}^2$ the limit below holds with probability 1, with the limit function g_{pl} as given.*

$$\lim_{n\to\infty} n^{-1} G_n(h) = g_{\text{pl}}(h) \equiv 1 + \frac{h_1 + h_2}{2} + \frac{1}{2}\sqrt{(h_1 - h_2)^2 + 4}. \tag{3.32}$$

PROOF. The limit is given in Theorem 2.4. The formula for g_{pl} comes from the duality (2.9) with g_{pp} given in (3.28). □

3.3. Shape theorem and the geodesic. Begin with this strengthening of the law of large numbers (3.4).

THEOREM 3.6. *Given $\varepsilon > 0$, there exists an almost surely finite random variable K such that*

$$|G_{0,x} - g_{\text{pp}}(x)| \leqslant \varepsilon |x|_1 \quad \textit{for all } x \in \mathbb{Z}^2_{\geqslant 0} \textit{ such that } |x_1| \geqslant K. \tag{3.33}$$

We postpone the proof of Theorem 3.6 towards the end of this section but apply it presently to prove the shape theorem. For real $t \geqslant 0$ define the randomly growing cluster in $\mathbb{R}^2_{\geqslant 0}$ as

$$A_t = \{\zeta \in \mathbb{R}^2_{\geqslant 0} : G_{0,\lfloor \zeta \rfloor} \leqslant t\}.$$

In other words, to get a continuum object, we attach a unit square to each integer point x such that $G_{0,x} \leqslant t$. The limit shape is defined by

$$\mathcal{A} = \{\zeta \in \mathbb{R}^2_{\geqslant 0} : g_{\text{pp}}(\zeta) \leqslant 1\}.$$

Since $g_{\text{pp}}(\zeta) \geqslant |\zeta|_1$, we have $\mathcal{A} \subset \{\zeta \in \mathbb{R}^2_{\geqslant 0} : |\zeta|_1 \leqslant 1\}$.

THEOREM 3.7 (Shape theorem). *Given $\varepsilon > 0$, there exists an almost surely finite random variable T such that*

$$t(1-\varepsilon)\mathcal{A} \subset A_t \subset t(1+\varepsilon)\mathcal{A} \qquad \textit{for all } t \geqslant T. \tag{3.34}$$

PROOF. With the random variable K given in Theorem 3.6, set

$$T = \max_{x \in \mathbb{Z}^2_{\geqslant 0}:\, |x|_1 \leqslant K} G_{0,x} \vee g_{\text{pp}}(x).$$

Consider $t \geqslant T$. We prove the two inclusions in turn.

Suppose $\zeta \in \mathbb{R}^2_{\geqslant 0}$ satisfies $g_{\text{pp}}(\zeta) \leqslant t(1-\varepsilon)$. Let $x = \lfloor \zeta \rfloor$, the closest integer point to the left and below ζ. Then also $|x|_1 \leqslant g_{\text{pp}}(x) \leqslant t(1-\varepsilon)$. To have $\zeta \in A_t$ we need to show that $x \in A_t$. If $|x|_1 \geqslant K$, Theorem 3.6 gives

$$G_{0,x} \leqslant g_{\text{pp}}(x) + \varepsilon |x|_1 \leqslant t(1-\varepsilon) + \varepsilon t(1-\varepsilon) \leqslant t.$$

While if $|x|_1 \leqslant K$, then the definition of T gives

$$G_{0,x} \leqslant T \leqslant t.$$

We have proved the first inclusion in (3.34).

For the second inclusion, suppose $G_{0,x} \leqslant t$. Then if $|x|_1 \geqslant K$,

$$G_{0,x} \geqslant g_{\text{pp}}(x) - \varepsilon |x|_1 \geqslant (1-\varepsilon)|x|_1,$$

and we conclude that $G_{0,x} \leqslant t$ implies

$$|x|_1 \leqslant K \vee \frac{t}{1-\varepsilon} = \frac{t}{1-\varepsilon}.$$

($T \geqslant K$ follows from the definition of T.) Then for $|x|_1 \geqslant K$,

$$g_{\text{pp}}(x) \leqslant G_{0,x} + \varepsilon|x|_1 \leqslant t + \frac{\varepsilon t}{1-\varepsilon}$$

while for $|x|_1 \leqslant K$, $g_{\text{pp}}(x) \leqslant T \leqslant t$ by the definition of T. Thus integer points $x \in A_t$ satisfy

$$g_{\text{pp}}(x) \leqslant t\big(1 + \tfrac{\varepsilon}{1-\varepsilon}\big).$$

Now, given $\zeta \in A_t$, let $x = \lfloor \zeta \rfloor$. Then, by the explicit formula (3.28) for g_{pp},

$$\begin{aligned} g_{\text{pp}}(\zeta) &\leqslant g_{\text{pp}}(x + e_1 + e_2) \leqslant g_{\text{pp}}(x) + 4 + 2\sqrt{x_1} + 2\sqrt{x_2} \\ &\leqslant t\big(1 + \tfrac{\varepsilon}{1-\varepsilon}\big) + 4 + 4\sqrt{\tfrac{t}{1-\varepsilon}} \\ &= t\Big(1 + \tfrac{\varepsilon}{1-\varepsilon} + \tfrac{4}{t} + \tfrac{4}{\sqrt{t(1-\varepsilon)}}\Big) \leqslant t(1+\varepsilon'), \end{aligned}$$

when $\varepsilon' > 0$ is given, and then $\varepsilon > 0$ is chosen small enough and T large enough. Thus we can conclude that $A_t \subset t(1+\varepsilon')\mathcal{A}$ for $t \geqslant T$. $\square$

Let $\pi_{\bullet}^{x,y} \in \Pi_{x,y}$ denote the (almost surely unique) maximizing path for $G_{x,y}$ defined by (3.1). That is, the last-passage value satisfies

$$G_{x,y} = \sum_{i=0}^{|y-x|_1} \omega_{\pi_i^{x,y}}. \tag{3.35}$$

Even though last-passage percolation seeks to maximize rather than minimize path weight, $\pi_{\bullet}^{x,y}$ is often called the *geodesic* from x to y. We measure the distance of the geodesic from the straight line segment between x and y with the quantity

$$D_{x,y} = \max_{0 \leqslant i \leqslant |y-x|_1} \left| \pi_i^{x,y} - \frac{|y-x|_1 - i}{|y-x|_1}\, x - \frac{i}{|y-x|_1}\, y \right|_1. \tag{3.36}$$

$D_{x,y}$ is larger than $\max_{0 \leqslant i \leqslant |y-x|_1} \operatorname{dist}(\pi_i^{x,y}, [x,y])$, the maximal Euclidean distance of a point $\pi_i^{x,y}$ of the geodesic from the line segment $[x,y]$. The reason is that the ℓ^1 norm dominates the Euclidean norm and the point $\frac{|y-x|_1 - i}{|y-x|_1}\, x + \frac{i}{|y-x|_1}\, y$ is not necessarily the orthogonal projection of $\pi_i^{x,y}$ onto the line segment $[x,y]$.

One can check that the shape function $g_{\text{pp}}(\xi) = \xi_1 + \xi_2 + 2(\xi_1\xi_2)^{1/2}$ has the following property: if $\xi, \zeta \neq 0$, then

$$g_{\text{pp}}(\xi + \zeta) > g_{\text{pp}}(\xi) + g_{\text{pp}}(\zeta)$$

unless ξ and ζ lie on the same ray from the origin, that is, $\zeta = c\xi$ for some $c > 0$. This tells us that, on the law of large numbers scale, the optimal path is always the straight line segment. In the next theorem we make this rigorous, by showing that the random geodesic to a point at distance N is within distance $o(N)$ of the straight line segment, with high probability.

THEOREM 3.8. *We have the limit*

$$\lim_{M\to\infty} \max_{x \in \mathbb{Z}^2_{\geqslant 0} :\, |x|_1 = M} \frac{D_{0,x}}{|x|_1} = 0 \quad \textit{in probability.} \tag{3.37}$$

We turn to the proofs of Theorems 3.6 and 3.8. Both proofs use an approximation in terms of a grid of points on the line segment $\mathcal{U} = [e_1, e_2]$, given in the next lemma.

LEMMA 3.9. *Fix $\delta > 0$ and $k \in \mathbb{Z}_{>0}$ such that $\frac{1}{k} < \delta$. Set $\xi_i = (\frac{i}{k}, \frac{k-i}{k}) \in \mathcal{U}$ for $0 \leqslant i \leqslant k$. Then for each $z = (z_1, z_2) \in \mathbb{R}^2_{\geqslant 0}$ such that*

$$1 + \delta \leqslant |z|_1 \leqslant 1 + 2\delta$$

there exists $i \in \{0, \ldots, k\}$ such that $\xi_i \leqslant z \leqslant \xi_i + 2\delta(e_1 + e_2)$.

PROOF. *Case 1.* Suppose $1 \leqslant z_2 \leqslant 1 + 2\delta$. Then for $i = 0$ we have

$$\xi_{0,2} = 1 \leqslant z_2 \leqslant 1 + 2\delta = \xi_{0,2} + 2\delta$$

and

$$\xi_{0,1} = 0 \leqslant z_1 \leqslant 1 + 2\delta - z_2 \leqslant 2\delta = \xi_{0,1} + 2\delta.$$

Case 2. For $0 \leqslant z_2 < 1$ choose $i \in \{1, \ldots, k\}$ such that

$$\xi_{i,2} = \tfrac{k-i}{k} \leqslant z_2 < \tfrac{k-i+1}{k}.$$

Then

$$z_1 \geqslant 1 + \delta - z_2 > 1 + \delta - 1 + \tfrac{i}{k} - \tfrac{1}{k} > \tfrac{i}{k} = \xi_{i,1}.$$

From the opposite side

$$z_1 \leqslant 1 + 2\delta - z_2 \leqslant 2\delta + \tfrac{i}{k} = \xi_{i,1} + 2\delta$$

and $z_2 < \frac{k-i}{k} + \frac{1}{k} \leqslant \frac{k-i}{k} + 2\delta = \xi_{i,2} + 2\delta$. □

PROOF OF THEOREM 3.6. Given $\varepsilon \in (0, 1)$, fix $\varepsilon_1 > 0$ small enough so that $17\sqrt{\varepsilon_1} < \varepsilon$. Fix an integer $k > \varepsilon_1^{-1}$. Let $\xi_i = (\frac{i}{k}, \frac{k-i}{k}) \in \mathcal{U}$ as in Lemma 3.9 and set $\xi_i' = \xi_i + 2\varepsilon_1(e_1 + e_2)$. An explicit calculation shows that

$$g_{\text{pp}}(\xi_i') - g_{\text{pp}}(\xi_i) \leqslant 16\sqrt{\varepsilon_1}.$$

Utilizing limit (3.28), let N_0 be an almost surely finite random variable such that, for all $\zeta \in \{\xi_1, \ldots, \xi_k, \xi_1', \ldots, \xi_k'\}$,

$$|\, G_{0,\lfloor N\zeta \rfloor} - N g_{\text{pp}}(\zeta) \,| \leqslant N\varepsilon_1 \qquad \text{for } N \geqslant N_0.$$

Increase N_0 if necessary to guarantee that

$$(N + 1)(1 + \varepsilon_1) < N(1 + 2\varepsilon_1) \qquad \text{for all } N \geqslant N_0.$$

Let $K = (1 + \varepsilon_1)N_0$. Given $x \in \mathbb{Z}^2_{\geqslant 0}$ such that $|x|_1 \geqslant K$, let $N \geqslant N_0$ be the integer such that

$$N(1 + \varepsilon_1) \leqslant |x|_1 < (N + 1)(1 + \varepsilon_1) < N(1 + 2\varepsilon_1).$$

Then by Lemma 3.9 we can fix j such that

$$\lfloor N\xi_j \rfloor \leqslant N\xi_j \leqslant x \leqslant \lfloor N\xi_j' \rfloor \leqslant N\xi_j'.$$

By monotonicity of G and monotonicity and homogeneity of g_{pp},

$$\begin{aligned} G_{0,x} - g_{\text{pp}}(x) &\geqslant G_{0,\lfloor N\xi_j \rfloor} - N g_{\text{pp}}(\xi_j') \\ &\geqslant G_{0,\lfloor N\xi_j \rfloor} - N g_{\text{pp}}(\xi_j) - 16N\sqrt{\varepsilon_1} \\ &\geqslant -N\varepsilon_1 - 16N\sqrt{\varepsilon_1} \;\geqslant\; -N\varepsilon \;\geqslant\; -\varepsilon|x|_1. \end{aligned}$$

From the other side,

$$\begin{aligned} G_{0,x} - g_{\text{pp}}(x) &\leqslant G_{0,\lfloor N\xi_j' \rfloor} - N g_{\text{pp}}(\xi_j) \\ &\leqslant G_{0,\lfloor N\xi_j' \rfloor} - N g_{\text{pp}}(\xi_j') + 16N\sqrt{\varepsilon_1} \\ &\leqslant N\varepsilon_1 + 16N\sqrt{\varepsilon_1} \;\leqslant\; N\varepsilon \;\leqslant\; \varepsilon|x|_1. \end{aligned}$$

This completes the proof of Theorem 3.6. □

PROOF OF THEOREM 3.8. Fix $0 < \varepsilon < 1$. We show that

$$\lim_{M\to\infty} \mathbb{P}\biggl\{ \max_{x\in\mathbb{Z}^2_{\geqslant 0}:\,|x|_1=M} \frac{D_{0,x}}{|x|_1} \geqslant \varepsilon \biggr\} = 0.$$

Consider M fixed but large until the end of the argument where we take it to ∞. Throughout the proof, x denotes a point in $\mathbb{Z}^2_{\geqslant 0}$ such that $|x|_1 = M$ and z denotes a point on the geodesic $\pi_\bullet^{0,x}$ that violates the condition $|z - (|z|_1/|x|_1)x|_1 < \varepsilon$.

With $0 < \varepsilon < 1$ given, we pick three small positive quantities δ_1, ε_1, and ε_2. Let $\varepsilon_1 > 0$ satisfy

$$0 < \varepsilon_1 < \varepsilon/8. \tag{3.38}$$

Then fix $\delta_1 > 0$ such that

$$0 < \delta_1 < \varepsilon - 6\varepsilon_1. \tag{3.39}$$

Given this $\delta_1 > 0$, let $\varepsilon_2 > 0$ be such that, for each $i = 0, \dots, k$,
(3.40)

$$g_{\text{pp}}(\xi_i') \geqslant g_{\text{pp}}(\zeta) + g_{\text{pp}}(\xi_i' - \zeta) + \varepsilon_2 \quad \text{for all } 0 \leqslant \zeta \leqslant \xi_i' \text{ such that } \left| \zeta - \frac{|\zeta|_1}{|\xi_i'|_1}\xi_i' \right|_1 \geqslant \delta_1.$$

(Of course in the cases $i = 0$ and $i = k$ there are no points ζ to consider.) Now shrink $\varepsilon_1 > 0$ further to ensure that

$$\varepsilon_2 > 3\varepsilon_1 + 16\sqrt{\varepsilon_1}. \tag{3.41}$$

Note that we can shrink ε_1 without violating (3.38) or (3.39), while keeping ε, δ_1 and ε_2 fixed.

As in Lemma 3.9 with $\delta = \varepsilon_1$, define $\xi_i = (\frac{i}{k}, \frac{k-i}{k})$ and $\xi_i' = \xi_i - 2\varepsilon_1(e_1 + e_2)$ for an integer $k > \varepsilon_1^{-1}$. Recall also

$$g_{\text{pp}}(\xi_i') - g_{\text{pp}}(\xi_i) \leqslant 16\sqrt{\varepsilon_1}. \tag{3.42}$$

Let N be the integer such that

$$N(1+\varepsilon_1) \leqslant M = |x|_1 < (N+1)(1+\varepsilon_1). \tag{3.43}$$

Increase M if necessary to guarantee that $(N+1)(1+\varepsilon_1) < N(1+2\varepsilon_1)$. Then by Lemma 3.9 for each x with $|x|_1 = M$ we can fix j such that

$$\lfloor N\xi_j \rfloor \leqslant N\xi_j \leqslant x \leqslant \lfloor N\xi_j' \rfloor \leqslant N\xi_j'. \tag{3.44}$$

We isolate two lemmas from the main body of the proof.

LEMMA 3.10. *Let z be a point that satisfies $0 \leqslant z \leqslant x$ and*

$$\left| z - \frac{|z|_1}{|x|_1}x \right|_1 \geqslant \varepsilon|x|_1. \tag{3.45}$$

Then

$$|z|_1 \geqslant \frac{\varepsilon}{2}|x|_1 \quad \text{and} \quad |z - x|_1 \geqslant \frac{\varepsilon}{2}|x|_1. \tag{3.46}$$

For j as in (3.44),

$$\left| \frac{z}{N} - \frac{|z|_1}{N|\xi'_j|_1}\xi'_j \right|_1 \geqslant \delta_1. \tag{3.47}$$

PROOF OF LEMMA 3.10. By the triangle inequality and $|z|_1 \leqslant |x|_1$,

$$\left| z - \frac{|z|_1}{|x|_1}x \right|_1 \leqslant 2|z|_1$$

and

$$\begin{aligned}\left| z - \frac{|z|_1}{|x|_1}x \right|_1 &\leqslant \left| \frac{|z|_1}{|x|_1}(z-x) + \frac{|x|_1 - |z|_1}{|x|_1}z \right|_1 \leqslant |z-x|_1 + |x|_1 - |z|_1 \\ &\leqslant 2|z-x|_1.\end{aligned}$$

Inequalities (3.46) have been verified.

For (3.47) we bound the distance $|x/|x|_1 - \xi'_j/|\xi'_j||$. Let $m \in \{1,2\}$ be the coordinate index. First an upper bound on $\frac{x_m}{|x|_1}$.

$$\frac{x_m}{|x|_1} \leqslant \frac{N\xi'_{j,m}}{N(1+\varepsilon_1)} = \frac{\xi'_{j,m}}{|\xi'_j|_1} \cdot \frac{1+4\varepsilon_1}{1+\varepsilon_1} \leqslant \frac{\xi'_{j,m}}{|\xi'_j|} + 3\varepsilon_1.$$

Then a lower bound.

$$\frac{x_m}{|x|_1} \geqslant \frac{N(\xi'_{j,m} - 2\varepsilon_1)}{N(1+2\varepsilon_1)} \geqslant \frac{\xi'_{j,m}}{|\xi'_j|_1} - 2\varepsilon_1.$$

From the last two displays

$$\left| \frac{x}{|x|_1} - \frac{\xi'_j}{|\xi'_j|_1} \right|_1 \leqslant 6\varepsilon_1. \tag{3.48}$$

On the other hand, from (3.45) and $|x|_1 \geqslant N(1+\varepsilon_1)$ we obtain

$$\left| \frac{z}{N} - \frac{|z|_1}{N} \cdot \frac{x}{|x|_1} \right|_1 \geqslant \varepsilon(1+\varepsilon_1).$$

By the triangle inequality

$$\begin{aligned}\left| \frac{z}{N} - \frac{|z|_1}{N|\xi'_j|_1}\xi'_j \right|_1 &\geqslant \left| \frac{z}{N} - \frac{|z|_1}{N} \cdot \frac{x}{|x|_1} \right|_1 - \frac{|z|_1}{N} \cdot \left| \frac{x}{|x|_1} - \frac{\xi'_j}{|\xi'_j|_1} \right|_1 \\ &\geqslant \varepsilon(1+\varepsilon_1) - 6\varepsilon_1(1+\varepsilon_1) \geqslant \delta_1.\end{aligned}$$

Claim (3.47) has been verified and Lemma 3.10 proved. □

For $N \in \mathbb{Z}_{>0}$ let B_N be the event on which the following statements hold:

$$G_{0,\lfloor N\xi_i \rfloor} \geqslant N g_{\mathrm{pp}}(\xi_i) - N\varepsilon_1 \qquad \text{for } i \in [k], \tag{3.49}$$

and whenever lattice points $0 \leqslant z \leqslant x$ and index j satisfy (3.43), (3.44), and (3.45), we have

$$G_{0,z} \leqslant g_{\mathrm{pp}}(z) + N\varepsilon_1 \tag{3.50}$$

and

$$G_{z,\lfloor N\xi'_j \rfloor} \leqslant g_{\mathrm{pp}}(N\xi'_j - z) + N\varepsilon_1. \tag{3.51}$$

LEMMA 3.11. *$\mathbb{P}(B_N) \to 1$ as $N \to \infty$.*

PROOF OF LEMMA 3.11. Both (3.49) and (3.50) are satisfed with high probability by Theorem 3.6. For (3.49) this is immediate. For (3.50) we need a few additional observations. By the assumptions on z and x and by (3.46),

$$|z|_1 \geqslant \frac{\varepsilon}{2}\,|x|_1 \geqslant \frac{N}{2}\varepsilon(1+\varepsilon_1),$$

and so $|z|_1$ grows large as N grows. Consequently (3.33) applies and gives

$$G_{0,z} \leqslant g_{\text{pp}}(z) + \varepsilon'|z|_1$$

for large enough N, for all z that satisfy (3.45), for any fixed $\varepsilon' > 0$. Then take $\varepsilon' < \varepsilon_1/(1+2\varepsilon_1)$ so that $\varepsilon'|z|_1 \leqslant \varepsilon'|x|_1 \leqslant N\varepsilon'(1+2\varepsilon_1) \leqslant N\varepsilon_1$.

To satisfy (3.51), consider N and j fixed, and let x and z vary. Translate and reflect the lattice to transform the process $\{G_{z,\lfloor N\xi'_j\rfloor}\}_{z\leqslant\lfloor N\xi'_j\rfloor}$ into the distributionally identical process $\{G_{0,\lfloor N\xi'_j\rfloor - z}\}_{z\leqslant\lfloor N\xi'_j\rfloor}$. We bound the norm of the endpoint above and below.

$$|\,\lfloor N\xi'_j\rfloor - z\,|_1 \leqslant |x-z|_1 + |\,\lfloor N\xi'_j\rfloor - x\,|_1 \leqslant |x|_1 + 4N\varepsilon_1 \leqslant N(1+6\varepsilon_1).$$

By the definition of B_M, the relevant points z satisfy condition (3.45), and hence by (3.46),

$$|\,\lfloor N\xi'_j\rfloor - z\,|_1 \geqslant |x-z|_1 - |\,\lfloor N\xi'_j\rfloor - x\,|_1 \geqslant \frac{\varepsilon}{2}\,|x|_1 - 4N\varepsilon_1 \geqslant N(\tfrac12\varepsilon(1+\varepsilon_1) - 4\varepsilon_1).$$

As N grows this last lower bound grows large, uniformly for the points z that satisfy (3.45). Hence for large enough N (3.33) applies and for these points z we have

$$G_{0,\lfloor N\xi'_j\rfloor - z} \leqslant g_{\text{pp}}(N\xi'_j - z) + \varepsilon'|\,\lfloor N\xi'_j\rfloor - z\,|_1 \leqslant g_{\text{pp}}(N\xi'_j - z) + N\varepsilon_1$$

once ε' is chosen small enough. Since we transformed the original process, we do not get (3.51) almost surely, but only with high probability. □

We conclude the proof of Theorem 3.8. Suppose that event B_N of (3.49)–(3.51) occurs and the geodesic $\pi_{\bullet}^{0,x}$ goes through a point z that satisfies (3.45). This gives $G_{0,x} \leqslant G_{0,z} + G_{z,x}$ (an inequality because the weight at z is counted twice). The following string of inequalities uses in turn (3.42), (3.49), monotonicity, (3.50)–(3.51), and finally (3.40) applied to $\zeta = z/N$:

$$\begin{aligned}
Ng_{\text{pp}}(\xi'_j) - N(\varepsilon_1 + 16\sqrt{\varepsilon_1}) &\leqslant Ng_{\text{pp}}(\xi_j) - N\varepsilon_1 \leqslant G_{0,\lfloor N\xi_j\rfloor} \leqslant G_{0,x} \\
&\leqslant G_{0,z} + G_{z,x} \leqslant G_{0,z} + G_{z,\lfloor N\xi'_j\rfloor} \\
&\leqslant N\Big[g_{\text{pp}}\Big(\frac{z}{N}\Big) + g_{\text{pp}}\Big(\xi'_j - \frac{z}{N}\Big)\Big] + 2N\varepsilon_1 \\
&\leqslant Ng_{\text{pp}}(\xi'_j) - N\varepsilon_2 + 2N\varepsilon_1.
\end{aligned}$$

By (3.41) we have a contradiction. Hence on the event B_N no geodesic to a point x that satisfies (3.43) can go through a point z that satisfies (3.45). Thus

$$\mathbb{P}\bigg\{ \frac{D_{0,x}}{|x|_1} \geqslant \varepsilon \text{ for some } x \text{ that satisfies (3.43)} \bigg\} \leqslant \mathbb{P}(B_N^{c}) \to 0.$$

Theorem 3.8 is proved. □

4. Busemann functions for the exponential CGM

In this section we prove the existence of the Busemann functions and show that they provide minimizers for the variational formulas. Section 4.1 states the results and Section 4.2 contains the proof of the main existence theorem.

4.1. Existence and properties of Busemann functions. We extend the constructions discussed in Section 3 to the full lattice $\mathbb{Z}^2$. As before a down-right path is a bi-infinite sequence $\mathcal{Y} = (y_k)_{k\in\mathbb{Z}}$ in $\mathbb{Z}^2$ such that $y_k - y_{k-1} \in \{e_1, -e_2\}$ for all $k \in \mathbb{Z}$. The lattice decomposes into a disjoint union $\mathbb{Z}^2 = \mathcal{G}_- \cup \mathcal{Y} \cup \mathcal{G}_+$ where the two regions are

$$\mathcal{G}_- = \{x \in \mathbb{Z}^2 : \exists j \in \mathbb{Z}_{>0} \text{ such that } x + j(e_1 + e_2) \in \mathcal{Y}\}$$

and

$$\mathcal{G}_+ = \{x \in \mathbb{Z}^2 : \exists j \in \mathbb{Z}_{>0} \text{ such that } x - j(e_1 + e_2) \in \mathcal{Y}\}.$$

It will be convenient to formalize the properties identified earlier in Theorem 3.1 in the following definition.

DEFINITION 4.1. Let $0 < \alpha < 1$. Let us say that a process

$$\{\eta_x, I_x, J_x, \check{\eta}_x : x \in \mathbb{Z}^2\} \tag{4.1}$$

is an *exponential-α last-passage percolation system* if the following properties hold.

(a) The process is stationary with marginal distributions

$$\eta_x, \check{\eta}_x \sim \text{Exp}(1), \quad I_x \sim \text{Exp}(1-\alpha), \quad \text{and} \quad J_x \sim \text{Exp}(\alpha). \tag{4.2}$$

For any down-right path $\mathcal{Y} = (y_k)_{k\in\mathbb{Z}}$ in $\mathbb{Z}^2$, the random variables

$$\{\check{\eta}_z : z \in \mathcal{G}_-\}, \quad \{t(\{y_{k-1}, y_k\}) : k \in \mathbb{Z}\}, \quad \text{and} \quad \{\eta_x : x \in \mathcal{G}_+\} \tag{4.3}$$

are all mutually independent, where the undirected edge variables $t(e)$ are defined as before in (3.16).

(b) The following equations are in force at all $x \in \mathbb{Z}^2$:

$$\check{\eta}_{x-e_1-e_2} = I_{x-e_2} \wedge J_{x-e_1} \tag{4.4}$$

$$I_x = \eta_x + (I_{x-e_2} - J_{x-e_1})^+ \tag{4.5}$$

$$J_x = \eta_x + (I_{x-e_2} - J_{x-e_1})^-. \tag{4.6}$$

Equations (4.4)–(4.6) above are the same as those in (3.7)–(3.9).

With the definition of the exponential-α LPP system we can state fairly succinctly the existence of Busemann functions in the next theorem. The setting is the following. $(\Omega, \mathfrak{S}, \mathbb{P})$ is a probability space whose generic sample point is denoted by ω. There is a group of measurable bijections $(\theta_x)_{x\in\mathbb{Z}^2}$ on Ω that preserve $\mathbb{P}$. $Y = (Y_x)_{x\in\mathbb{Z}^2}$ are i.i.d. Exp(1) weights defined on Ω that satisfy $Y_x(\omega) = Y_0(\theta_x\omega)$. The LPP process is defined in terms of the weights Y:

$$G_{x,y} = \max_{x_\bullet \in \Pi_{x,y}} \sum_{k=0}^{|y-x|_1} Y_{x_k}. \tag{4.7}$$

The canonical choice $\Omega = \mathbb{R}_+^{\mathbb{Z}^2}$ with translations $(\theta_x\omega)_y = \omega_{x+y}$, an i.i.d. product measure $\mathbb{P}$ and the coordinate process $Y_x(\omega) = \omega_x$ is one possibility. The additional generality is meaningful for the uniqueness in Theorem 4.3.

THEOREM 4.2. *For each* $0 < \alpha < 1$ *there exist a stationary cocycle* B^α *and a family of random weights* $\{X_x^\alpha\}_{x\in\mathbb{Z}^2}$ *on* $(\Omega, \mathfrak{S}, \mathbb{P})$ *with the following properties.*

(i) *For each* $0 < \alpha < 1$, *process*

$$\{X_x^\alpha,\, B^\alpha_{x-e_1,x},\, B^\alpha_{x-e_2,x},\, Y_x : x \in \mathbb{Z}^2\}$$

is an exponential-α *last-passage system as described in Definition* 4.1.

(ii) *There exists a single event* Ω_2 *of full probability such that for all* $\omega \in \Omega_2$, *all* $x \in \mathbb{Z}^2$ *and all* $\lambda < \rho$ *in* $(0, 1)$ *we have the inequalities*

$$B^\lambda_{x,x+e_1}(\omega) \leqslant B^\rho_{x,x+e_1}(\omega) \quad \text{and} \quad B^\lambda_{x,x+e_2}(\omega) \geqslant B^\rho_{x,x+e_2}(\omega). \tag{4.8}$$

Furthermore, for all $\omega \in \Omega_2$ *and* $x, y \in \mathbb{Z}^2$, *the function* $\lambda \to B^\lambda_{x,y}(\omega)$ *is right-continuous with left limits.*

(iii) *For each fixed* $0 < \alpha < 1$ *there exists an event* $\Omega_2^{(\alpha)}$ *of full probability such that the following holds: for each* $\omega \in \Omega_2^{(\alpha)}$ *and any sequence* $v_n \in \mathbb{Z}^2$ *such that* $|v_n|_1 \to \infty$ *and*

$$\lim_{n\to\infty} \frac{v_n}{|v_n|_1} = \xi(\alpha) = \left(\frac{(1-\alpha)^2}{(1-\alpha)^2+\alpha^2}, \frac{\alpha^2}{(1-\alpha)^2+\alpha^2}\right). \tag{4.9}$$

we have the limits

$$B^\alpha_{x,y}(\omega) = \lim_{n\to\infty} [G_{x,v_n}(\omega) - G_{y,v_n}(\omega)] \qquad \forall x, y \in \mathbb{Z}^2. \tag{4.10}$$

The LPP process $G_{x,y}$ *is now defined by* (4.7). *Furthermore, for all* $\omega \in \Omega_2^{(\alpha)}$ *and* $x, y \in \mathbb{Z}^2$,

$$\lim_{\lambda\to\alpha} B^\lambda_{x,y}(\omega) = B^\alpha_{x,y}(\omega). \tag{4.11}$$

Part (i) of Theorem 4.2 together with (4.4) implies

$$Y_x = B^\alpha_{x,\,x+e_1} \wedge B^\alpha_{x,\,x+e_2}. \tag{4.12}$$

In other words, cocycle B^α is adapted to the potential of the corner growth model in the sense of (2.17). In the opposite lattice direction we have the identity

$$X_x^\alpha = B^\alpha_{x-e_1,\,x} \wedge B^\alpha_{x-e_2,\,x}, \tag{4.13}$$

which follows from (4.5)–(4.6). (*Note that in this construction the given* Y*-weights are now playing the role of the* $\check{\omega}$*-weights of Theorem* 3.1, *and the constructed weights* X^α *play the role of the* ω*-weights of Theorem* 3.1.)

From the explicit formula for the shape function (Theorem 3.4),

$$g_{\text{pp}}(s,t) = s + t + 2\sqrt{st}$$

and from this we derive

$$\nabla g_{\text{pp}}(s,t) = \big(1 + \sqrt{t/s}\,, 1 + \sqrt{s/t}\,\big).$$

By the exponential distributions of B^α-increments given in part (i) of Theorem 4.2 we obtain the relation

$$\big(\mathbb{E}[B^\alpha_{0,e_1}], \mathbb{E}[B^\alpha_{0,e_2}]\big) = \Big(\frac{1}{1-\alpha}, \frac{1}{\alpha}\Big) = \nabla g_{\text{pp}}(\xi(\alpha)). \tag{4.14}$$

This is natural since the limit (4.10) represents B^α as a microscopic gradient of passage time.

The next theorem gives a uniqueness statement.

THEOREM 4.3. *Assume the setting of Theorem 4.2. Fix $0 < \rho < 1$.*

(i) *Suppose that, indexed by the nearest-neighbor edges in the quadrant with lower left corner $u = (u_1, u_2) \in \mathbb{Z}^2$, there are random variables*

$$(A_{x,\,x+e_1}, A_{x,\,x+e_2})_{x \in u+\mathbb{Z}^2_{\geqslant 0}}$$

on $(\Omega, \mathfrak{S}, \mathbb{P})$ with properties (a) *and* (b)*:*

(a) $A_{x,\,x+e_1} \sim \mathrm{Exp}(1-\rho)$ *and* $A_{x,\,x+e_2} \sim \mathrm{Exp}(\rho)$ *for all $x \in u + \mathbb{Z}^2_{\geqslant 0}$. For any $v = (v_1, v_2) \in u + \mathbb{Z}^2_{>0}$, the random variables*

$$(A_{v-ie_1,\,v-(i-1)e_1}, A_{v-je_2,\,v-(j-1)e_2})_{1 \leqslant i \leqslant v_1-u_1,\, 1 \leqslant j \leqslant v_2-u_2}$$

are independent.

(b) *For each $x \in u + \mathbb{Z}^2_{\geqslant 0}$ we have $\mathbb{P}$-almost surely additivity around the unit square with lower left corner at x:*

$$A_{x,\,x+e_1} + A_{x+e_1,\,x+e_1+e_2} = A_{x,\,x+e_2} + A_{x+e_2,\,x+e_1+e_2},$$

and recovery:

$$Y_x = A_{x,\,x+e_1} \wedge A_{x,\,x+e_2}.$$

Then $A_{x,\,x+e_r} = B^{\rho}_{x,\,x+e_r}$ for all $x \in u + \mathbb{Z}^2_{\geqslant 0}$ and $r \in \{1,2\}$, $\mathbb{P}$-almost surely.

(ii) *In particular, suppose there are random variables $(U_x, A_{x-e_1,x}, A_{x-e_2,x})_{x\in\mathbb{Z}^2}$ on $(\Omega, \mathfrak{S}, \mathbb{P})$ such that $\{U_x, A_{x-e_1,x}, A_{x-e_2,x}, Y_x : x \in \mathbb{Z}^2\}$ is an exponential-ρ last-passage system as described in Definition 4.1. Then $U_x = X^{\rho}_x$, $A_{x-e_1,x} = B^{\rho}_{x-e_1,x}$ and $A_{x-e_2,x} = B^{\alpha}_{x-e_2,x}$ for all x, $\mathbb{P}$-almost surely.*

The second part of assumption (a) above says that the edge variables A are independent on the north and east boundaries of any rectangle $[u, v]$. Part (ii) about the uniqueness of an exponential-ρ last-passage system that includes the given weights Y is a consequence of part (i) of the theorem because assumptions (a) and (b) are true for an exponential-ρ last-passage system.

Before turning to the proofs of Theorems 4.2 and 4.3 in Section 4.2, we revisit previous themes.

Busemann function as a stationary last-passage percolation process. Fix $0 < \rho < 1$. The Busemann function B^{ρ} can be viewed as an increment-stationary LPP process in two ways, depending on whether we use the weights Y or X^{ρ}. Fix a reference point $v \in \mathbb{Z}^2$. We construct two LPP processes with boundary given by B^{ρ}. $\{G^{NE}_{u,v}\}_{u \leqslant v}$ has boundary conditions on the north and east and admissible paths from v proceed down-left using steps $\{-e_1, -e_2\}$, while $\{G^{SW}_{v,w}\}_{w \geqslant v}$ has boundary conditions on the south and west and admissible paths are of the same up-right kind as before. The definitions are

$$G^{NE}_{v-ke_r,v} = B^{\rho}_{v-ke_r,v} \qquad \text{for } r \in \{1,2\} \text{ and } k \in \mathbb{Z}_{\geqslant 0}$$

and in the bulk

$$G^{NE}_{u,v} = Y_u + G^{NE}_{u+e_1,v} \vee G^{NE}_{u+e_2,v} \qquad \text{for } u \leqslant v - e_1 - e_2$$

and then

$$G^{SW}_{v,\,v+ke_r} = B^{\rho}_{v,\,v+ke_r} \qquad \text{for } r \in \{1,2\} \text{ and } k \in \mathbb{Z}_{\geqslant 0}$$

and in the bulk

$$G^{SW}_{v,w} = X^{\rho}_w + G^{SW}_{v,\,w-e_1} \vee G^{SW}_{v,\,w-e_2} \qquad \text{for } w \geqslant v + e_1 + e_2.$$

Inductive arguments using (4.12) and (4.13) and the additivity of B^ρ prove the next theorem.

THEOREM 4.4. *For all $u \leqslant v$ and all $w \geqslant v$, $G^{NE}_{u,v} = B^\rho_{u,v}$ and $G^{SW}_{v,w} = B^\rho_{v,w}$.*

The LPP processes constructed here have the same distribution as G^ρ constructed in (3.13)–(3.14) (modulo a lattice reversal for G^{NE}): with v fixed,

$$\{B^\rho_{v-x,\,v}\}_{x\in\mathbb{Z}^2_{\geqslant 0}} \overset{d}{=} \{G^\rho_{v,\,v+x}\}_{x\in\mathbb{Z}^2_{\geqslant 0}} \quad \text{and} \quad \{B^\rho_{v,\,v+x}\}_{x\in\mathbb{Z}^2_{\geqslant 0}} \overset{d}{=} \{G^\rho_{v,\,v+x}\}_{x\in\mathbb{Z}^2_{\geqslant 0}}. \tag{4.15}$$

This follows because processes $\{G^{NE}_{v-x,\,v}\}_{x\in\mathbb{Z}^2_{\geqslant 0}}$, $\{G^{SW}_{v,\,v+x}\}_{x\in\mathbb{Z}^2_{\geqslant 0}}$ and $\{G^\rho_{v,\,v+x}\}_{x\in\mathbb{Z}^2_{\geqslant 0}}$ are constructed with identical procedures from distributionally identical inputs.

Busemann functions as extremals in the variational formulas. Recall the variational formula for the point-to-point limit shape function, specialized to the two-dimensional corner growth model.

$$g_{\text{pp}}(\xi) = \inf_{B\in\mathcal{K}} \mathbb{P}\text{-}\operatorname{ess\,sup} \max_{i=1,2} \{Y_0 - B(0,e_i) - h(B)\cdot\xi\}. \tag{4.16}$$

Above $h(B)$ is the negative of the mean vector:

$$h(B) = -\big(\mathbb{E}[B(0,e_1)], \mathbb{E}[B(0,e_2)]\big). \tag{4.17}$$

The next theorem shows that cocycle B^α from Theorem 4.2 is a minimizer in (4.16) for the characteristic direction $\xi(\alpha)$.

THEOREM 4.5. *Continue with the setting of Theorem 4.2. The following hold for each $0<\alpha<1$.*

(i) *The characteristic direction $\xi(\alpha)$ and the vector $h(B^\alpha)$ are dual to each other in the sense that*

$$g_{\text{pl}}(h(B^\alpha)) = g_{\text{pp}}(\xi(\alpha)) + h(B^\alpha)\cdot\xi(\alpha). \tag{4.18}$$

(ii) *We have the identities*

$$\begin{aligned} g_{\text{pp}}(\xi(\alpha)) &= -h(B^\alpha)\cdot\xi(\alpha) \\ &= \max_{i=1,2}\big[\,Y_0(\omega) - B^\alpha_{0,e_i}(\omega) - h(B^\alpha)\cdot\xi(\alpha)\,\big] \quad \mathbb{P}\text{-almost surely.} \end{aligned} \tag{4.19}$$

In other words, B^α minimizes in the variational formula (4.16) in the characteristic direction $\xi(\alpha)$, without the essential supremum.

PROOF. Part (i). From (4.14) we have

$$-h(B^\alpha) = \nabla g_{\text{pp}}(\xi(\alpha))$$

and hence by the strict concavity of g_{pp}, $\xi(\alpha)$ is the unique maximizer in the duality

$$g_{\text{pl}}(h(B^\alpha)) = \sup_{\xi\in\mathcal{U}}[\,g_{\text{pp}}(\xi) + h(B^\alpha)\cdot\xi\,].$$

Part (ii). By the duality (4.18), by $g_{\text{pl}}(h(B^\alpha)) = 0$ from Lemma 2.14, and by the adaptedness (4.12),

$$\begin{aligned} g_{\text{pp}}(\xi(\alpha)) &= g_{\text{pl}}(h(B^\alpha)) - h(B^\alpha)\cdot\xi(\alpha) = -h(B^\alpha)\cdot\xi(\alpha) \\ &= Y_0 - B^\alpha_{0,e_1}\wedge B^\alpha_{0,e_2} - h(B^\alpha)\cdot\xi(\alpha) \\ &= \max_{i=1,2}\big[\,Y_0 - B^\alpha_{0,e_i} - h(B^\alpha)\cdot\xi(\alpha)\,\big]. \end{aligned}$$

□

4.2. Proof of the existence of Busemann functions. This section proves the existence and properties of the limiting Busemann functions. The proof is an adaptation of the one given in [**31**] for the exactly solvable log-gamma (or inverse gamma) polymer model.

Before specializing to the processes we study, we state and prove general inequalities for planar last-passage increments. Early appearances of these types of inequalities in first-passage percolation can be found in [**3**, **4**]. Let real weights $\{\widetilde{Y}_x\}_{x\in\mathbb{Z}^2}$ be given. Define last-passage times

$$\widetilde{G}_{x,y} = \max_{x_{0,n}} \sum_{k=0}^{n} \widetilde{Y}_{x_k} \tag{4.20}$$

where the maximum is over up-right paths from $x_0 = x$ to $x_n = y$ with $n = |y-x|_1$ and the admissible steps are as before $x_k - x_{k-1} \in \{e_1, e_2\}$. The convention is $\widetilde{G}_{v,v} = 0$. For $x \leqslant v - e_1$ and $y \leqslant v - e_2$ denote increments by

$$\widetilde{I}_{x,v} = \widetilde{G}_{x,v} - \widetilde{G}_{x+e_1,v} \qquad \text{and} \qquad \widetilde{J}_{y,v} = \widetilde{G}_{y,v} - \widetilde{G}_{y+e_2,v}\,.$$

For the precise statement of the lemma below it is important that the sum in (4.20) includes the first weight $\widetilde{Y}_x$, but the increments $\widetilde{I}$ and $\widetilde{J}$ are not sensitive to whether the last weight $\widetilde{Y}_y$ is included or excluded.

LEMMA 4.6. *For $x \leqslant v - e_1$ and $y \leqslant v - e_2$*

$$\widetilde{I}_{x,v+e_2} \geqslant \widetilde{I}_{x,v} \geqslant \widetilde{I}_{x,v+e_1} \qquad \textit{and} \qquad \widetilde{J}_{y,v+e_2} \leqslant \widetilde{J}_{y,v} \leqslant \widetilde{J}_{y,v+e_1}\,. \tag{4.21}$$

PROOF. Let $v = (m,n)$. The proof goes by an induction argument that starts from the north and east boundaries. On the north, for $x = (k,n)$ for some $k < m$,

$$\begin{aligned}
\widetilde{I}_{(k,n),(m,n+1)} &= \widetilde{G}_{(k,n),(m,n+1)} - \widetilde{G}_{(k+1,n),(m,n+1)} \\
&= \widetilde{Y}_{k,n} + \widetilde{G}_{(k+1,n),(m,n+1)} \vee \widetilde{G}_{(k,n+1),(m,n+1)} - \widetilde{G}_{(k+1,n),(m,n+1)} \\
&\geqslant \widetilde{Y}_{k,n} = \widetilde{G}_{(k,n),(m,n)} - \widetilde{G}_{(k+1,n),(m,n)} = \widetilde{I}_{(k,n),(m,n)}\,.
\end{aligned}$$

A similar argument (or the above inequality applied to transposed lattice points $(a',b') = (b,a)$) gives, for $y = (m,\ell)$ for some $\ell < n$,

$$\widetilde{J}_{(m,\ell),(m+1,n)} \geqslant \widetilde{J}_{(m,\ell),(m,n)}.$$

We also have the equalities, first for $y = (m,\ell)$ for some $\ell < n$

$$\begin{aligned}
\widetilde{J}_{(m,\ell),(m,n+1)} &= \widetilde{G}_{(m,\ell),(m,n+1)} - \widetilde{G}_{(m,\ell+1),(m,n+1)} \\
&= \widetilde{Y}_{m,\ell} = \widetilde{G}_{(m,\ell),(m,n)} - \widetilde{G}_{(m,\ell+1),(m,n)} = \widetilde{J}_{(m,\ell),(m,n)}
\end{aligned}$$

and similarly also

$$\widetilde{I}_{(k,n),(m+1,n)} = \widetilde{I}_{(k,n),(m,n)}.$$

These inequalities start the induction. Now let $u \leqslant v - e_1 - e_2$. Assume by induction that (4.21) holds for $x = u + e_2$ and $y = u + e_1$. We prove the first inequalities of (4.21) for $x = u$.

$$\begin{aligned}
\widetilde{I}_{u,v+e_2} &= \widetilde{G}_{u,v+e_2} - \widetilde{G}_{u+e_1,v+e_2} = \widetilde{Y}_u + (\widetilde{G}_{u+e_2,v+e_2} - \widetilde{G}_{u+e_1,v+e_2})^+ \\
&= \widetilde{Y}_u + (\widetilde{I}_{u+e_2,v+e_2} - \widetilde{J}_{u+e_1,v+e_2})^+ \\
&\geqslant \widetilde{Y}_u + (\widetilde{I}_{u+e_2,v} - \widetilde{J}_{u+e_1,v})^+ = \widetilde{I}_{u,v}\,.
\end{aligned}$$

The last equality comes by repeating the first three equalities with v instead of $v + e_2$.

Replacing the pair $(v + e_2, v)$ with $(v, v + e_1)$ in the argument above gives $\widetilde{I}_{u,v} \geqslant \widetilde{I}_{u,v+e_1}$. A symmetric argument works for the $\widetilde{J}$ inequalities. □

We introduce the following general notational device which we illustrate in the context of (4.20). If Λ is a subset of admissible paths from x to y, then

$$\widetilde{G}_{x,y}(\Lambda) = \max_{x_{0,n} \in \Lambda} \sum_{k=0}^{n} \widetilde{Y}_{x_k} \tag{4.22}$$

is the last-passage value obtained when the maximum is restricted to paths $x_{0,n}$ in Λ.

Return to the context of the CGM with exponential weights. Fix an origin $u \in \mathbb{Z}^2$ and a parameter $0 < \rho < 1$ for the moment. Assume given independent variables

$$\{\sigma_x,\ I_{u+ie_1},\ J_{u+je_2} : x \in u + \mathbb{Z}^2_{>0},\ i, j \in \mathbb{Z}_{>0}\} \tag{4.23}$$

as in (3.5), with marginal distributions (3.6). (We use σ for these auxiliary Exp(1) weights instead of the ω of (3.5) to avoid confusion with the sample point ω of Theorem 4.2.) Let the variables

$$\{\sigma_x,\ I_{x-e_2},\ J_{x-e_1},\ \check{\sigma}_{x-e_1-e_2} : x \in u + \mathbb{Z}^2_{>0}\} \tag{4.24}$$

be defined via equations (3.7)–(3.9) from initial variables (4.23), so that the properties given in Theorem 3.1 are satisfied, with σ replacing ω.

Utilizing the edge weights I_x and J_x and the vertex weights $\check{\sigma}_x$ we define several last-passage percolation processes. First a process with i.i.d. weights:

$$\check{G}_{x,y} = \max_{x_\bullet \in \Pi_{x,y}} \sum_{k=0}^{|y-x|_1} \check{\sigma}_{x_k} \qquad \text{for } y \geqslant x \geqslant u. \tag{4.25}$$

Its increments are defined by

$$\begin{aligned} &\check{I}_{x,v} = \check{G}_{x,v} - \check{G}_{x+e_1,v} \quad \text{for } u \leqslant x \leqslant v - e_1 \\ \text{and} \quad &\check{J}_{y,v} = \check{G}_{y,v} - \check{G}_{y+e_2,v} \quad \text{for } u \leqslant y \leqslant v - e_2. \end{aligned}$$

Then we introduce an auxiliary increment-stationary last-passage process $\check{G}^{NE}_{x,v}$ for $u \leqslant x \leqslant v$, with boundary edge weights on the north and east borders. First set $\check{G}^{NE}_{v,v} = 0$. On the north and east boundaries emanating from v in the negative directions,

$$\check{G}^{NE}_{v-ke_1,v} = \sum_{i=0}^{k-1} I_{v-ie_1} \quad \text{and} \quad \check{G}^{NE}_{v-\ell e_2,v} = \sum_{j=0}^{\ell-1} J_{v-je_2} \qquad \text{for } k, \ell \geqslant 1. \tag{4.26}$$

In the bulk for $u \leqslant x \leqslant v - e_1 - e_2$ we define,

$$\begin{aligned} \check{G}^{NE}_{x,v} &= \check{\sigma}_x + \check{G}^{NE}_{x+e_1,v} \vee \check{G}^{NE}_{x+e_2,v} \\ &= \max_{1 \leqslant k \leqslant (v-x)\cdot e_1} \Big\{ \check{G}_{x,v-ke_1-e_2} + \sum_{i=0}^{k-1} I_{v-ie_1} \Big\} \\ &\qquad \bigvee \max_{1 \leqslant \ell \leqslant (v-x)\cdot e_2} \Big\{ \check{G}_{x,v-e_1-\ell e_2} + \sum_{j=0}^{\ell-1} J_{v-je_2} \Big\}. \end{aligned} \tag{4.27}$$

In the next lemma we check that the increments of the $\check{G}^{NE}$ process are in fact the I and J variables already given in (4.24).

LEMMA 4.7. *For $u \leqslant x \leqslant v - e_1$ and $u \leqslant y \leqslant v - e_2$,*

$$I_{x+e_1} = \check{G}^{NE}_{x,v} - \check{G}^{NE}_{x+e_1,v} \qquad \text{and} \qquad J_{y+e_2} = \check{G}^{NE}_{y,v} - \check{G}^{NE}_{y+e_2,v}. \tag{4.28}$$

PROOF. The claim is true for $x = v - ke_1$ and $y = v - \ell e_2$ by definition (4.26). Here is the induction step for the edge $(x, x + e_1)$, assuming that (4.28) has been proved for edges $(x + e_2, x + e_1 + e_2)$ and $(x + e_1, x + e_1 + e_2)$.

$$\begin{aligned} \check{G}^{NE}_{x,v} - \check{G}^{NE}_{x+e_1,v} &= \check{\sigma}_x + \bigl(\check{G}^{NE}_{x+e_2,v} - \check{G}^{NE}_{x+e_1,v}\bigr)^+ = \check{\sigma}_x + \bigl(I_{x+e_1+e_2} - J_{x+e_1+e_2}\bigr)^+ \\ &= I_{x+e_1} \wedge J_{x+e_2} + \bigl(I_{x+e_1} - J_{x+e_2}\bigr)^+ = I_{x+e_1}. \end{aligned}$$

The second last equality used (3.7) and the cocycle property $I_{x+e_1} + J_{x+e_1+e_2} = J_{x+e_2} + I_{x+e_1+e_2}$. A similar argument extends (4.28) to the edge $(x, x + e_2)$. ☐

Next we consider restricted $\check{G}^{NE}$ last-passage values utilizing the notation introduced in (4.22). In particular, we consider last-passage values of the kind

$$\check{G}^{NE}_{x,\,v}(v - e_1 \in x_\bullet) = \max_{1 \leqslant k \leqslant (v-x)\cdot e_1} \Bigl\{ \check{G}_{x, v-ke_1-e_2} + \sum_{i=0}^{k-1} I_{v-ie_1} \Bigr\} \tag{4.29}$$

where the condition $v - e_1 \in x_\bullet$ means that the path goes through the point $v - e_1$, which is equivalent to saying that the last step of the path goes from $v - e_1$ to v. We show that the asymptotics of the restricted $\check{G}^{NE}$ are the expected ones and calculate the limits.

LEMMA 4.8. *Fix a point $a \in u + \mathbb{Z}^2_{\geqslant 0}$ and reals $0 < s, t < \infty$. Let $v_n \in u + \mathbb{Z}^2_{\geqslant 0}$ be such that $|v_n|_1 \to \infty$ and $v_n/|v_n|_1 \to (s,t)/(s+t)$ as $n \to \infty$. Then we have the following almost sure limits.*

$$\begin{aligned} |v_n|_1^{-1}\, \check{G}^{NE}_{a,\,v_n}(v_n - e_1 \in x_\bullet) \underset{n\to\infty}{\longrightarrow} (s+t)^{-1} \sup_{0 \leqslant \tau \leqslant s} \Bigl\{ \frac{\tau}{1-\rho} + g_{\rm pp}(s-\tau, t) \Bigr\} \\ = \begin{cases} g_{\rm pp}(s,t) = \bigl(\sqrt{s} + \sqrt{t}\bigr)^2, & \dfrac{s}{t} \leqslant \Bigl(\dfrac{1-\rho}{\rho}\Bigr)^2 \\[2ex] \dfrac{s}{1-\rho} + \dfrac{t}{\rho}, & \dfrac{s}{t} \geqslant \Bigl(\dfrac{1-\rho}{\rho}\Bigr)^2 \end{cases} \end{aligned} \tag{4.30}$$

and

$$\begin{aligned} |v_n|_1^{-1}\, \check{G}^{NE}_{a,\,v_n}(v_n - e_2 \in x_\bullet) \underset{n\to\infty}{\longrightarrow} (s+t)^{-1} \sup_{0 \leqslant \tau \leqslant t} \Bigl\{ \frac{\tau}{\rho} + g_{\rm pp}(s, t-\tau) \Bigr\} \\ = \begin{cases} g_{\rm pp}(s,t) = \bigl(\sqrt{s} + \sqrt{t}\bigr)^2, & \dfrac{s}{t} \geqslant \Bigl(\dfrac{1-\rho}{\rho}\Bigr)^2 \\[2ex] \dfrac{s}{1-\rho} + \dfrac{t}{\rho}, & \dfrac{s}{t} \leqslant \Bigl(\dfrac{1-\rho}{\rho}\Bigr)^2. \end{cases} \end{aligned} \tag{4.31}$$

PROOF. We prove the limit in (4.30). Finding the supremum is calculus. The proof of (4.31) being entirely analogous. Fix $\varepsilon > 0$, let $M = \lfloor \varepsilon^{-1} \rfloor$, and

$$q^n_j = j\Bigl\lfloor \frac{\varepsilon |v_n|_1 s}{s+t} \Bigr\rfloor \text{ for } 0 \leqslant j \leqslant M-1, \text{ and } q^n_M = (v_n - a)\cdot e_1.$$

For large enough n it is the case that $q^n_M - C\varepsilon < q^n_{M-1} < q^n_M$.

Suppose a maximal path for $\check G^{NE}_{a,\,v_n}(v_n - e_1 \in x_\bullet)$ enters the north boundary from the bulk at the point $v_n - (\ell, 0)$ with $q^n_j < \ell \leqslant q^n_{j+1}$. By nonnegativity of the weights,

$$\begin{aligned}\check G^{NE}_{a,\,v_n}(v_n - e_1 \in x_\bullet) &= \check G_{a,\,v_n-(\ell,1)} + \sum_{i=0}^{\ell-1} I_{v_n - ie_1} \\ &\leqslant \check G_{a,\,v_n-(q^n_j,1)} + \frac{q^n_j}{1-\rho} + \sum_{i=0}^{q^n_{j+1}-1} \bigl(I_{v_n-ie_1} - \tfrac{1}{1-\rho}\bigr) + \frac{q^n_{j+1}-q^n_j}{1-\rho}.\end{aligned}$$

Collect the bounds for all the intervals $(q^n_j, q^n_{j+1}]$:

$$\begin{aligned}\check G^{NE}_{a,\,v_n}(v_n - e_1 \in x_\bullet) \leqslant \max_{0\leqslant j\leqslant M-1} \Bigl\{ &\check G_{a,\,v_n-(q^n_j,1)} + \frac{q^n_j}{1-\rho} \\ &+ \sum_{i=0}^{q^n_{j+1}-1} \bigl(I_{v_n-ie_1} - \tfrac{1}{1-\rho}\bigr) + \frac{q^n_{j+1}-q^n_j}{1-\rho} \Bigr\}\end{aligned} \tag{4.32}$$

Divide through by $|v_n|_1$ and let $n \to \infty$. On the right-hand side above, Theorem 3.6 and the homogeneity of $g_{\rm pp}$ give

$$\frac{\check G_{a,\,v_n-(q^n_j,1)}}{|v_n|_1} = g_{\rm pp}\Bigl(\frac{v_n}{|v_n|_1} - \frac{q^n_j}{|v_n|_1}e_1 + \frac{O(1)}{|v_n|_1}\Bigr) + o(1) \longrightarrow \frac{g_{\rm pp}(s - sj\varepsilon, t)}{s+t}.$$

The mean zero i.i.d. sum satisfies

$$\frac{1}{|v_n|_1} \sum_{i=0}^{q^n_{j+1}-1} \bigl(I_{v_n-ie_1} - \tfrac{1}{1-\rho}\bigr) \longrightarrow 0.$$

We get the upper bound

$$\begin{aligned}\varlimsup_{n\to\infty} |v_n|_1^{-1} \check G^{NE}_{0,\,v_n}(v_n - e_1 \in x_\bullet) &\leqslant (s+t)^{-1} \max_{0\leqslant j\leqslant M-1} \Bigl[g_{\rm pp}(s - sj\varepsilon, t) + \frac{sj\varepsilon}{1-\rho} + C\varepsilon\Bigr] \\ &\leqslant (s+t)^{-1} \sup_{0\leqslant\tau\leqslant s} \Bigl[\frac{\tau}{1-\rho} + g_{\rm pp}(s-\tau, t)\Bigr] + C\varepsilon.\end{aligned}$$

Let $\varepsilon \searrow 0$ to complete the proof of the upper bound.

To get the matching lower bound let the supremum

$$\sup_{\tau\in[0,s]} \bigl\{\tfrac{\tau}{1-\rho} + g_{\rm pp}(s-\tau, t)\bigr\}$$

be attained at $\tau^* \in [0, s]$. With $m_n = |v_n|_1/(s+t)$ we have

$$\check G^{NE}_{a,\,v_n}(v_n - e_1 \in x_\bullet) \geqslant \check G_{a, v_n - (\lfloor m_n\tau^*\rfloor \vee 1, 1)} + \sum_{i=0}^{(\lfloor m_n\tau^*\rfloor - 1)^+} I_{v_n - ie_1}.$$

Let $n \to \infty$ to get

$$\varliminf_{n\to\infty} |v_n|_1^{-1} \check G^{NE}_{0,\,v_n}(v_n - e_1 \in x_\bullet) \geqslant (s+t)^{-1}\Bigl[g_{\rm pp}(s-\tau^*, t) + \frac{\tau^*}{1-\rho}\Bigr].$$

This completes the proof of the limit in (4.30). □

As a consequence we record the expected asymptotics for the unrestricted process with edge weights on the north and east:

(4.33)

$$\begin{aligned}
&\lim_{n\to\infty} |v_n|_1^{-1}\, \check G^{NE}_{a,\,v_n} \\
&= \lim_{n\to\infty} |v_n|_1^{-1}\, \check G^{NE}_{a,\,v_n}(v_n - e_1 \in x_\bullet) \bigvee \check G^{NE}_{a,\,v_n}(v_n - e_2 \in x_\bullet) \\
&= (s+t)^{-1} \sup_{0\leqslant\tau\leqslant s} \Big\{ \frac{\tau}{1-\rho} + g_{\rm pp}(s-\tau, t) \Big\} \bigvee \sup_{0\leqslant\tau\leqslant t} \Big\{ \frac{\tau}{\rho} + g_{\rm pp}(s, t-\tau) \Big\} \\
&= (s+t)^{-1} \sup_{0\leqslant\tau\leqslant s} \Big\{ \frac{\tau}{1-\rho} + \big(\sqrt{s-\tau} + \sqrt{t}\,\big)^2 \Big\} \bigvee \sup_{0\leqslant\tau\leqslant t} \Big\{ \frac{\tau}{\rho} + \big(\sqrt{s} + \sqrt{t-\tau}\,\big)^2 \Big\} \\
&= \frac{s}{1-\rho} + \frac{t}{\rho}.
\end{aligned}$$

In the next lemma we derive bounds on the limiting local gradients of the last-passage values $\check G$ defined in (4.25) in terms of the i.i.d. $\check\sigma$ weights. Recall definition (3.30) of the characteristic direction $\xi(\rho) = \big(\frac{(1-\rho)^2}{(1-\rho)^2+\rho^2}, \frac{\rho^2}{(1-\rho)^2+\rho^2}\big)$.

LEMMA 4.9. *Consider two sequences $\{v_n\}$ and $\{w_n\}$ in $a+\mathbb{Z}^2_{\geqslant 0}$ such that $|v_n|_1 \wedge |w_n|_1 \to \infty$ and*

$$\lim_{n\to\infty} \frac{v_n}{|v_n|_1} = (s, 1-s) \quad \textit{and} \quad \lim_{n\to\infty} \frac{w_n}{|w_n|_1} = (t, 1-t).$$

Assume that

$$s < \xi_1(\rho) = \frac{(1-\rho)^2}{(1-\rho)^2+\rho^2} < t.$$

Then we have the following statements.

(i) *Almost surely*

$$\check G^{NE}_{a,v_n}(v_n - e_2 \in x_\bullet) = \check G^{NE}_{a,v_n} \qquad \textit{and} \qquad \check G^{NE}_{a,w_n}(w_n - e_1 \in x_\bullet) = \check G^{NE}_{a,w_n} \tag{4.34}$$

for all large enough n.

(ii) *The following inequalities hold almost surely:*

$$\varlimsup_{n\to\infty} [\check G_{a,w_n} - \check G_{a+e_1,w_n}] \leqslant I_{a+e_1} \leqslant \varliminf_{n\to\infty} [\check G_{a,v_n} - \check G_{a+e_1,v_n}] \tag{4.35}$$

and

$$\varlimsup_{n\to\infty} [\check G_{a,v_n} - \check G_{a+e_2,v_n}] \leqslant J_{a+e_2} \leqslant \varliminf_{n\to\infty} [\check G_{a,w_n} - \check G_{a+e_2,w_n}]. \tag{4.36}$$

PROOF. Part (i). We prove the second statement of (4.34). The maximizing path to w_n comes through either $w_n - e_1$ or $w_n - e_2$. So to get a contradiction we can assume that $\mathbb{P}(A) > 0$ for the event A on which $\check G^{NE}_{a,w_n}(w_n - e_2 \in x_\bullet) = \check G^{NE}_{a,w_n}$ happens for infinitely many n. On the event A we can take limits (4.31) and (4.33) to get

$$\sup_{0\leqslant\tau\leqslant 1-t} \Big\{ \frac{\tau}{\rho} + \big(\sqrt{t} + \sqrt{1-t-\tau}\,\big)^2 \Big\} = \frac{t}{1-\rho} + \frac{1-t}{\rho}.$$

But by (4.31), $\frac{t}{1-t} > \big(\frac{1-\rho}{\rho}\big)^2$ implies that the supremum on the left equals $\big(\sqrt{t} + \sqrt{1-t}\,\big)^2$ which is strictly less than the right-hand side by (3.31). Thus $\mathbb{P}(A) > 0$ is not possible.

Part (ii). We prove the statements for w_n. By a combination of developments from above we derive the sequence of inequalities and equalities in (4.37) below. The steps are justified one by one after the inequalities. In the first step, $\check{G}^N_{a,w_n+e_2}$ denotes a last-passage process in the rectangle $\{x : a \leqslant x \leqslant w_n + e_2\}$ that uses $\check{\sigma}$ weights on the horizontal lines below the top one, and the I weights on the top horizontal line $x \cdot e_2 = w_n \cdot e_2 + 1$ (north boundary of the rectangle), with an irrelevant zero weight assigned at the top right corner $w_n + e_2$. The third step below is valid almost surely for large n. This proves the first inequality of (4.35).

$$\begin{aligned}
&\check{G}_{a,w_n} - \check{G}_{a+e_1,w_n} \\
&\qquad \leqslant \check{G}^N_{a,\,w_n+e_2} - \check{G}^N_{a+e_1,\,w_n+e_2} \\
&\qquad = \check{G}^{NE}_{a,\,w_n+e_1+e_2}(w_n + e_2 \in x_\bullet) - \check{G}^{NE}_{a+e_1,\,w_n+e_1+e_2}(w_n + e_2 \in x_\bullet) \\
&\qquad = \check{G}^{NE}_{a,\,w_n+e_1+e_2} - \check{G}^{NE}_{a+e_1,\,w_n+e_1+e_2} \\
&\qquad = I_{a+e_1}.
\end{aligned} \tag{4.37}$$

The first inequality in (4.37) above is a special case of the first inequality of (4.21), applied to the situation where the weights in (4.20) are given by

$$\widetilde{Y}_x = \begin{cases} 0 & \text{if } x = w_n + e_2, \\ I_{x+e_1} & \text{if } x = w_n + e_2 - ie_1 \text{ for some } i \geqslant 1, \\ \check{\sigma}_x & \text{if } x \leqslant w_n \end{cases}$$

In the first equality in (4.37) we move the upper right corner from $w_n + e_2$ one step to the right to $w_n + e_1 + e_2$ so that we can include the boundary weights both on the north and east boundaries. This is exactly the definition of $\check{G}^{NE}$ in (4.26) and (4.27). To preserve the equality we force the paths to go through $w_n + e_2$.

The second equality in (4.37) is valid almost surely for large enough n, by the already proved (4.34). The last equality comes from (4.28).

Similarly we reason for the e_2 increment:

$$\begin{aligned}
&\check{G}_{a,w_n} - \check{G}_{a+e_2,w_n} \\
&\qquad \geqslant \check{G}^N_{a,\,w_n+e_2} - \check{G}^N_{a+e_2,\,w_n+e_2} \\
&\qquad = \check{G}^{NE}_{a,\,w_n+e_1+e_2}(w_n + e_2 \in x_\bullet) - \check{G}^{NE}_{a+e_2,\,w_n+e_1+e_2}(w_n + e_2 \in x_\bullet) \\
&\qquad = \check{G}^{NE}_{a,\,w_n+e_1+e_2} - \check{G}^{NE}_{a+e_2,\,w_n+e_1+e_2} \\
&\qquad = J_{a+e_2}.
\end{aligned} \tag{4.38}$$

This proves the last inequality of (4.36). □

Next we use the estimates above to build an exponential-α last-passage system from limits of local gradients of last-passage values. Denote the Exp(λ) cumulative distribution function by

$$F_\lambda(s) = \begin{cases} 0, & s < 0 \\ 1 - e^{-\lambda s}, & s \geqslant 0. \end{cases}$$

LEMMA 4.10. *Let i.i.d.* Exp(1) *weights* $Y = (Y_x)_{x\in\mathbb{Z}^2}$ *be given and define the point-to-point last-passage process* $\{G_{x,y}\}$ *by* (4.7). *Fix* $0 < \alpha < 1$ *and a sequence* $v_n \in \mathbb{Z}^2$ *such that* $|v_n|_1 \to \infty$ *and*

$$\lim_{n\to\infty} \frac{v_n}{|v_n|_1} = \xi(\alpha) = \left(\frac{(1-\alpha)^2}{(1-\alpha)^2+\alpha^2}, \frac{\alpha^2}{(1-\alpha)^2+\alpha^2} \right) \in \text{ri}\,\mathcal{U}. \tag{4.39}$$

(i) *The limits*

$$B^\alpha_{x,y} = \lim_{n\to\infty} [G_{x,v_n} - G_{y,v_n}] \tag{4.40}$$

exist $\mathbb{P}$*-almost surely for all* $x, y \in \mathbb{Z}^2$ *and satisfy additivity* $B^\alpha_{x,y} + B^\alpha_{y,z} = B^\alpha_{x,z}$.

(ii) *Define*

$$X^\alpha_x = B^\alpha_{x-e_1,x} \wedge B^\alpha_{x-e_2,x} \qquad \text{for } x \in \mathbb{Z}^2. \tag{4.41}$$

Then the process

$$\{X^\alpha_x,\, B^\alpha_{x-e_1,x},\, B^\alpha_{x-e_2,x},\, Y_x : x \in \mathbb{Z}^2\}$$

is an exponential-α *last-passage system as described in Definition* 4.1.

PROOF. Part (i). Fix $a \in \mathbb{Z}^2$ and let

$$\overline{B} = \varlimsup_{n\to\infty} [G_{a,v_n} - G_{a+e_1,v_n}] \quad \text{and} \quad \underline{B} = \varliminf_{n\to\infty} [G_{a,v_n} - G_{a+e_1,v_n}]. \tag{4.42}$$

To get control of the distributions of $\overline{B}$ and $\underline{B}$, we realize the process $\{G_{a,y}\}$ on another probability space as an instance of $\{\breve{G}_{a,y}\}$. Then we can apply bounds (4.35)–(4.36).

Let $0 < \lambda < \alpha < \rho < 1$. This implies

$$\frac{(1-\rho)^2}{(1-\rho)^2+\rho^2} < \frac{(1-\alpha)^2}{(1-\alpha)^2+\alpha^2} < \frac{(1-\lambda)^2}{(1-\lambda)^2+\lambda^2}. \tag{4.43}$$

Take any lattice point $u \leqslant a$ as an origin. Suppose on some arbitrary probability space we have mutually independent variables $\sigma = (\sigma_x)_{x\in\mathbb{Z}^2}$, $(I^\lambda_{u+ie_1})_{i\geqslant 1}$, $(J^\lambda_{u+je_2})_{j\geqslant 1}$, $(I^\rho_{u+ie_1})_{i\geqslant 1}$, and $(J^\rho_{u+je_2})_{j\geqslant 1}$ with marginal distributions

$$\begin{aligned} \sigma_x \sim \text{Exp}(1), \quad I^\lambda_{u+ie_1} &\sim \text{Exp}(1-\lambda), \quad J^\lambda_{u+je_2} \sim \text{Exp}(\lambda), \\ I^\rho_{u+ie_1} &\sim \text{Exp}(1-\rho), \quad \text{and} \quad J^\rho_{u+je_2} \sim \text{Exp}(\rho). \end{aligned} \tag{4.44}$$

In other words, for parameters λ and ρ we have initial conditions of the kind described in (3.5) and (3.6). Iterating with equations (3.7)–(3.9), on the quadrant $u + \mathbb{Z}^2_{\geqslant 0}$, construct two processes of the type (3.12): one denoted by

$$\{\sigma_x,\, I^\lambda_{x-e_2},\, J^\lambda_{x-e_1},\, \breve{\sigma}^{[\lambda]}_{x-e_1-e_2} : x \in u + \mathbb{Z}^2_{>0}\} \tag{4.45}$$

with parameter λ, and the other denoted by

$$\{\sigma_x,\, I^\rho_{x-e_2},\, J^\rho_{x-e_1},\, \breve{\sigma}^{[\rho]}_{x-e_1-e_2} : x \in u + \mathbb{Z}^2_{>0}\} \tag{4.46}$$

with parameter ρ. By construction these processes have the properties given in Theorem 3.1, process (4.45) with parameter λ and with configuration $\breve{\sigma}^{[\lambda]}$ replacing $\breve{\omega}$, and process (4.46) with parameter ρ (as stated in Theorem 3.1) but with $\breve{\sigma}^{[\rho]}$ replacing $\breve{\omega}$.

We stipulated that the initial edge weights in (4.44) on the axes $\{u + ie_k : i \geqslant 1, k = 1, 2\}$ for the λ and ρ systems were independent. In fact the coupling between processes (4.45) and (4.46) is immaterial because the two processes will not be used jointly.

The key point is that both $\{\breve\sigma^{[\lambda]}\}_{x\in u+\mathbb{Z}^2_{\geqslant 0}}$ and $\{\breve\sigma^{[\rho]}\}_{x\in u+\mathbb{Z}^2_{\geqslant 0}}$ are i.i.d. Exp(1) variables. The superscripts $[\lambda]$ and $[\rho]$ simply remind us that these variables were constructed from edge variables with parameters λ and ρ, respectively. As in (4.25), define last-passage processes $\breve G^{[\lambda]}$ and $\breve G^{[\rho]}$ that use i.i.d. Exp(1) weights $\breve\sigma^{[\lambda]}$ and $\breve\sigma^{[\rho]}$, respectively:

$$\breve G^{[\lambda]}_{x,y} = \max_{x_\bullet\,\in\,\Pi_{x,y}} \sum_{k=0}^{|y-x|_1} \breve\sigma^{[\lambda]}_{x_k} \quad\text{and}\quad \breve G^{[\rho]}_{x,y} = \max_{x_\bullet\,\in\,\Pi_{x,y}} \sum_{k=0}^{|y-x|_1} \breve\sigma^{[\rho]}_{x_k} \quad\text{for } y \geqslant x \geqslant u. \tag{4.47}$$

The distributional equality of last-passage processes

$$\{\breve G^{[\lambda]}_{x,y} : y \geqslant x \geqslant u\} \overset{d}{=} \{\breve G^{[\rho]}_{x,y} : y \geqslant x \geqslant u\} \overset{d}{=} \{G_{x,y} : y \geqslant x \geqslant u\} \tag{4.48}$$

follows because they are defined the same way from i.i.d. Exp(1) weights.

We derive bounds for the distribution functions of $\overline{B}$ and $\underline{B}$.

$$\begin{aligned}\mathbb{P}\{\overline{B} \leqslant s\} &= \mathbb{P}\{\,\varlimsup_{n\to\infty}(G_{a,v_n} - G_{a+e_1,v_n}) \leqslant s\} = \mathbb{P}\{\,\varlimsup_{n\to\infty}(\breve G^{[\rho]}_{a,v_n} - \breve G^{[\rho]}_{a+e_1,v_n}) \leqslant s\}\\ &\geqslant \mathbb{P}\{I^{\rho}_{a+e_1} \leqslant s\} = F_{1-\rho}(s)\end{aligned}$$

and

$$\begin{aligned}\mathbb{P}\{\underline{B} \leqslant s\} &= \mathbb{P}\{\,\varliminf_{n\to\infty}(G_{a,v_n} - G_{a+e_1,v_n}) \leqslant s\} = \mathbb{P}\{\,\varliminf_{n\to\infty}(\breve G^{[\lambda]}_{a,v_n} - \breve G^{[\lambda]}_{a-e_1,v_n}) \leqslant s\}\\ &\leqslant \mathbb{P}\{I^{\lambda}_{a+e_1} \leqslant s\} = F_{1-\lambda}(s).\end{aligned}$$

Above we first replaced the weights ω with $\breve\sigma^{[\lambda]}$ and $\breve\sigma^{[\rho]}$, respectively. Then we applied (4.35), as justified by (4.43). Last we used the known distributions of the I increment variables. Since $\overline{B} \geqslant \underline{B}$ always, we have deduced that

$$F_{1-\rho}(s) \leqslant \mathbb{P}\{\overline{B} \leqslant s\} \leqslant \mathbb{P}\{\underline{B} \leqslant s\} \leqslant F_{1-\lambda}(s) \qquad \text{for all } \lambda, \rho \text{ such that } \lambda < \alpha < \rho.$$

Letting $\lambda \nearrow \alpha$ and $\rho \searrow \alpha$ allows us to conclude that $\overline{B} = \underline{B} \sim \text{Exp}(1-\alpha)$. This proves the limit in (4.40) for $(x,y) = (a, a+e_1)$. Proof of the limit for $(x,y) = (a, a+e_2)$ proceeds analogously. Since $a \in \mathbb{Z}^2$ was arbitrary, we have the limit in (4.40) for all nearest-neighbor pairs x, y.

An arbitrary increment $y-x$ can be decomposed into a sum of nearest-neighbor increments, and then the limit follows for all pairs x, y by the additivity on the right-hand side of (4.40). Along the way one also derives the additivity $B^\alpha_{x,y} + B^\alpha_{y,z} = B^\alpha_{x,z}$.

Part (ii). We need to verify the properties in Definition 4.1. We begin with the joint distribution of B^α-increments along a down-right path.

Consider the joint distribution of $k+\ell$ nearest-neighbor increments $B^\alpha_{x_i,x_i+e_1}$ and $B^\alpha_{y_j,y_j+e_2}$ for $1 \leqslant i \leqslant k$ and $1 \leqslant j \leqslant \ell$. Fix an origin u such that all x_i and y_j lie in the quadrant $u + \mathbb{Z}^2_{\geqslant 0}$. By the distributional equality (4.48), part (i) gives the existence of the almost sure limits

$$\breve B^{[\lambda]}_{x,y} = \lim_{n\to\infty}(G^{[\lambda]}_{x,v_n} - G^{[\lambda]}_{y,v_n}) \quad\text{and}\quad \breve B^{[\rho]}_{x,y} = \lim_{n\to\infty}(G^{[\rho]}_{x,v_n} - G^{[\rho]}_{y,v_n}) \quad \forall x, y \in u + \mathbb{Z}^2_{\geqslant 0}.$$

Then, first by distributional equality of processes,

$$\begin{aligned}&\mathbb{P}\{B^{\alpha}_{x_i,\,x_i+e_1}\leqslant s_i,\ B^{\alpha}_{y_j,\,y_j+e_2}>t_j\ \ \forall i\in[k],j\in[\ell]\,\}\\&=\mathbb{P}\{\breve{B}^{[\lambda]}_{x_i,\,x_i+e_1}\leqslant s_i,\ \breve{B}^{[\lambda]}_{y_j,\,y_j+e_2}>t_j\ \ \forall i\in[k],j\in[\ell]\,\}\\&\leqslant\mathbb{P}\{I^{\lambda}_{x_i+e_1}\leqslant s_i,\ J^{\lambda}_{y_j+e_2}>t_j\ \ \forall i\in[k],j\in[\ell]\,\}\end{aligned}$$

The last step came from the inequalities (4.35) and (4.36) (the case of v_n of those inequalities is the right one to look at). Similarly, using the remaining two inequalities of (4.35) and (4.36), we deduce

$$\begin{aligned}&\mathbb{P}\{B^{\alpha}_{x_i,\,x_i+e_1}\leqslant s_i,\ B^{\alpha}_{y_j,\,y_j+e_2}>t_j\ \ \forall i\in[k],j\in[\ell]\,\}\\&=\mathbb{P}\{\breve{B}^{[\rho]}_{x_i,\,x_i+e_1}\leqslant s_i,\ \breve{B}^{[\rho]}_{y_j,\,y_j+e_2}>t_j\ \ \forall i\in[k],j\in[\ell]\,\}\\&\geqslant\mathbb{P}\{I^{\rho}_{x_i+e_1}\leqslant s_i,\ J^{\rho}_{y_j+e_2}>t_j\ \ \forall i\in[k],j\in[\ell]\,\}.\end{aligned}$$

Assume in particular now that the edges $\{x_i, x_i+e_1\}$ and $\{y_j, y_j+e_2\}$ lie on a given down-right path. Then, part (i) of Theorem 3.1 applied to the processes (4.45) and (4.46) turns the bounds above into

$$\begin{aligned}&\prod_{i\in[k]}F_{1-\rho}(s_i)\cdot\prod_{j\in[\ell]}(1-F_\rho(t_j))\\&\qquad\leqslant\ \mathbb{P}\{B^{\alpha}_{x_i,\,x_i+e_1}\leqslant s_i,\ B^{\alpha}_{y_j,\,y_j+e_2}>t_j\ \ \forall i\in[k],j\in[\ell]\,\}\\&\qquad\leqslant\prod_{i\in[k]}F_{1-\lambda}(s_i)\cdot\prod_{j\in[\ell]}(1-F_\lambda(t_j)).\end{aligned}$$

Letting again $\lambda\nearrow\alpha$ and $\rho\searrow\alpha$ shows that along a down-right path, the variables $B^{\alpha}_{x,\,x+e_1}$ and $B^{\alpha}_{y,\,y+e_2}$ are independent with distributions $\text{Exp}(1-\alpha)$ and $\text{Exp}(\alpha)$, respectively, as required by part (a) of Definition 4.1.

Fix a down-right path $\mathcal{Y}$ in $\mathbb{Z}^2$. We verify the distributional properties on $\mathcal{G}_-$, $\mathcal{Y}$ and $\mathcal{G}_+$ inside an arbitrarily large rectangle.

Consider a large rectangle $\mathcal{D}=\{x\in\mathbb{Z}^2:(M_0,N_0)\leqslant x\leqslant(M_1,N_1)\}$ whose lower left and upper right corners are (M_0,N_0) and (M_1,N_1). The e_1-edge variables $B^{\alpha}_{(i,N_1),(i+1,N_1)}$ for $M_0\leqslant i\leqslant M_1-1$ on the north boundary, the e_2-edge variables $B^{\alpha}_{(M_1,j),(M_1,j+1)}$ for $N_0\leqslant j\leqslant N_1-1$ on the east boundary, and the bulk variables Y_x for $(M_0,N_0)\leqslant x\leqslant(M_1-1,N_1-1)$ are mutually independent. (The Y-variables are independent of the B^{α}-variables to their north and east because limit (4.40) constructs a B^{α}-variable in terms of Y-weights to its north and east.)

In other words, the B^{α}-increments on the north and east boundaries and the Y-weights in the bulk of the rectangle satisfy the properties of the I,J boundary weights and $\breve{\omega}$-bulk weights in Theorem 3.1. Next we show by a south and westward induction that the joint distribution of (X^{α},B,Y) is the correct one.

We claim that the variables satisfy the equations

$$\begin{aligned}X^{\alpha}_{x+e_1+e_2}&=B^{\alpha}_{x+e_2,\,x+e_1+e_2}\wedge B^{\alpha}_{x+e_1,\,x+e_1+e_2}\\B^{\alpha}_{x,\,x+e_1}&=Y_x+(B^{\alpha}_{x+e_2,\,x+e_1+e_2}-B^{\alpha}_{x+e_1,\,x+e_1+e_2})^{+}\\B^{\alpha}_{x,\,x+e_2}&=Y_x+(B^{\alpha}_{x+e_2,\,x+e_1+e_2}-B^{\alpha}_{x+e_1,\,x+e_1+e_2})^{-}.\end{aligned}\tag{4.49}$$

In contrast with the iteration (3.7)–(3.9), the equations above proceed to the south and west.

The first equation in (4.49) is definition (4.41). The second and third come from the limits (4.40). For example,

$$\begin{aligned} B^{\alpha}_{x,\,x+e_1} &= \lim_{n\to\infty}[G_{x,v_n} - G_{x+e_1,\,v_n}] = \lim_{n\to\infty}[Y_x + G_{x+e_1,\,v_n} \vee G_{x+e_2,\,v_n} - G_{x+e_1,\,v_n}] \\ &= Y_x + \lim_{n\to\infty}[G_{x+e_2,\,v_n} - G_{x+e_1,\,v_n}]^+ \\ &= Y_x + \lim_{n\to\infty}\big[(G_{x+e_2,\,v_n} - G_{x+e_1+e_2,\,v_n}) - (G_{x+e_1,\,v_n} - G_{x+e_1+e_2,\,v_n})\big]^+ \\ &= Y_x + [B^{\alpha}_{x+e_2,\,x+e_1+e_2} - B^{\alpha}_{x+e_1,\,x+e_1+e_2}]^+. \end{aligned}$$

By Lemma 3.2 and induction, for any down-right path from the upper left corner (M_0, N_1) to the lower right corner (M_1, N_0), inside the rectangle $\mathcal{D}$, the B^α-increments on the path, the Y-weights below and to the left of the path, and the X^α weights above and to the right of the path, are all independent with the correct marginal distributions stipulated in (4.2). Part (a) of Definition 4.1 has been verified.

It remains to check that equations (3.7)–(3.9) are satisfied.

$$\begin{aligned} B^{\alpha}_{x,x+e_1} \wedge B^{\alpha}_{x,x+e_2} &= \lim_{n\to\infty}[G_{x,v_n} - G_{x+e_1,v_n}] \wedge [G_{x,v_n} - G_{x+e_2,v_n}] \\ &= \lim_{n\to\infty}[Y_x + G_{x+e_1,v_n} \vee G_{x+e_2,v_n} - G_{x+e_1,v_n}] \\ &\qquad\qquad \wedge [Y_x + G_{x+e_1,v_n} \vee G_{x+e_2,v_n} - G_{x+e_2,v_n}] \\ &= Y_x + [G_{x+e_2,v_n} - G_{x+e_1,v_n}]^+ \wedge [G_{x+e_1,v_n} - G_{x+e_2,v_n}]^+ \\ &= Y_x. \end{aligned}$$

This verifies (3.7) (at x instead of at $x - e_1 - e_2$).

By definition (4.41) and the additivity of B^α,

$$\begin{aligned} B^{\alpha}_{x-e_1,\,x} &= B^{\alpha}_{x-e_1,\,x} \wedge B^{\alpha}_{x-e_2,\,x} + (B^{\alpha}_{x-e_1,\,x} - B^{\alpha}_{x-e_2,\,x})^+ \\ &= X^{\alpha}_x + (B^{\alpha}_{x-e_1-e_2,\,x-e_2} - B^{\alpha}_{x-e_1-e_2,\,x-e_1})^+. \end{aligned}$$

This is (3.8). Equation (3.9) is verified in a similar manner. □

LEMMA 4.11. *Fix a countable dense subset $D \subset (0,1)$. Then there exists an event Ω_0 of full probability such that the following holds for each $\omega \in \Omega_0$.*

(i) *For each $\rho \in D$ the process $\{B^{\rho}_{x,y}(\omega)\}_{x,y\in\mathbb{Z}^2}$ is well-defined by the limits in (4.40) for the specific sequence $v_n = \lfloor n\xi(\rho)\rfloor$. These processes satisfy the following inequalities:*

$$\begin{aligned} &B^{\lambda}_{x,x+e_1}(\omega) \leqslant B^{\rho}_{x,x+e_1}(\omega) \\ \text{and}\quad &B^{\lambda}_{x,x+e_2}(\omega) \geqslant B^{\rho}_{x,x+e_2}(\omega) \qquad \text{for all } x \in \mathbb{Z}^2 \text{ and } \lambda < \rho \text{ in } D. \end{aligned} \tag{4.50}$$

(ii) *For each $\alpha \in (0,1)$, and any sequence u_n in $\mathbb{Z}^2$ such that $|u_n|_1 \to \infty$ and*

$$\lim_{n\to\infty}\frac{u_n}{|u_n|_1} = \xi(\alpha) = \left(\frac{(1-\alpha)^2}{(1-\alpha)^2+\alpha^2}, \frac{\alpha^2}{(1-\alpha)^2+\alpha^2}\right), \tag{4.51}$$

we have these bounds:

$$\begin{aligned} \sup_{\lambda\in D:\lambda<\alpha} B^{\lambda}_{x,x+e_1} &\leqslant \varliminf_{n\to\infty}[G_{x,u_n} - G_{x+e_1,u_n}] \\ &\leqslant \varlimsup_{n\to\infty}[G_{x,u_n} - G_{x+e_1,u_n}] \leqslant \inf_{\rho\in D:\rho>\alpha} B^{\rho}_{x,x+e_1} \end{aligned} \tag{4.52}$$

and

$$\begin{aligned}\sup_{\rho\in D:\rho>\alpha} B^{\alpha}_{x,x+e_2} &\leqslant \varliminf_{n\to\infty}\,[G_{x,u_n} - G_{x+e_2,u_n}] \\ &\leqslant \varlimsup_{n\to\infty}\,[G_{x,u_n} - G_{x+e_2,u_n}] \leqslant \inf_{\lambda\in D:\lambda<\alpha} B^{\alpha}_{x,x+e_2}.\end{aligned} \tag{4.53}$$

PROOF. Define Ω_0 to be the event on which the limits in (4.40) hold for each $\rho \in D$, for the specific sequence $v_n = \lfloor n\xi(\rho)\rfloor$. Then fix $\omega \in \Omega_0$. Let $0 < \lambda < \alpha < \rho < 1$ be such that $\lambda, \rho \in D$. Let u_n be any sequence in $\mathbb{Z}^2$ such that $|u_n|_1 \to \infty$ and (4.51) holds. Let $b_n = |u_n|_1$. Then (4.43) implies that for large n,

$$\begin{aligned} &\lfloor b_n\xi(\rho)\rfloor \cdot e_1 < u_n \cdot e_1 < \lfloor b_n\xi(\lambda)\rfloor \cdot e_1 \\ \text{and}\quad &\lfloor b_n\xi(\lambda)\rfloor \cdot e_2 < u_n \cdot e_2 < \lfloor b_n\xi(\rho)\rfloor \cdot e_2.\end{aligned} \tag{4.54}$$

Then Lemma 4.6 gives the bounds

$$G_{x,\lfloor b_n\xi(\lambda)\rfloor} - G_{x+e_1,\lfloor b_n\xi(\lambda)\rfloor} \leqslant G_{x,u_n} - G_{x+e_1,u_n} \leqslant G_{x,\lfloor b_n\xi(\rho)\rfloor} - G_{x+e_1,\lfloor b_n\xi(\rho)\rfloor}$$

and

$$G_{x,\lfloor b_n\xi(\rho)\rfloor} - G_{x+e_2,\lfloor b_n\xi(\rho)\rfloor} \leqslant G_{x,u_n} - G_{x+e_2,u_n} \leqslant G_{x,\lfloor b_n\xi(\lambda)\rfloor} - G_{x+e_2,\lfloor b_n\xi(\lambda)\rfloor}.$$

All the inequalities claimed follow by taking $n \to \infty$. □

PROOF OF THEOREM 4.2. We start with a countable dense subset $D \subset (0,1)$, the processes B^λ for $\lambda \in D$ defined by the limits

$$B^{\lambda}_{x,y} = \lim_{n\to\infty}\,[G_{x,\lfloor n\xi(\lambda)\rfloor} - G_{y,\lfloor n\xi(\lambda)\rfloor}\,] \tag{4.55}$$

on the event Ω_0 of full probability given in Lemma 4.11. For each $\lambda \in D$ we have the additivity

$$B^{\lambda}_{x,y} + B^{\lambda}_{y,z} = B^{\lambda}_{x,z} \tag{4.56}$$

and from Lemma 4.10 we know that, with $X^\lambda_x = B^\lambda_{x-e_1,x} \wedge B^\lambda_{x-e_2,x}$, the process

$$\{X^\lambda_x, B^\lambda_{x-e_1,x}, B^\lambda_{x-e_2,x}, Y_x : x \in \mathbb{Z}^2\}$$

is an exponential-λ last-passage system as described in Definition 4.1.

Let Ω_1 be the subset of Ω_0 on which

$$\begin{aligned} &\sup_{\lambda\in D:\lambda<\gamma} B^{\lambda}_{x,x+e_1} = B^{\gamma}_{x,x+e_1} = \inf_{\rho\in D:\rho>\gamma} B^{\rho}_{x,x+e_1} \\ \text{and}\quad &\sup_{\rho\in D:\rho>\gamma} B^{\rho}_{x,x+e_2} = B^{\gamma}_{x,x+e_2} = \inf_{\lambda\in D:\lambda<\gamma} B^{\lambda}_{x,x+e_2} \qquad \text{for all } \gamma\in D.\end{aligned} \tag{4.57}$$

Event Ω_1 has full probability because of the monotonicity and control of distributions: for example, for the first equality in (4.57) reason as follows: by (4.50)

$$\sup_{\lambda\in D:\lambda<\gamma} B^{\lambda}_{x,x+e_1} = \lim_{D\ni\lambda\nearrow\gamma} B^{\lambda}_{x,x+e_1} \leqslant B^{\gamma}_{x,x+e_1},$$

but by Lemma 4.10(b) and the convergence of distributions, both $\lim_{D\ni\lambda\nearrow\gamma} B^{\lambda}_{x,x+e_1}$ and $B^{\gamma}_{x,x+e_1}$ have Exp$(1-\gamma)$ distribution. Hence they agree almost surely. From (4.57), the monotonicity in (4.50), and additivity to extend from nearest-neighbor increments, we get continuity of the process on D:

$$\lim_{D\ni\lambda\to\gamma} B^{\lambda}_{x,y} = B^{\gamma}_{x,y} \qquad \text{for all } \gamma\in D, \text{ on the event } \Omega_1. \tag{4.58}$$

By discarding another zero probability event we can assume that Ω_1 is invariant under translations.

In order to prove that B^γ is a stationary cocycle (Definition 2.6) for $\gamma \in D$, it remains to check the stationarity $B^\gamma_{x,y}(\theta_z\omega) = B^\gamma_{z+x,\,z+y}(\omega)$. We apply bounds (4.52)–(4.53) to the sequence $u_n = \lfloor n\xi(\gamma)\rfloor + z$ that satisfies (4.51) with limit $\xi(\gamma)$. Together with (4.57) these bounds give (with an extension by additivity) almost sure equalities

$$B^\gamma_{x,y} = \lim_{n\to\infty}[G_{x,\,\lfloor n\xi(\gamma)\rfloor+z} - G_{y,\,\lfloor n\xi(\gamma)\rfloor+z}]$$

for any fixed z and all x, y. Consequently

$$\begin{aligned} B^\gamma_{x+z,y+z} &= \lim_{n\to\infty}[G_{x+z,\,\lfloor n\xi(\gamma)\rfloor+z} - G_{y+z,\,\lfloor n\xi(\gamma)\rfloor+z}] \\ &= \lim_{n\to\infty}[G_{x,\,\lfloor n\xi(\gamma)\rfloor} - G_{y,\,\lfloor n\xi(\gamma)\rfloor}] \circ \theta_z \\ &= B^\gamma_{x,y} \circ \theta_z. \end{aligned}$$

We have now checked that B^γ is a cocycle for each $\gamma \in D$.

Next, we take the step to general $\alpha \in (0,1)$. For each $\alpha \in (0,1)$ define processes X^α and B^α by taking right limits from values in D: set for each $\omega \in \Omega_1$ and $x, y \in \mathbb{Z}^2$

$$(X^\alpha_x(\omega), B^\alpha_{x,y}(\omega)) = \lim_{D\ni\lambda\searrow\alpha}(X^\lambda_x(\omega), B^\lambda_{x,y}(\omega)). \tag{4.59}$$

These limits exist for nearest-neighbor pairs x, y by the monotonicity in (4.50), and extend to all pairs x, y by additivity on the right. The limit for $X^\alpha_x(\omega)$ comes along as a function by virtue of the definition $X^\lambda_x(\omega) = B^\lambda_{x-e_1,x}(\omega) \wedge B^\lambda_{x-e_2,x}(\omega)$ which is then also preserved to the limit.

Extend these functions in some arbitrary way outside Ω_1. By the arguments given above, we have not altered these functions on Ω_1 if α happens to lie in D. The properties of both Definitions 2.6 and 4.1 are preserved by the limits: B^α is a stationary cocycle and $\{X^\alpha_x, B^\alpha_{x-e_1,x}, B^\alpha_{x-e_2,x}, Y_x : x \in \mathbb{Z}^2\}$ is an exponential-α last-passage system. We have verified part (i) of the theorem.

Inequalities (4.8) are valid on the event Ω_1 simultaneously for all $\lambda, \rho \in D$ and preserved by the limit in (4.59). (4.58), (4.59) and the monotonicity and additivity together give the cadlag paths $\lambda \mapsto B^\lambda_{x,y}(\omega)$ for $\omega \in \Omega_1$. Part (ii) is proved.

For part (iii), fix $0 < \alpha < 1$ and let $\Omega_2^{(\alpha)}$ be the intersection of the event Ω_1 above (which is contained in the event Ω_0 of Lemma 4.11) with the event on which

$$\begin{aligned} &\sup_{\lambda\in D:\lambda<\alpha} B^\lambda_{x,x+e_1} = \inf_{\rho\in D:\rho>\alpha} B^\rho_{x,x+e_1} \\ \text{and} \quad &\sup_{\rho\in D:\rho>\alpha} B^\alpha_{x,x+e_2} = \inf_{\lambda\in D:\lambda<\alpha} B^\alpha_{x,x+e_2}. \end{aligned} \tag{4.60}$$

The equalities above hold with probability 1 by the argument used already above. First, by monotonicity inequality $\leqslant$ holds in both equalities above. Then the suprema and infima are limits, and the left- and right-hand sides of the equalities above are equal in distribution. Hence the left- and right-hand sides agree almost surely. As a consequence we get limit (4.11).

The coincidence of the lower and upper bounds in (4.52)–(4.53) imply that the claimed limit in (4.10) holds for nearest-neighbor pairs x, y. Extend to all x, y by additivity.

This completes the proof of Theorem 4.2. □

PROOF OF THEOREM 4.3. As already observed, part (ii) of the theorem follows from part (i). Assumptions (a) and (b) of part (i) of Theorem 4.3 enable us to repeat the development of Lemmas 4.7–4.9 for the LPP processes constructed as in (4.26)–(4.27) but this time with the inputs $I_x = A_{x-e_1,x}$, $J_x = A_{x-e_2,x}$ and $\breve{\sigma}_x = Y_x$.

The conclusion of part (i) comes from an application of Lemma 4.9. Given ρ, let $0 < \lambda_1 < \rho < \lambda_2 < 1$. Then the characteristic directions satisfy $\xi_1(\lambda_1) > \xi_1(\rho) > \xi_1(\lambda_2)$. Let $\{v_n\}$ and $\{w_n\}$ be sequences in $a + \mathbb{Z}^2_{\geqslant 0}$ such that $|v_n|_1 \wedge |w_n|_1 \to \infty$ and

$$\lim_{n\to\infty} \frac{v_n}{|v_n|_1} = \xi(\lambda_2) \quad \text{and} \quad \lim_{n\to\infty} \frac{w_n}{|w_n|_1} = \xi(\lambda_1).$$

The hypotheses of Lemma 4.9 are satisfied. Since we have the Busemann limits in (4.10) for the weights $\breve{\sigma} = Y$, conclusions (4.35)–(4.36) give

$$B^{\lambda_1}_{a,\,a+e_1} \leqslant A_{a,\,a+e_1} \leqslant B^{\lambda_2}_{a,\,a+e_1} \quad \mathbb{P}\text{-a.s.}$$

with the opposite inequalities for the edge $(a, a+e_2)$. Let $\lambda_1, \lambda_2 \to \rho$ and apply limit (4.11) to conclude that $A_{a,a+e_i} = B^{\rho}_{a,a+e_i}$ a.s. Then from the properties of an exponential system, $U_x = A_{x-e_1,x} \wedge A_{x-e_2,x} = B^{\rho}_{x-e_1,x} \wedge B^{\rho}_{x-e_2,x} = X^{\rho}_x$. □

4.3. Midpoint problem. Let again $\pi^{0,x}_{\bullet}$ denote the (almost surely unique) maximizing path or geodesic for $G_{0,x}$ defined by (4.7). The *midpoint problem* asks whether the probability that the geodesic goes through some particular point roughly midway between 0 and x converges to zero, as the distance to x grows. We prove this in the next theorem for the case where x has a limiting direction. Since Theorem 3.8 already showed that the geodesic stays within distance $o(|x|_1)$ of the ray to x, we consider intermediate points in this narrow range.

In undirected first-passage percolation the midpoint problem was solved under a differentiability assumption on the limit shape by [**20**] and without any such unverifiable assumptions by [**1**].

For $m \in \mathbb{Z}_{\geqslant 0}$ let

$$\mathbb{L}_m = \{x \in \mathbb{Z}^2_{\geqslant 0} : |x|_1 = m\} = \{(j, m-j) : 0 \leqslant j \leqslant m\}$$

denote the set of lattice points on level m in the first quadrant. Fix a parameter $\rho \in (0,1)$ and for $n \in \mathbb{Z}_{>0}$ let $v_n \in \mathbb{L}_n$ be a sequence of lattice points that escapes in the characteristic direction of ρ: $v_n/n \to \xi(\rho)$. Then fix $t \in (0,1)$ to select an intermediate level. Let $z_n \in \mathbb{L}_{\lfloor nt \rfloor}$ be a sequence of lattice points on successive levels $\lfloor nt \rfloor$ such that $z_n/n \to t\xi(\rho)$.

THEOREM 4.12. *With assumptions as in the paragraph above,* $\lim_{n\to\infty} \mathbb{P}\{\pi^{0,v_n}_{\lfloor nt \rfloor} = z_n\} = 0.$

PROOF. Let ${}^{o}G_{x,y}$ denote the last-passage value without the first weight, that is,

$${}^{o}G_{x,y} = \max_{x_{\bullet} \in \Pi_{x,y}} \sum_{k=1}^{|y-x|_1} Y_{x_k}$$

where a path from x to y is given by $x_{\bullet} = \{x = x_0, x_1, \ldots, x_{|y-x|_1} = y\}$.

Let R denote reflection of the lattice across the origin that acts on weights by $(RY)_x = Y_{-x}$. Then define $\widetilde{B}^{\alpha}_{x,y} = B^{\alpha}_{-x,-y} \circ R$. By Theorem 4.2, there exists an event $\Omega_3^{(\alpha)}$ of full probability on which the following holds: for any sequence $u_n \in \mathbb{Z}^2$ such that $|u_n|_1 \to \infty$ and

$$\lim_{n\to\infty} \frac{u_n}{|u_n|_1} = -\xi(\alpha) = -\left(\frac{(1-\alpha)^2}{(1-\alpha)^2+\alpha^2}, \frac{\alpha^2}{(1-\alpha)^2+\alpha^2}\right),$$

we have the limits

$$\widetilde{B}^{\alpha}_{x,y} = \lim_{n\to\infty} (G_{u_n,x} - G_{u_n,y}) \qquad \forall x, y \in \mathbb{Z}^2.$$

In other words, $\widetilde{B}^{\alpha}$ is a Busemann function in the southwest direction. For fixed x, y the distributional relationship is $\widetilde{B}^{\alpha}_{x,y} \overset{d}{=} B^{\alpha}_{y,x}$.

Observe that $\pi^{0,v_n}_{\lfloor nt\rfloor} = z_n$ if and only if for all $y \in \mathbb{L}_{\lfloor nt\rfloor}\backslash\{z_n\}$,

$$\begin{aligned} G_{0,z_n} + {}^oG_{z_n,v_n} &> G_{0,y} + {}^oG_{y,v_n} \\ \text{if and only if} \quad G_{0,z_n} - G_{0,y} &> {}^oG_{y,v_n} - {}^oG_{z_n,v_n}. \end{aligned}$$

The left and right-hand sides of the last inequality above are independent.

In order to replace ${}^oG$ above with G so that we can directly apply Theorem 4.2, we augment the probability space with an independent collection of i.i.d. Exp(1) weights $\{Y'_x\}_{x\in\mathbb{Z}^2}$. Let G' denote last-passage values that use Y' weights.

Consider points y of the form $y = z_n + (k, -k)$ for $1 \leqslant k \leqslant m$. Below replace the weight Y_x with Y'_x for lattice points x strictly to the right of $\mathbb{L}_{\lfloor nt\rfloor}$. Then shift the origin to z_n.

$$\begin{aligned}
&\mathbb{P}\{\pi^{0,v_n}_{\lfloor nt\rfloor} = z_n\} \\
&\leqslant \mathbb{P}\big\{G_{0,z_n} - G_{0,z_n+(k,-k)} > {}^oG_{z_n+(k,-k),v_n} - {}^oG_{z_n,v_n} \ \ \forall k \in [m]\big\} \\
&\leqslant \mathbb{P}\big\{G_{0,z_n} - G_{0,z_n+(k,-k)} > G'_{z_n+(k,-k),v_n} - G'_{z_n,v_n} - Y'_{z_n+(k,-k)} + Y'_{z_n} \ \ \forall k \in [m]\big\} \\
&\leqslant \mathbb{P}\big\{G_{-z_n,0} - G_{-z_n,(k,-k)} > G'_{(k,-k),v_n-z_n} - G'_{0,v_n-z_n} - Y'_{(k,-k)} + Y'_0 \ \ \forall k \in [m]\big\} \\
&= \mathbb{P}\bigg\{ \sum_{i=1}^{k} [G_{-z_n,(i-1,-i+1)} - G_{-z_n,(i,-i)}] \\
&\qquad\qquad > \sum_{i=1}^{k} [G'_{(i,-i),v_n-z_n} - G'_{(i-1,-i+1),v_n-z_n}] - Y'_{(k,-k)} + Y'_0 \ \ \forall k \in [m]\bigg\} \\
&\underset{n\to\infty}{\longrightarrow} \mathbb{P}\bigg\{ \sum_{i=1}^{k} \widetilde{B}^{\rho}_{(i-1,-i+1),(i,-i)} > \sum_{i=1}^{k} B^{\rho}_{(i,-i),(i-1,-i+1)} - Y'_{(k,-k)} + Y'_0 \ \ \forall k \in [m]\bigg\} \\
&= \mathbb{P}\big\{S_k > -Y'_{(k,-k)} + Y'_0 \ \ \forall k \in [m]\big\} \\
&\leqslant \mathbb{P}\big\{S_k > -Y'_{(k,-k)} \ \ \forall k \in [m]\big\}.
\end{aligned}$$

After the limit above, B^{ρ} is the Busemann function for the weights Y', $\widetilde{B}^{\rho}$ is the southwest Busemann function for the weights Y, and these two Busemann functions are independent. We defined the symmetric random walk

$$S_k = \sum_{i=1}^{k} [\,\widetilde{B}^{\rho}_{(i-1,-i+1),(i,-i)} - B^{\rho}_{(i,-i),(i-1,-i+1)}\,]. \tag{4.61}$$

Note that $S_{\bullet}$ is not independent of $Y'_{(k,-k)}$. We continue to bound the last probability as follows.

$$\begin{aligned}
&\mathbb{P}\{S_k > -Y'_{(k,-k)} \quad \forall k \in [m]\} \\
&\leqslant \mathbb{P}\{\max_{1\leqslant k\leqslant m} Y'_{(k,-k)} > m^{1/4}\} + \mathbb{P}\{S_k > -m^{1/4} \quad \forall k \in [m]\} \\
&\leqslant me^{-m^{1/4}} + \mathbb{P}\left\{\inf_{0\leqslant s\leqslant 1} \frac{S_{\lfloor sm\rfloor}}{\sqrt{m}} > -m^{-1/4}\right\}.
\end{aligned}$$

The last line converges to zero as $m \to \infty$ because $\inf_{0\leqslant s\leqslant 1} m^{-1/2}S_{\lfloor sm\rfloor}$ converges in distribution to $\inf_{0\leqslant s\leqslant 1} \sigma B_s$ where σ^2 is the variance of a term in (4.61) and $B_{\bullet}$ is standard Brownian motion. □

5. Fluctuation exponents

Before specializing to the two-dimensional corner growth model with exponential weights, we discuss the general setting in the context of the Kardar-Parisi-Zhang (KPZ) class of models, in arbitrary dimension.

5.1. Heuristics in the KPZ class. Stochastic models in the KPZ class are expected to have common values of two *fluctuation exponents* χ and ζ that depend only on the dimension $d \geqslant 2$. In the context of the last-passage percolation process $G_{0,\lfloor N\mathbf{v}\rfloor}$ on $\mathbb{Z}^d$ discussed in Section 2.1, in a fixed direction $\mathbf{v} \in \mathbb{R}^d_{>0}$, these exponents are defined informally as follows:

(a) the fluctuations of the last-passage value $G_{0,\lfloor N\mathbf{v}\rfloor}$ are expected to have order of magnitude N^{χ};

(b) the optimal path from 0 to $\lfloor N\mathbf{v}\rfloor$ is expected to fluctuate on the scale N^{ζ} around the straight line segment from 0 to $N\mathbf{v}$.

Certain properties are expected to hold for these exponents. We state two below and give heuristic justification.

The *KPZ scaling relation* states that

$$\chi = 2\zeta - 1. \tag{5.1}$$

It is expected to hold in broad generality in all dimensions. A version of it has been proved by Chatterjee [**15**] for first-passage percolation in two dimensions. The proof was simplified and generalized by Auffinger and Damron [**6**] who also extended the result to positive temperature polymer models [**5**].

The direction $\chi \geqslant 2\zeta - 1$ is the easier of the two inequalities. We give a nonrigorous argument for this inequality as a consequence of the limiting shape function g_{pp} having quadratic curvature. The idea is simple: if the path wanders too much, it picks up inadmissibly deviant values of g_{pp}, by the assumed curvature of g_{pp}.

Consider the setting of the d-dimensional CGM with admissible steps $\mathcal{R} = \{e_1, \dots, e_d\}$ and an i.i.d. environment $\omega = (\omega_x)_{x\in\mathbb{Z}^d}$ with $\mathbb{E}(|\omega_x|^{d+\varepsilon}) < \infty$ for some $\varepsilon > 0$. Then the continuous, concave, homogeneous shape function

$$g_{\text{pp}}(\mathbf{u}) = \lim_{N\to\infty} N^{-1}G_{0,\lfloor N\mathbf{u}\rfloor} \tag{5.2}$$

exists as an almost sure limit for all $\mathbf{u} \in \mathbb{R}^d_{\geqslant 0}$.

Assume that for some fixed $\mathbf{v} \in \mathbb{R}^d_{>0}$,

$$G_{0,\lfloor N\mathbf{v}\rfloor} = Ng_{\text{pp}}(\mathbf{v}) + \Theta_P(N^{\chi}) \tag{5.3}$$

where $Y_N = \Theta_P(N^\chi)$ means that $N^{-\chi} Y_N$ is tight but does not converge to zero. Suppose that the second derivative of $g_{\rm pp}$ at $\mathbf{v}$ is strictly negative definite: for any vector $\mathbf{w}$,

$$D^2_{\mathbf{w}}\, g_{\rm pp}(\mathbf{v}) = \frac{\partial^2}{\partial\theta^2} g_{\rm pp}(\mathbf{v}+\theta\mathbf{w})\Big|_{\theta=0} = \mathbf{w}^T D^2 g_{\rm pp}(\mathbf{v})\, \mathbf{w} < 0.$$

(The shape function is concave so a nonvanishing second derivative is negative.)

If maximizing paths are to fluctuate on the scale N^ζ, then let us assume that the maximizing path for $G_{0,\,2N\mathbf{v}}$ goes through the point $N\mathbf{v} + N^\zeta \mathbf{w}_N$ for some nondegenerate, tight random vector $\mathbf{w}_N$. (We drop integer parts to lighten up the notation. In any case the argument presented below is not rigorous.) Then

$$\begin{aligned} 2Ng_{\rm pp}(\mathbf{v}) + \Theta_P(N^\chi) &= G_{0,\,2N\mathbf{v}} = G_{0,\,N\mathbf{v}+N^\zeta\mathbf{w}_N} + G_{N\mathbf{v}+N^\zeta\mathbf{w}_N,\,2N\mathbf{v}} \\ &= Ng_{\rm pp}(\mathbf{v}+N^{\zeta-1}\mathbf{w}_N) + Ng_{\rm pp}(\mathbf{v}-N^{\zeta-1}\mathbf{w}_N) + \Theta_P(N^\chi) \\ &\approx 2Ng_{\rm pp}(\mathbf{v}) + N^{2\zeta-1} D^2_{\mathbf{w}_N}\, g_{\rm pp}(\mathbf{v}) + \Theta_P(N^\chi). \end{aligned}$$

Cancelling the constant $2Ng_{\rm pp}(\mathbf{v})$ leaves

$$\Theta_P(N^\chi) \approx N^{2\zeta-1} D^2_{\mathbf{w}_N}\, g_{\rm pp}(\mathbf{v}) + \Theta_P(N^\chi).$$

This forces $2\zeta - 1 \leqslant \chi$.

In two dimensions it is also expected that

$$\chi = \zeta/2. \tag{5.4}$$

We give a heuristic justification for this relation under some natural assumptions. Assume we have a Busemann function in direction $\mathbf{u}$:

$$B^{\mathbf{u}}_{x,y} = \lim_{M\to\infty} \big(G_{x,M\mathbf{u}} - G_{y,M\mathbf{u}}\big). \tag{5.5}$$

Assume that in some direction $\mathbf{v}$,

$$B^{\mathbf{u}}_{0,m\mathbf{v}} = \nabla g_{\rm pp}(\mathbf{u}) \cdot m\mathbf{v} + \Theta_P(m^{1/2}). \tag{5.6}$$

This says that in direction $\mathbf{v}$, $B^{\mathbf{u}}$ has fluctuation exponent $\frac{1}{2}$, in other words, diffusive fluctuations. (These assumptions are satisfied in the exponential CGM, for example for $\mathbf{v} = e_1$ or e_2, as verified by Theorem 4.2 and (4.14).)

A natural *planar* assumption is that geodesics (maximizing paths) from different points in a particular direction eventually coalesce. Since maximizing paths fluctuate on the scale N^ζ, it is reasonable to assume that if the initial points x, y in (5.5) are separated by a distance of order N^ζ, then once M is much larger than N, the maximizing paths from x and y to $M\mathbf{u}$ have merged at $a_N N\mathbf{u}$ for some (tight random) constant a_N. Then the limit value has been reached:

$$B^{\mathbf{u}}_{0,N^\zeta\mathbf{v}} = G_{0,\,a_N N\mathbf{u}} - G_{N^\zeta\mathbf{v},\,a_N N\mathbf{u}}.$$

Apply (5.3) and (5.6) to the two sides above :

$$\begin{aligned} \nabla g_{\rm pp}(\mathbf{u}) \cdot N^\zeta\mathbf{v} + \Theta_P(N^{\zeta/2}) &= Ng_{\rm pp}(a_N\mathbf{u}) - Ng_{\rm pp}(a_N\mathbf{u} - N^{\zeta-1}\mathbf{v}) + \Theta_P(N^\chi) \\ &\approx \nabla g_{\rm pp}(a_N\mathbf{u}) \cdot N^\zeta\mathbf{v} + \Theta_P(N^\chi). \end{aligned}$$

By homogeneity $\nabla g_{\rm pp}(\mathbf{u}) = \nabla g_{\rm pp}(a_N\mathbf{u})$ and hence the above gives $\Theta_P(N^{\zeta/2}) \approx \Theta_P(N^\chi)$. This gives a heuristic justification of (5.4).

5.2. Exponents for the stationary exponential CGM. We derive the KPZ exponents $\chi = \frac{1}{3}$ and $\zeta = \frac{2}{3}$ for the stationary exponential CGM in 1+1 dimensions. Recall the setting from Section 3. The parameter $0 < \rho < 1$ of the boundary weights is fixed. We are given mutually independent random variables

$$\{\omega_x, I_{ie_1}, J_{je_2} : x \in \mathbb{Z}^2_{>0}, i, j \in \mathbb{Z}_{>0}\} \tag{5.7}$$

with marginal distributions

$$\omega_x \sim \text{Exp}(1), \quad I_{ie_1} \sim \text{Exp}(1-\rho), \quad \text{and} \quad J_{je_2} \sim \text{Exp}(\rho). \tag{5.8}$$

The last-passage process $G^\rho_{0,x}$ is defined for $x \geqslant 0$ by $G^\rho_{0,0} = 0$,

$$G^\rho_{0,\,me_1} = \sum_{i=1}^m I_{ie_1} \quad \text{and} \quad G^\rho_{0,\,ne_2} = \sum_{j=1}^n J_{je_2}, \tag{5.9}$$

and then for $(m,n) \in \mathbb{Z}^2_{>0}$,

$$\begin{aligned} G^\rho_{0,\,(m,n)} = \max_{1\leqslant k\leqslant m} &\Big\{ \sum_{i=1}^k I_{ie_1} + G_{ke_1+e_2,\,(m,n)} \Big\} \\ &\bigvee \max_{1\leqslant \ell\leqslant n} \Big\{ \sum_{j=1}^\ell J_{je_2} + G_{\ell e_2+e_1,\,(m,n)} \Big\}. \end{aligned} \tag{5.10}$$

For $a \in \mathbb{Z}^2_{>0}$ the quantity $G_{a,(m,n)}$ inside the braces is the last-passage value defined in (3.1) for i.i.d. Exp(1) weights ω.

We do not have a closed form expression for $\mathbb{V}\text{ar}[G^\rho_{0,(m,n)}]$ but we can access it well enough to show that it obeys the fluctuation exponent 1/3 characteristic of the KPZ class. However, there in an extra twist. Notice in (5.8) that the boundary weights ω_{ie_1} and ω_{je_2} are larger on average than the bulk weights $\{\omega_x\}_{x\in\mathbb{Z}^2_{>0}}$. This implies that the boundaries are attractive to the maximizing path. It turns out that only when we take the point x to infinity in the characteristic direction $((1-\rho)^2, \rho^2)$, the pull of the boundaries balances out and $G^\rho_{0,x}$ obeys KPZ fluctuations. Otherwise the boundaries swamp the effects of the percolation and $G^\rho_{0,x}$ obeys the classical central limit theorem.

Let N be a scaling parameter that increases to ∞. We consider the point-to-point last-passage percolation from 0 to a point $(m,n) = (m(N), n(N))$ that is taken to infinity as $N \to \infty$. Let κ_N denote the deviation of (m,n) from the characteristic direction:

$$\kappa_N = |\,m - N(1-\rho)^2\,| \vee |\,n - N\rho^2\,|. \tag{5.11}$$

THEOREM 5.1. *Assume weight distributions* (5.8) *and let* a_0 *be a fixed positive constant. Then there exists a constant* $0 < C = C(\rho, a_0) < \infty$ *such that*

$$C^{-1}N^{2/3} \leqslant \mathbb{V}\text{ar}[G^\rho_{0,(m,n)}] \leqslant CN^{2/3} \tag{5.12}$$

holds for all $(m,n) \in \mathbb{Z}^2_{>0}$ *and* $N \geqslant 1$ *for which* κ_N *in* (5.11) *satisfies* $\kappa_N \leqslant a_0 N^{2/3}$.

The proof gives the following dependence of the bounds on a_0:

$$c_1(\rho)e^{-C_2(\rho)a_0^3}N^{2/3} \leqslant \mathbb{V}\text{ar}[G^\rho_{0,(m,n)}] \leqslant C_3(\rho)(1+a_0)N^{2/3} \tag{5.13}$$

where $c_1(\rho)$, $C_2(\rho)$ and $C_3(\rho)$ are functions of ρ alone.

As a fairly immediate corollary we obtain the behavior in off-characteristic directions. For concreteness, we state the result for the case where the horizontal direction is abnormally large.

COROLLARY 5.2. *Assume weight distributions* (5.8). *Suppose* $m, n \to \infty$. *Define parameter* N *by* $n = N\rho^2$, *and assume that*

$$N^{-\alpha}\big(m - N(1-\rho)^2\big) \to c_1 > 0 \quad \text{as } m, n \to \infty$$

for some $\alpha > 2/3$. *Then as* $m, n \to \infty$,

$$N^{-\alpha/2}\{G^{\rho}_{0,(m,n)} - \mathbb{E}[G^{\rho}_{0,(m,n)}]\}$$

converges in distribution to a centered normal distribution with variance $c_1(1-\rho)^{-2}$.

PROOF. Recall that overline means centering of a random variable.

$$N^{-\alpha/2}\,\overline{G}^{\rho}_{0,(m,n)} = N^{-\alpha/2}\,\overline{G}^{\rho}_{0,(\lfloor N(1-\rho)^2\rfloor,\,\lfloor N\rho^2\rfloor)} + N^{-\alpha/2}\sum_{i=\lfloor N(1-\rho)^2\rfloor+1}^{m} \bar{I}_{(i,n)}$$

The mean square of the first term on the right is of order $N^{-\alpha}\cdot N^{2/3}$ and hence in the limit vanishes in L^2 and in probability. The second term is a sum of approximately $c_1 N^{\alpha}$ mean zero i.i.d. terms with variance $\mathbb{E}[\bar{I}^2_{(i,n)}] = (1-\rho)^{-2}$. This sum gives the normal limit, by the standard central limit theorem. □

As discussed in Section 3.3, for a given endpoint (m, n) the last-passage problem (5.10) has an almost surely unique maximizing path or geodesic $\pi^{0,(m,n)}_{\bullet} = (\pi^{0,(m,n)}_k)_{k=0}^{m+n}$ from $\pi^{0,(m,n)}_0 = 0$ to $\pi^{0,(m,n)}_{m+n} = (m, n)$ that satisfies

$$G^{\rho}_{0,(m,n)} = \sum_{k=1}^{m+n} \omega_{\pi^{0,(m,n)}_k},$$

where we utilized the notational convention on the axes that $\omega_{ke_1} = I_{ke_1}$ and $\omega_{\ell e_2} = J_{\ell e_2}$.

The theorem below quantifies the fluctuations of the maximizing path $\pi^{0,(m,n)}_{\bullet}$ around a fixed point on the straight line from 0 to (m, n) when (m, n) points in the characteristic direction of the parameter ρ. As before, the endpoint (m, n) is of order N. Qualitatively speaking, the theorem says that $\pi^{0,(m,n)}_{\bullet}$ hits a square centered on the diagonal with high probability if the square is large enough on the scale $N^{2/3}$, and $\pi^{0,(m,n)}_{\bullet}$ misses the square with at least a fixed positive probability if the square is small enough on the scale $N^{2/3}$.

Let

$$A_r(x) = \{y \in \mathbb{R}^2 : x_i \leqslant y_i \leqslant x_i + r \text{ for } i = 1, 2\}$$

denote the closed square with side length r and lower left corner at $x = (x_1, x_2) \in \mathbb{R}^2$. $\mathbb{P}^{\rho}_{0,(m,n)}$ denotes the probability measure of the point-to-point LPP process in the rectangle $[0, (m, n)]$.

THEOREM 5.3. *Fix $0 < \rho < 1$. Assume weight distributions* (5.8) *and let*

$$(m, n) = (\lfloor N(1-\rho)^2 \rfloor, \lfloor N\rho^2 \rfloor).$$

(a) *For each $0 < \delta < 1$ we have this upper bound, with a constant $C = C(\rho)$:*

$$\sup_{0 \leqslant t \leqslant 1-\delta} \mathbb{P}^{\rho}_{0,(m,n)} \Big\{ \pi^{0,(m,n)}_{\bullet} \textit{ does not intersect } A_{rN^{2/3}}(tm, tn) \Big\} \leqslant \frac{C}{r^3} \tag{5.14}$$

for $r \geqslant 1$ and $N \geqslant \frac{1}{\delta}$.

(b) *Fix $0 < t < 1$. There exist $\delta, \varepsilon > 0$ that depend on ρ and $N_0 = N_0(\rho, t) < \infty$ such that*

$$\mathbb{P}^{\rho}_{0,(m,n)} \Big\{ \pi^{0,(m,n)}_{\bullet} \textit{ does not intersect } A_{\delta N^{2/3}}(tm, tn) \Big\} \geqslant \varepsilon \qquad \textit{for } N \geqslant N_0. \tag{5.15}$$

Estimate (5.14) gives the upper bound $\zeta \leqslant \frac{2}{3}$ and (5.15) the lower bound $\zeta \geqslant \frac{2}{3}$ for the path exponent ζ. Since $\pi^{0,(m,n)}_0 = 0$, for $t = 0$ the upper bound (5.14) holds trivially while the lower bound (5.15) fails trivially.

REMARK 5.4 (Exponents for LPP with i.i.d. exponential weights). After these theorems, the natural next question concerns the exponents for the last-passage value and maximizing path of the LPP process $G_{0,y}$ of (3.1) with i.i.d. Exp(1) weights ω. Presently the coupling approach of this article gives the exponent $\chi = \frac{1}{3}$ and the upper bound $\zeta \leqslant \frac{2}{3}$ for the i.i.d. exponential case. Results to this effect can be found in [**9**] for the exponential CGM, and in [**42, 53**] for positive temperature polymer models. △

REMARK 5.5 (Fluctuation exponent for the Busemann function). Through the distributional equality (4.15), Theorem 5.1 gives the following corollary for the fluctuations of B^{α} in the characteristic direction $\xi(\alpha)$: for each $0 < \alpha < 1$ there exists a finite constant $C = C(\alpha)$ such that

$$C^{-1} N^{2/3} \leqslant \mathbb{V}\mathrm{ar}[B^{\alpha}_{0,\lfloor N\xi(\alpha)\rfloor}] \leqslant C N^{2/3} \tag{5.16}$$

for N large enough so that $\lfloor N\xi(\alpha)\rfloor \neq 0$. △

We turn to proving Theorems 5.1 and 5.3. The remainder of this section is divided into four parts: a variance formula, an upper bound proof for Theorem 5.1, a lower bound proof for Theorem 5.1, and the proof of Theorem 5.3.

5.3. Variance formula. The first step towards the proof of Theorem 5.1 is an explicit formula tying together $\mathbb{V}\mathrm{ar}[G^{\rho}_{0,(m,n)}]$ and the amount of weight the maximizing path $\pi^{0,(m,n)}_{\bullet}$ collects on the boundary. For $r = 1, 2$, define the exit time (or exit point) of the maximizing path from the e_r axis by

$$\tau_r = \max\{k \geqslant 0 : \pi^{0,(m,n)}_k \cdot e_{3-r} = 0\}, \qquad r = 1, 2. \tag{5.17}$$

If the first step of the path $\pi^{0,(m,n)}_{\bullet}$ from the origin is e_r, then $1 \leqslant \tau_r \leqslant (m,n) \cdot e_r$ and $\tau_{3-r} = 0$. In other words, almost surely exactly one of τ_1 and τ_2 is positive (but which one is positive varies with the realization of the weights ω).

Further, introduce the sums of weights along the axes:

$$S_{1,k} = \sum_{i=1}^{k} I_{ie_1} \quad \text{and} \quad S_{2,\ell} = \sum_{j=1}^{\ell} J_{je_2}.$$

Then S_{r,τ_r} is the amount of weight that the maximizing path collects on the e_r-axis. Again, for each weight configuration ω, exactly one of S_{1,τ_1} and S_{2,τ_2} is positive and the other one zero.

Notational comment. The joint distribution of the random variables associated with the increment-stationary point-to-point LPP process is determined by these parameters: the planar rectangle $[0,(m,n)]$ and the parameter $\rho \in (0,1)$ of the boundary edge weights in (5.7)–(5.8). The probability measure for this situation will be denoted by $\mathbb{P}^{\rho}_{0,(m,n)}$. When processes with different values of ρ are coupled in the same rectangle, the variables themselves will be adorned with superscripts, as in τ^{ρ}_r and $S^{\rho}_{r,k}$. When the parameters ρ or (m,n) are understood from the context or already present in the notation for the random variables, as for example in $G^{\rho}_{0,(m,n)}$ in (5.12), we omit the parameters from $\mathbb{P}$. △

Next we state the variance formula for the last-passage value in the increment-stationary CGM. Formulas of this type for exactly solvable models go back to the work of Ferrari and Fontes [**25**] on the asymmetric simple exclusion process. In the particle system context the formulas relate the variance of the current to a expectation of a second-class particle.

THEOREM 5.6. *Assume weight distributions* (5.8). *Then for* $(m,n) \in \mathbb{Z}^2_{>0}$,

$$\mathbb{V}\mathrm{ar}[G^{\rho}_{0,(m,n)}] = -\frac{m}{(1-\rho)^2} + \frac{n}{\rho^2} + \frac{2}{1-\rho}\,\mathbb{E}^{\rho}_{0,(m,n)}[S_{1,\tau_1}] \tag{5.18}$$

$$= \frac{m}{(1-\rho)^2} - \frac{n}{\rho^2} + \frac{2}{\rho}\,\mathbb{E}^{\rho}_{0,(m,n)}[S_{2,\tau_2}]. \tag{5.19}$$

PROOF. We prove (5.18). The second line (5.19) follows for example by transposition. Begin with a covariance manipulation. For this proof , abbreviate the G^{ρ}-increments on the sides of the rectangle $[0,(m,n)]$ as follows:

$$S_{\mathcal{N}} = G^{\rho}_{0,(m,n)} - G^{\rho}_{0,(0,n)}, \quad S_{\mathcal{S}} = G^{\rho}_{0,(m,0)},$$
$$S_{\mathcal{E}} = G^{\rho}_{0,(m,n)} - G^{\rho}_{0,(m,0)}, \quad S_{\mathcal{W}} = G^{\rho}_{0,(0,n)}.$$

Then

$$\begin{aligned}\mathbb{V}\mathrm{ar}\big[G^{\rho}_{0,(m,n)}\big] &= \mathbb{V}\mathrm{ar}(S_{\mathcal{W}} + S_{\mathcal{N}}) = \mathbb{V}\mathrm{ar}(S_{\mathcal{W}}) + \mathbb{V}\mathrm{ar}(S_{\mathcal{N}}) + 2\,\mathbb{C}\mathrm{ov}(S_{\mathcal{W}}, S_{\mathcal{N}})\\ &= \mathbb{V}\mathrm{ar}(S_{\mathcal{W}}) + \mathbb{V}\mathrm{ar}(S_{\mathcal{N}}) + 2\,\mathbb{C}\mathrm{ov}(S_{\mathcal{S}} + S_{\mathcal{E}} - S_{\mathcal{N}}, S_{\mathcal{N}})\\ &= \mathbb{V}\mathrm{ar}(S_{\mathcal{W}}) - \mathbb{V}\mathrm{ar}(S_{\mathcal{N}}) + 2\,\mathbb{C}\mathrm{ov}(S_{\mathcal{S}}, S_{\mathcal{N}})\\ &= \frac{n}{\rho^2} - \frac{m}{(1-\rho)^2} + 2\,\mathbb{C}\mathrm{ov}(S_{\mathcal{S}}, S_{\mathcal{N}}).\end{aligned} \tag{5.20}$$

The two last equalities came from Theorem 3.1(i) applied to the down-right path that gives the north and east boundaries of the rectangle $[0,(m,n)]$: first the independence of $S_{\mathcal{E}}$ and $S_{\mathcal{N}}$ and then the marginal distributions of $S_{\mathcal{W}}$ and $S_{\mathcal{N}}$ as sums of i.i.d. exponentials.

To prove (5.18) it remains to compute $\mathbb{C}\mathrm{ov}(S_{\mathcal{S}}, S_{\mathcal{N}})$. In order to differentiate with respect to the parameter of the weights $I_{(i,0)}$ on the x-axis (term $S_{\mathcal{S}}$) while fixing everything else, consider a system with two independent parameters α and ρ on the boundaries and with initial weight distributions (for $i,j \in \mathbb{Z}_{>0}$)

$$\omega_{(i,j)} \sim \mathrm{Exp}(1), \quad I_{ie_1} \sim \mathrm{Exp}(\alpha), \quad \text{and} \quad J_{je_2} \sim \mathrm{Exp}(\rho). \tag{5.21}$$

Use the superscript α,ρ on $G^{\alpha,\rho}_{0,(m,n)}$, $\mathbb{P}^{\alpha,\rho}$, and $\mathbb{E}^{\alpha,\rho}$ for the LPP process with these weights. If $\alpha \neq 1-\rho$ this is not a stationary LPP process, so we do not know the distribution of $S_{\mathcal{N}}$. We can still derive the identity

$$\mathbb{C}\mathrm{ov}^{\alpha,\rho}(S_{\mathcal{S}}, S_{\mathcal{N}}) = -\frac{\partial}{\partial\alpha}\mathbb{E}^{\alpha,\rho}(S_{\mathcal{N}}). \tag{5.22}$$

The sum $S_{\mathcal{S}} = \sum_{i=1}^m I_{ie_1}$ has the gamma density

$$f_{m,\alpha}(s) = \frac{\alpha^m}{(m-1)!}s^{m-1}e^{-\alpha s}.$$

By changing variables in the integrals, with test functions φ and ψ,

$$\mathbb{E}^{\alpha,\rho}[\,\varphi(I_{e_1},\dots,I_{me_1})\,\psi(S_{\mathcal{S}})\,] = \int_{\mathbb{R}^m_{>0}} \varphi(x_1,\dots,x_m)\,\psi(\textstyle\sum_1^m x_i)\,\alpha^m\,e^{-\alpha\sum_1^m x_i}\,dx_{1,m}$$

$$= \int_0^\infty ds\,\tfrac{\alpha^m}{(m-1)!}s^{m-1}e^{-\alpha s}\,\psi(s)$$

$$\times \int_{\mathbb{R}^{m-1}_{>0}} \mathbf{1}_{\{\sum_1^{m-1}x_i<s\}}\,\varphi\big(x_1,\dots,x_{m-1},s-\textstyle\sum_1^{m-1}x_i\big)\,\tfrac{(m-1)!}{s^{m-1}}\,dx_{1,m-1}$$

we see that the conditional distribution of $(I_{e_1},\dots,I_{me_1})$, given $S_{\mathcal{S}}$, is independent of the parameter α.

We verify (5.22).

$$\begin{aligned}
-\frac{\partial}{\partial\alpha}\mathbb{E}^{\alpha,\rho}(S_{\mathcal{N}}) &= -\frac{\partial}{\partial\alpha}\int_0^\infty \mathbb{E}^{\alpha,\rho}(S_{\mathcal{N}}\,|\,S_{\mathcal{S}}=s)\,f_{m,\alpha}(s)\,ds \\
&= \int_0^\infty \mathbb{E}^{\alpha,\rho}(S_{\mathcal{N}}\,|\,S_{\mathcal{S}}=s)\Big(-\frac{\partial}{\partial\alpha}f_{m,\alpha}(s)\Big)\,ds \\
&= \int_0^\infty \mathbb{E}^{\alpha,\rho}(S_{\mathcal{N}}\,|\,S_{\mathcal{S}}=s)\Big(s-\frac{m}{\alpha}\Big)f_{m,\alpha}(s)\,ds \\
&= \mathbb{E}^{\alpha,\rho}[S_{\mathcal{N}}S_{\mathcal{S}}] - \mathbb{E}^{\alpha,\rho}[S_{\mathcal{N}}]\cdot\frac{m}{\alpha} = \mathbb{C}\mathrm{ov}^{\alpha,\rho}(S_{\mathcal{S}},S_{\mathcal{N}}).
\end{aligned}$$

On the other hand, we calculate the derivative in (5.22) by putting the α-dependence directly on the weights, by realizing them as functions of uniform random variables. Let $U_1,\dots,U_m$ be i.i.d. Unif$(0,1)$ random variables. The inverse function of the Exp(α) distribution function $F_\alpha(x) = 1-e^{-\alpha x}$ is $F_\alpha^{-1}(u) = -\alpha^{-1}\log(1-u)$ and this satisfies $\frac{\partial}{\partial\alpha}F_\alpha^{-1}(u) = -\alpha^{-1}F_\alpha^{-1}(u)$. For the Exp$(\alpha)$ weights on the x-axis take $I_{ie_1} = F_\alpha^{-1}(U_i)$. Let $\widetilde{\mathbb{E}}$ denote the expectation over $U_1,\dots,U_m$ and the other weights J_{je_2} and $\omega_{(i,j)}$ from (5.21). $\widetilde{\mathbb{E}}$ does not depend on α. τ_1^α is the exit point of the maximizing path from the x-axis when the edge weights on the x-axis are $I_{ie_1} = F_\alpha^{-1}(U_i)$.

$$\begin{aligned}
\frac{\partial}{\partial\alpha}\mathbb{E}^{\alpha,\rho}(S_{\mathcal{N}}) &= \frac{\partial}{\partial\alpha}\mathbb{E}^{\alpha,\rho}[G_{0,(m,n)} - G_{0,(0,n)}] = \frac{\partial}{\partial\alpha}\mathbb{E}^{\alpha,\rho}[G_{0,(m,n)}] \\
&= \frac{\partial}{\partial\alpha}\widetilde{\mathbb{E}}\Big[\sum_{i=1}^{\tau_1^\alpha}F_\alpha^{-1}(U_i) + G_{(\tau_1^\alpha,1),(m,n)}\Big] = \widetilde{\mathbb{E}}\Big[\sum_{i=1}^{\tau_1^\alpha}\frac{\partial}{\partial\alpha}F_\alpha^{-1}(U_i)\Big] \\
&= -\alpha^{-1}\widetilde{\mathbb{E}}\Big[\sum_{i=1}^{\tau_1^\alpha}F_\alpha^{-1}(U_i)\Big] = -\alpha^{-1}\mathbb{E}^{\alpha,\rho}[S_{1,\tau_1}].
\end{aligned} \tag{5.23}$$

An intuitive justification of the fourth equality above is that, with probability one, if α' is close enough to α, then $\tau_1^{\alpha'} = \tau_1^{\alpha}$. A detailed justification of (5.23) follows.

Let $\varepsilon > 0$ be small. Since $F_{\alpha-\varepsilon}^{-1}(U_i) > F_{\alpha}^{-1}(U_i)$, replacing α with $\alpha-\varepsilon$ increases weights on the x-axis without altering other weights. Consequently $\tau_1^{\alpha-\varepsilon} \geqslant \tau_1^{\alpha}$ and $G_{0,(m,n)}^{\alpha-\varepsilon,\rho} \geqslant G_{0,(m,n)}^{\alpha,\rho}$. Now consider the differences between the expectations above at $\alpha-\varepsilon$ and α.

$$\widetilde{\mathbb{E}}\Big[\sum_{i=1}^{\tau_1^{\alpha-\varepsilon}} F_{\alpha-\varepsilon}^{-1}(U_i) + G_{(\tau_1^{\alpha-\varepsilon},1),(m,n)}\Big] - \widetilde{\mathbb{E}}\Big[\sum_{i=1}^{\tau_1^{\alpha}} F_{\alpha}^{-1}(U_i) + G_{(\tau_1^{\alpha},1),(m,n)}\Big]$$

$$= \widetilde{\mathbb{E}}\Big[\sum_{i=1}^{\tau_1^{\alpha}} \big(F_{\alpha-\varepsilon}^{-1}(U_i) - F_{\alpha}^{-1}(U_i)\big)\Big] \tag{5.24}$$

$$+ \widetilde{\mathbb{E}}\Big[\sum_{i=\tau_1^{\alpha}+1}^{\tau_1^{\alpha-\varepsilon}} F_{\alpha-\varepsilon}^{-1}(U_i) + G_{(\tau_1^{\alpha-\varepsilon},1),(m,n)} - G_{(\tau_1^{\alpha},1),(m,n)}\Big]. \tag{5.25}$$

The expectation on line (5.24) equals

$$\frac{\varepsilon}{\alpha-\varepsilon}\,\widetilde{\mathbb{E}}\Big[\sum_{i=1}^{\tau_1^{\alpha}} F_{\alpha}^{-1}(U_i)\Big] = \big(\varepsilon + O(\varepsilon^2)\big)\alpha^{-1}\mathbb{E}^{\alpha,\rho}[S_{1,\tau_1}].$$

We show that the expectation on line (5.25) is of order $o(\varepsilon)$. The integrand inside the expectation vanishes if $\tau_1^{\alpha-\varepsilon} = \tau_1^{\alpha}$. Hence we may multiply the integrand by the indicator of the complementary event and then bound it as follows:

$$\begin{aligned}
&\Big(\sum_{i=\tau_1^{\alpha}+1}^{\tau_1^{\alpha-\varepsilon}} F_{\alpha-\varepsilon}^{-1}(U_i) + G_{(\tau_1^{\alpha-\varepsilon},1),(m,n)} - G_{(\tau_1^{\alpha},1),(m,n)}\Big)\cdot \mathbf{1}_{\{\tau_1^{\alpha-\varepsilon} > \tau_1^{\alpha}\}}\\
&\leqslant \Big(\sum_{i=1}^{\tau_1^{\alpha-\varepsilon}} F_{\alpha-\varepsilon}^{-1}(U_i) + G_{(\tau_1^{\alpha-\varepsilon},1),(m,n)} - \sum_{i=1}^{\tau_1^{\alpha}} F_{\alpha}^{-1}(U_i) - G_{(\tau_1^{\alpha},1),(m,n)}\Big)\cdot \mathbf{1}_{\{\tau_1^{\alpha-\varepsilon} > \tau_1^{\alpha}\}}\\
&= \big(G_{0,(m,n)}^{\alpha-\varepsilon,\rho} - G_{0,(m,n)}^{\alpha,\rho}\big)\cdot \mathbf{1}_{\{\tau_1^{\alpha-\varepsilon} > \tau_1^{\alpha}\}} \leqslant \big(S_{1,m}^{\alpha-\varepsilon} - S_{1,m}^{\alpha}\big)\cdot \mathbf{1}_{\{\tau_1^{\alpha-\varepsilon} > \tau_1^{\alpha}\}}.
\end{aligned}$$

The last inequality comes from the observation that, since the weights away from the south boundary do not change as α changes, the increase $G_{0,(m,n)}^{\alpha-\varepsilon,\rho} - G_{0,(m,n)}^{\alpha,\rho}$ is bounded by the total increase $S_{1,m}^{\alpha-\varepsilon} - S_{1,m}^{\alpha}$ in the weight on the south boundary. We bound the expectation on line (5.25):

$$\begin{aligned}
&\widetilde{\mathbb{E}}\Big[\sum_{i=\tau_1^{\alpha}+1}^{\tau_1^{\alpha-\varepsilon}} F_{\alpha-\varepsilon}^{-1}(U_i) + G_{(\tau_1^{\alpha-\varepsilon},1),(m,n)} - G_{(\tau_1^{\alpha},1),(m,n)}\Big]\\
&\qquad \leqslant \widetilde{\mathbb{E}}\big[S_{1,m}^{\alpha-\varepsilon} - S_{1,m}^{\alpha}\,,\ \tau_1^{\alpha-\varepsilon} > \tau_1^{\alpha}\big]\\
&\qquad \leqslant \big\{\widetilde{\mathbb{E}}\big[(S_{1,m}^{\alpha-\varepsilon} - S_{1,m}^{\alpha})^2\big]\big\}^{1/2}\big(\widetilde{\mathbb{P}}\{\tau_1^{\alpha-\varepsilon} > \tau_1^{\alpha}\}\big)^{1/2}\\
&\qquad = \frac{\varepsilon}{\alpha-\varepsilon}\big\{\widetilde{\mathbb{E}}\big[(S_{1,m}^{\alpha})^2\big]\big\}^{1/2}\big(\widetilde{\mathbb{P}}\{\tau_1^{\alpha-\varepsilon} > \tau_1^{\alpha}\}\big)^{1/2}\\
&\qquad = C(m,\alpha)\,\frac{\varepsilon}{\alpha-\varepsilon}\big(\widetilde{\mathbb{P}}\{\tau_1^{\alpha-\varepsilon} > \tau_1^{\alpha}\}\big)^{1/2}
\end{aligned}$$

for a constant $C(m,\alpha)$. It remains to argue that

$$\lim_{\varepsilon\to 0}\widetilde{\mathbb{P}}\{\tau_1^{\alpha-\varepsilon} > \tau_1^{\alpha}\} = 0.$$

This comes from dominated convergence as follows. For almost every fixed weights, there exists $\delta > 0$ such that the maximizing path $x_\bullet^{\alpha,\rho *}$ for $G^{\alpha,\rho}_{0,(m,n)}$ has weight at least δ above the weight of the next best path. Thus, once ε is small enough so that $S^{\alpha-\varepsilon}_{1,m} < S^{\alpha}_{1,m} + \delta$, path $x_\bullet^{\alpha,\rho *}$ is maximal also for $G^{\alpha-\varepsilon,\rho}_{0,(m,n)}$ and then $\tau_1^{\alpha-\varepsilon} = \tau_1^{\alpha}$. Thus $\lim_{\varepsilon\to 0}\mathbf{1}_{\{\tau_1^{\alpha-\varepsilon} > \tau_1^{\alpha}\}} = 0$ almost surely. We have verified that the expectation on line (5.25) is of order $o(\varepsilon)$.

Now return to (5.24)–(5.25) to collect the steps to compute the derivative:

$$\begin{aligned}
&\frac{\partial}{\partial\alpha}\widetilde{\mathbb{E}}\Big[\sum_{i=1}^{\tau_1^{\alpha}} F_{\alpha}^{-1}(U_i) + G_{(\tau_1^{\alpha},1),(m,n)}\Big] \\
&= -\lim_{\varepsilon\to 0}\frac{1}{\varepsilon}\bigg\{\widetilde{\mathbb{E}}\Big[\sum_{i=1}^{\tau_1^{\alpha-\varepsilon}} F_{\alpha-\varepsilon}^{-1}(U_i) + G_{(\tau_1^{\alpha-\varepsilon},1),(m,n)}\Big] \\
&\qquad\qquad - \widetilde{\mathbb{E}}\Big[\sum_{i=1}^{\tau_1^{\alpha}} F_{\alpha}^{-1}(U_i) + G_{(\tau_1^{\alpha},1),(m,n)}\Big]\bigg\} \\
&= -\lim_{\varepsilon\to 0}\frac{1}{\varepsilon}\Big\{\big(\varepsilon + O(\varepsilon^2)\big)\alpha^{-1}\mathbb{E}^{\alpha,\rho}[S_{1,\tau_1}] + o(\varepsilon)\Big\} \;=\; -\alpha^{-1}\mathbb{E}^{\alpha,\rho}[S_{1,\tau_1}].
\end{aligned}$$

Calculation (5.23) has been justified.

To finish the proof, return to treat the last term on the last line of (5.20):

$$\begin{aligned}
\mathbb{C}\mathrm{ov}(S_{\mathcal{S}}, S_{\mathcal{N}}) &= \mathbb{C}\mathrm{ov}^{1-\rho,\rho}(S_{\mathcal{S}}, S_{\mathcal{N}}) = -\frac{\partial}{\partial\alpha}\mathbb{E}^{\alpha,\rho}(S_{\mathcal{N}})\big|_{\alpha=1-\rho} \\
&= \frac{1}{1-\rho}\mathbb{E}^{1-\rho,\rho}[S_{1,\tau_1}] = \frac{1}{1-\rho}\mathbb{E}^{\rho}[S_{1,\tau_1}].
\end{aligned}$$

This completes the proof of identity (5.18). ☐

5.4. Upper bound for the passage time exponent. This upper bound proof is an improved version of the proof originally given in [**9**]. Article [**9**] was a corner growth model adaptation of the arguments of the earlier work [**13**] on increasing sequences among planar Poisson points.

We couple the boundary variables for two different parameters $0 < \rho < \lambda < 1$ as follows:

$$I^{\lambda}_{ie_1} = \frac{1-\rho}{1-\lambda} I^{\rho}_{ie_1} > I^{\rho}_{ie_1} \quad\text{and}\quad J^{\lambda}_{je_2} = \frac{\rho}{\lambda}J^{\rho}_{je_2} < J^{\rho}_{je_2}. \tag{5.26}$$

From this follows for example that $\tau_1^{\lambda} \geqslant \tau_1^{\rho}$ and $\tau_2^{\lambda} \leqslant \tau_2^{\rho}$, and also

$$S^{\lambda}_{1,\ell} - S^{\rho}_{1,\ell} \leqslant S^{\lambda}_{1,k} - S^{\rho}_{1,k} \qquad \text{for } 0 \leqslant \ell \leqslant k. \tag{5.27}$$

We begin with auxiliary lemmas.

LEMMA 5.7. *Let $0 < \varepsilon < 1$. Then there exists a constant $C = C(\varepsilon)$ such that, for $\varepsilon \leqslant \rho < \lambda \leqslant 1-\varepsilon$,*

$$\mathbb{V}\mathrm{ar}[G^{\lambda}_{0,(m,n)}] \leqslant \mathbb{V}\mathrm{ar}[G^{\rho}_{0,(m,n)}] + Cm(\lambda-\rho).$$

PROOF. From (5.26)

$$S^{\lambda}_{2,\tau_2} = \sum_{j=1}^{\tau_2^{\lambda}} J^{\lambda}_{je_2} \leqslant \sum_{j=1}^{\tau_2^{\rho}} J^{\lambda}_{je_2} = \frac{\rho}{\lambda}\sum_{j=1}^{\tau_2^{\rho}} J^{\rho}_{je_2} = \frac{\rho}{\lambda} S^{\rho}_{2,\tau_2}.$$

Using the second line of (5.18),

$$\begin{aligned}
\mathbb{V}\mathrm{ar}[G^{\lambda}_{0,(m,n)}] &= \frac{m}{(1-\lambda)^2} - \frac{n}{\lambda^2} + \frac{2}{\lambda}\mathbb{E}[S^{\lambda}_{2,\tau_2}] \\
&\leqslant \frac{(1-\rho)^2}{(1-\lambda)^2}\cdot\frac{m}{(1-\rho)^2} - \frac{\rho^2}{\lambda^2}\cdot\frac{n}{\rho^2} + \frac{\rho^2}{\lambda^2}\cdot\frac{2}{\rho}\mathbb{E}[S^{\rho}_{2,\tau_2}] \\
&= \frac{\rho^2}{\lambda^2}\cdot \mathbb{V}\mathrm{ar}[G^{\rho}_{0,(m,n)}] + \frac{m}{(1-\rho)^2}\Big(\frac{(1-\rho)^2}{(1-\lambda)^2} - \frac{\rho^2}{\lambda^2}\Big) \\
&\leqslant \mathbb{V}\mathrm{ar}[G^{\rho}_{0,(m,n)}] + Cm(\lambda-\rho).
\end{aligned}$$

□

LEMMA 5.8. *Let $0 < \varepsilon < 1$. Then there exists a constant $C = C(\varepsilon)$ such that, for $\varepsilon \leqslant \rho \leqslant 1-\varepsilon$,*

$$\mathbb{E}^{\rho}_{0,(m,n)}[S_{1,\tau_1}] \leqslant C\big(\mathbb{E}^{\rho}_{0,(m,n)}[\tau_1] + 1\big). \tag{5.28}$$

PROOF. Recall that $I_{ie_1} \sim \mathrm{Exp}(1-\rho)$. The steps of the derivation are self-explanatory.

$$\begin{aligned}
\mathbb{E}[S_{1,\tau_1}] &= \mathbb{E}\Big[\sum_{i=1}^{\tau_1} I_{ie_1}\Big] = \mathbb{E}\Big[\sum_{i=1}^{\tau_1} \overline{I}_{ie_1}\Big] + \frac{\mathbb{E}[\tau_1]}{1-\rho} = \sum_{k=1}^{m}\mathbb{E}[\,\overline{S}_{1,k}\,, \tau_1 = k] + \frac{\mathbb{E}[\tau_1]}{1-\rho} \\
&\leqslant \sum_{k=1}^{m}\mathbb{E}[\,\overline{S}_{1,k}\,, \overline{S}_{1,k} \leqslant k,\ \tau_1 = k] + \sum_{k=1}^{m}\mathbb{E}[\,\overline{S}_{1,k}\,, \overline{S}_{1,k} > k,\ \tau_1 = k] + \frac{\mathbb{E}[\tau_1]}{1-\rho} \\
&\leqslant \sum_{k=1}^{m} k\,\mathbb{P}(\tau_1 = k) + \sum_{k=1}^{m}\mathbb{E}[\,\overline{S}_{1,k}\,, \overline{S}_{1,k} > k] + \frac{\mathbb{E}[\tau_1]}{1-\rho} \\
&\leqslant \mathbb{E}[\tau_1]\frac{2-\rho}{1-\rho} + \sum_{k=1}^{m}\big(\mathbb{E}[\,\overline{S}^2_{1,k}\,]\big)^{1/2}\big(P\{\overline{S}_{1,k} > k\}\big)^{1/2} \\
&\leqslant \mathbb{E}[\tau_1]\frac{2-\rho}{1-\rho} + C(\rho)\sum_{k=1}^{\infty} k^{1/2}e^{-c(\rho)k} \leqslant C(\rho)\big(\mathbb{E}[\tau_1] + 1\big).
\end{aligned}$$

□

The main estimate for the upper bound in (5.12) is contained in the next proposition.

PROPOSITION 5.9. *Consider the increment-stationary CGM in the rectangle $[0,(m,n)]$ for $(m,n) \in \mathbb{Z}^2_{>0}$ and with weight distributions (5.8) for a given $0 < \rho < 1$. Let κ_N be defined by (5.11). Three positive constants a_0, a_1 and N_0 are given and the assumption is that*

$$\kappa_N \leqslant a_0 N^{2/3} \quad \textit{and} \quad m \leqslant a_1 N \quad \textit{for} \quad N \geqslant N_0. \tag{5.29}$$

Then there exist constants $c_2, c_3 < \infty$ such that the following two bounds hold:

$$\mathbb{P}^{\rho}_{0,(m,n)}\{\tau_1 \geqslant \ell\} \leqslant c_3\Big(\frac{N^2}{\ell^3} + (1+a_0)\frac{N^{8/3}}{\ell^4}\Big) \tag{5.30}$$

for $N \geqslant N_0$ and $1 \vee c_2\kappa_N \leqslant \ell \leqslant m$ and

$$\mathbb{E}^{\rho}_{0,(m,n)}[(\tau_1)^q] \leqslant \Big(c_2 a_0^q + \frac{c_3}{3-q}\Big)N^{2q/3} \tag{5.31}$$

for $N \geqslant N_0$ and $1 \leqslant q < 3$. The functional dependencies of the constants c_2, c_3 on the parameters is as follows:

$$c_2 = c_2(\rho) \quad \text{and} \quad c_3 = c_3(a_1, \rho). \tag{5.32}$$

Furthermore, c_2 and c_3 are locally bounded functions of their arguments.

The upper variance bound in (5.12) follows from a combination of (5.18), assumption (5.11), (5.28), and (5.31) for $q = 1$.

Proof of Proposition 5.9. Consider $N \geqslant N_0$ so that the assumptions are in force. Assume that, for some $0 < c_2 < \infty$, the integer ℓ satisfies

$$1 \vee c_2\kappa_N \leqslant \ell \leqslant m \leqslant a_1 N. \tag{5.33}$$

The proof will choose $c_2 = c_2(\rho)$ large enough. Let $0 < r < 1$ be a constant that will be set small enough in the proof. Let

$$\lambda = \rho + \frac{r\ell}{N}. \tag{5.34}$$

We take $r = r(a_1, \rho)$ at least small enough so that $ra_1 < \frac{1}{2}(1-\rho)$. This guarantees that for $N \geqslant N_0$, $\lambda \in (\rho, \frac{1+\rho}{2})$ is also a legitimate parameter for an increment-stationary CGM.

In the first inequality below use $S^{\lambda}_{1,k} + G_{(k,1),(m,n)} \leqslant G^{\lambda}_{0,(m,n)}$ and then (5.27). Recall that $\overline{X} = X - \mathbb{E}X$ denotes a centered random variable.

$$\begin{aligned}
\mathbb{P}\{\tau_1^{\rho} \geqslant \ell\} &= \mathbb{P}\{\,\exists k \geqslant \ell :\, S^{\rho}_{1,k} + G_{(k,1),(m,n)} = G^{\rho}_{0,(m,n)}\,\} \\
&\leqslant \mathbb{P}\{\,\exists k \geqslant \ell :\, S^{\lambda}_{1,k} - S^{\rho}_{1,k} \leqslant G^{\lambda}_{0,(m,n)} - G^{\rho}_{0,(m,n)}\,\} \\
&= \mathbb{P}\{\,S^{\lambda}_{1,\ell} - S^{\rho}_{1,\ell} \leqslant G^{\lambda}_{0,(m,n)} - G^{\rho}_{0,(m,n)}\,\} \\
&= \mathbb{P}\Big\{\,\overline{S^{\lambda}_{1,\ell}} - \overline{S^{\rho}_{1,\ell}} \leqslant \overline{G^{\lambda}_{0,(m,n)}} - \overline{G^{\rho}_{0,(m,n)}} \\
&\qquad\qquad - \big(\,\mathbb{E}[S^{\lambda}_{1,\ell} - S^{\rho}_{1,\ell}] - \mathbb{E}[G^{\lambda}_{0,(m,n)} - G^{\rho}_{0,(m,n)}]\big)\Big\}
\end{aligned} \tag{5.35}$$

Next we compute and bound the means of the random variables in the last probability above. This will show that the event in the probability is a deviation, and not a typical event. First

$$\begin{aligned}
\mathbb{E}[S^{\lambda}_{1,\ell} - S^{\rho}_{1,\ell}] &= \ell\Big(\frac{1}{1-\lambda} - \frac{1}{1-\rho}\Big) = \frac{\ell}{(1-\lambda)(1-\rho)}(\lambda - \rho) \\
&= \frac{1}{(1-\lambda)(1-\rho)} \cdot \frac{r\ell^2}{N}
\end{aligned} \tag{5.36}$$

Introduce the quantities

$$\kappa^1_N = m - N(1-\rho)^2 \quad \text{and} \quad \kappa^2_N = n - N\rho^2 \tag{5.37}$$

that satisfy (with κ_N as in (5.11))

$$|\kappa^1_N| \vee |\kappa^2_N| \leqslant \kappa_N.$$

Then the LPP values.

$$
\begin{aligned}
\mathbb{E}[G^{\lambda}_{0,(m,n)} - G^{\rho}_{0,(m,n)}] &= m\Big(\frac{1}{1-\lambda} - \frac{1}{1-\rho}\Big) + n\Big(\frac{1}{\lambda} - \frac{1}{\rho}\Big) \\
&= \Big(\frac{m}{(1-\lambda)(1-\rho)} - \frac{n}{\lambda\rho}\Big)(\lambda-\rho) \\
&= N\Big(\frac{1-\rho}{1-\lambda} - \frac{\rho}{\lambda}\Big)(\lambda-\rho) + \Big(\frac{\kappa^1_N}{(1-\lambda)(1-\rho)} - \frac{\kappa^2_N}{\lambda\rho}\Big)(\lambda-\rho) \\
&= \frac{N}{\lambda(1-\lambda)}(\lambda-\rho)^2 + \Big(\frac{\kappa^1_N}{(1-\lambda)(1-\rho)} - \frac{\kappa^2_N}{\lambda\rho}\Big)(\lambda-\rho) \\
&= \frac{r^2\ell^2}{\lambda(1-\lambda)N} + \Big(\frac{\kappa^1_N}{(1-\lambda)(1-\rho)} - \frac{\kappa^2_N}{\lambda\rho}\Big)\frac{r\ell}{N} \\
&\leqslant \frac{r^2\ell^2}{\lambda(1-\lambda)N} + \frac{1}{(1-\lambda)(1-\rho)\wedge\lambda\rho}\cdot\frac{r\ell^2}{c_2N}
\end{aligned} \tag{5.38}
$$

The last inequality used the assumption $\kappa_N \leqslant \ell/c_2$ from (5.33).

The key step is the appearance of the quadratic $(\lambda-\rho)^2$ in the middle of the development above. This is how curvature enters this upper bound proof. The quadratic comes because $\lambda \mapsto \mathbb{E}[G^{\lambda}_{0,(m,n)}]$ is minimized at $\lambda = \rho$ by virtue of the choice of (m,n) as the (approximate) characteristic direction for ρ.

On the last line above the first term is the important one and the second an error that came from the κ_N discrepancies. Comparison of the means in (5.36) and (5.38) shows that if we choose c_2 large enough and then r small enough, both as functions of (λ,ρ), then for a large enough constant $c_3 = c_3(\lambda,\rho)$,

$$
\mathbb{E}[S^{\lambda}_{1,\ell} - S^{\rho}_{1,\ell}] > \mathbb{E}[G^{\lambda}_{0,(m,n)} - G^{\rho}_{0,(m,n)}] + \frac{r\ell^2}{c_3N}. \tag{5.39}
$$

Since the range $\lambda \in (\rho, \frac{1+\rho}{2})$ is determined by ρ, the dependence on λ can be dropped and we have $c_2 = c_2(\rho)$, $r = r(a_1,\rho)$ and $c_3 = c_3(\rho)$.

We continue the bound on $\mathbb{P}\{\tau^{\rho}_1 \geqslant \ell\}$ from line (5.35) and apply (5.39). Below we subsume (r, a_1, ρ, λ)-dependent factors into a constant $C = C(a_1,\rho)$. Along the way we use Lemma 5.7, Theorem 5.6, (5.37), $\kappa_N \leqslant c_2^{-1}\ell$, $m \leqslant a_1N$, and Lemma 5.8.

$$
\begin{aligned}
\mathbb{P}\{\tau^{\rho}_1 \geqslant \ell\} &\leqslant \mathbb{P}\Big\{\overline{S^{\lambda}_{1,\ell}} - \overline{S^{\rho}_{1,\ell}} \leqslant \overline{G^{\lambda}_{0,(m,n)}} - \overline{G^{\rho}_{0,(m,n)}} - \frac{r\ell^2}{c_3N}\Big\} \\
&\leqslant \mathbb{P}\Big\{\overline{S^{\lambda}_{1,\ell}} - \overline{S^{\rho}_{1,\ell}} \leqslant -\frac{r\ell^2}{2c_3N}\Big\} + \mathbb{P}\Big\{\overline{G^{\lambda}_{0,(m,n)}} - \overline{G^{\rho}_{0,(m,n)}} \geqslant \frac{r\ell^2}{2c_3N}\Big\} \\
&\leqslant \frac{CN^2}{\ell^4}\mathbb{V}\mathrm{ar}[S^{\lambda}_{1,\ell} - S^{\rho}_{1,\ell}] + \frac{CN^2}{\ell^4}\mathbb{V}\mathrm{ar}[G^{\lambda}_{0,(m,n)} - G^{\rho}_{0,(m,n)}] \\
&\leqslant \frac{CN^2}{\ell^3} + \frac{CN^2}{\ell^4}\big(\mathbb{V}\mathrm{ar}[G^{\lambda}_{0,(m,n)}] + \mathbb{V}\mathrm{ar}[G^{\rho}_{0,(m,n)}]\big) \\
&\leqslant \frac{CN^2}{\ell^3} + \frac{CN^2}{\ell^4}\big(\mathbb{V}\mathrm{ar}[G^{\rho}_{0,(m,n)}] + m(\lambda-\rho)\big) \\
&= \frac{CN^2}{\ell^3} + \frac{CN^2}{\ell^4}\Big(-\frac{m}{(1-\rho)^2} + \frac{n}{\rho^2} + \frac{2}{1-\rho}\mathbb{E}[S^{\rho}_{1,\tau_1}] + a_1N\cdot\frac{r\ell}{N}\Big) \\
&\leqslant \frac{CN^2}{\ell^3} + \frac{CN^2}{\ell^4}\big(\mathbb{E}[\tau^{\rho}_1] + \ell\big) \leqslant \frac{CN^2}{\ell^3} + \frac{CN^2}{\ell^4}\mathbb{E}[\tau^{\rho}_1].
\end{aligned} \tag{5.40}
$$

The fourth inequality above used the fact that the boundary variables have diffusive fluctuations, that is, the variance of $S^{\lambda}_{1,\ell} - S^{\rho}_{1,\ell}$ is of order ℓ.

Define a constant

$$b = c_2 a_0 + 1 + C,$$

with c_2 from (5.33) and $C = C(a_1, \rho)$ from line (5.40) above. By the assumption $\kappa_N \leqslant a_0 N^{2/3}$ this ensures $bN^{2/3} \geqslant c_2 \kappa_N$, which lets us use bound (5.40) for integers $\ell \geqslant bN^{2/3}$. By adjusting the constant C in the front of (5.40) we can apply the bound to all real $\ell \geqslant bN^{2/3}$. Note first below that $\tau^{\rho}_1 \leqslant m$ because we are considering the point-to-point LPP process in the rectangle $[0, (m,n)]$. Then get an upper bound by increasing the upper integration limit to ∞.

$$\begin{aligned}\mathbb{E}[\tau^{\rho}_1] &= \int_0^m \mathbb{P}(\tau^{\rho}_1 \geqslant s)\, ds \;\leqslant\; bN^{2/3} + C\int_{bN^{2/3}}^{\infty} \Bigl(\frac{N^2}{s^3} + \frac{N^2}{s^4}\mathbb{E}[\tau^{\rho}_1]\Bigr)\, ds \\ &= bN^{2/3} + \frac{CN^{2/3}}{2b^2} + \frac{C}{3b^3}\mathbb{E}[\tau^{\rho}_1] \;\leqslant\; bN^{2/3} + \tfrac{1}{2}N^{2/3} + \tfrac{1}{3}\mathbb{E}[\tau^{\rho}_1].\end{aligned}$$

From this we obtain the bound

$$\mathbb{E}[\tau^{\rho}_1] \leqslant \bigl(c_2(\rho)a_0 + C_1(a_1,\rho)\bigr)N^{2/3}$$

and thereby (5.31) has been proved for $q = 1$. Substituting this bound back into line (5.40) gives

$$\mathbb{P}\{\tau^{\rho}_1 \geqslant \ell\} \leqslant C_2\Bigl(\frac{N^2}{\ell^3} + (1+a_0)\frac{N^{8/3}}{\ell^4}\Bigr)$$

for a constant $C_2 = C_2(a_1,\rho)$, verifying (5.30). Another integration with $b = c_2(\rho)a_0 + 1 + C_2$ proves (5.31) for $1 < q < 3$:

$$\begin{aligned}\mathbb{E}[(\tau^{\rho}_1)^q] &= \int_0^{\infty} \mathbb{P}(\tau^{\rho}_1 \geqslant s)\, qs^{q-1}\, ds \\ &\leqslant b^q N^{2q/3} + C_2\int_{bN^{2/3}}^{\infty} \bigl(N^2 s^{q-4} + (1+a_0)N^{8/3}s^{q-5}\bigr)\, ds \\ &= b^q N^{2q/3} + \frac{C_2 b^{q-3}}{3-q}N^{2q/3} + \frac{C_2}{b^{4-q}}(1+a_0)N^{2q/3} \\ &\leqslant \Bigl(c_2(\rho)a_0^q + \frac{C_3}{3-q}\Bigr)N^{2q/3}\end{aligned}$$

where we summarized the (a_1,ρ)-dependent constants into $C_3 = C_3(a_1,\rho)$. Note that a_0 acquired a power q because it is contained in b.

This completes the proof of Proposition 5.9, with c_3 defined as the constant that appears in front of the right-hand sides of (5.30)–(5.31). □

5.5. Lower bound for the passage time exponent. This lower bound proof is adapted to the corner growth model from the one given in [**11**] for the asymmetric simple exclusion process (ASEP).

The parameter $0 < \rho < 1$ of the increment-stationary LPP process is fixed. Let N be the scaling parameter that is sent to infinity and (m,n) the endpoint of the point-to-point LPP process. As before, we measure deviation from the characteristic direction of ρ by $\kappa_N = |m - N(1-\rho)^2| \vee |n - N\rho^2|$. We prove the lower bounds on the right and left tail stated in the theorem below.

THEOREM 5.10. *Assume that for some constants $a_0, k_0 > 0$, we have $\kappa_N \leqslant a_0 N^{2/3}$ for $N \geqslant k_0$. Then we have the conclusions below.*

There exist constants $1 < a_1(\rho), a_2(\rho), N_0(\rho, s) < \infty$ such that, for $s \geqslant a_2(\rho)(1+a_0^2)$ and $N \geqslant N_0(\rho, s) \vee k_0$,

$$\mathbb{P}\{\omega : G^{\rho}_{0,(m,n)} \geqslant \mathbb{E}[G^{\rho}_{0,(m,n)}] + sN^{1/3}\} \geqslant e^{-a_1(\rho)s^{3/2}}. \tag{5.41}$$

Furthermore, there exist constants $0 < t = t(\rho, a_0), a_3(\rho, a_0), N_1(\rho) < \infty$ such that for $N \geqslant N_1(\rho) \vee k_0$,

$$\mathbb{P}\{\omega : G^{\rho}_{0,(m,n)} \leqslant \mathbb{E}[G^{\rho}_{0,(m,n)}] - tN^{1/3}\} \geqslant a_3(\rho, a_0)t^2. \tag{5.42}$$

From (5.41) we derive the lower variance bound in (5.12):

$$\mathbb{V}\text{ar}[G^{\rho}_{0,(m,n)}] = \mathbb{E}\big[\big(G_{0,(m,n)} - \mathbb{E}^{\rho}[G_{0,(m,n)}]\big)^2\,\big] \geqslant s^2 N^{2/3} \cdot e^{-a_1(\rho)s^{3/2}}.$$

Setting $s = a_2(\rho)(1 + a_0^2)$ gives the lower bound in (5.13).

In this proof also we perturb the parameter of the boundary weights. Introduce a quantity $r > 0$ which, in the end, will be a constant multiple of $s^{1/2}$ in (5.41). Define another parameter for the increment-stationary CGM by

$$\lambda = \rho + \frac{r}{N^{1/3}}.$$

To guarantee that $\lambda \in (\rho, \frac{1+\rho}{2})$ we assume that

$$N \geqslant N_0 = N_0(\rho, r) = 8\big(\tfrac{r}{1-\rho}\big)^3.$$

N_0 can increase during the proof, but remains a function of ρ and r.

Notational comment. In this section we argue by switching measures on the space of weights, in addition to coupling different weight configurations. Hence it is convenient to attach the parameters ρ and λ to the measure $\mathbb{P}$ and the expectation $\mathbb{E}$ and variance $\mathbb{V}$ar to indicate which distribution is placed on the weights. We denote all the weights by ω_x. So for example under the probability measure $\mathbb{P}^{\rho}_{0,(m,n)}$ the weights on the rectangle $[0, (m, n)]$ have distributions as in (5.8) but without the I and J notation, namely

$$\begin{aligned} &\omega_x \sim \text{Exp}(1) \text{ for bulk vertices } x \in \mathbb{Z}^2_{>0}, \\ &\omega_{ie_1} \sim \text{Exp}(1-\rho), \text{ and } \omega_{je_2} \sim \text{Exp}(\rho). \end{aligned} \tag{5.43}$$

The point-to-point LPP process is defined by

$$G_{0,(m,n)} = \max_{x_{0,m+n} \in \Pi_{0,(m,n)}} \sum_{k=1}^{m+n} \omega_{x_k} \tag{5.44}$$

with maximum over paths that satisfy $x_0 = 0$ and $x_{m+n} = (m, n)$. △

For $N \in \mathbb{Z}_{>0}$ and $r > 0$ define the event

$$A_{N,r} = \big\{(1-\rho)rN^{2/3} \leqslant \tau_1 \leqslant 4\rho^{-1}rN^{2/3}\big\}. \tag{5.45}$$

Variable τ_1 is the exit point from the e_1-axis of the maximizing path from 0 to (m, n) in (5.44), defined as in (5.17). We develop a lower bound for the probability of $A_{N,r}$ under $\mathbb{P}^{\lambda}_{0,(m,n)}$, that is, for the increment-stationary process with parameter λ, restricted to the rectangle $[0, (m, n)]$. Note that this rectangle is *not* of the characteristic shape for λ. We take advantage of this in the proof.

LEMMA 5.11. *There exists a constant $C_1 = C_1(\rho)$ such that for*

$$r \geqslant 1 \vee \frac{2a_0}{\rho \wedge (1-\rho)} \tag{5.46}$$

and for $N \geqslant N_0(\rho, r)$ we have this bound:

$$\mathbb{P}^{\lambda}_{0,(m,n)}(A_{N,r}) \geqslant 1 - C_1 r^{-3}. \tag{5.47}$$

PROOF. We derive first an upper bound for $\mathbb{P}^{\lambda}_{0,(m,n)}\{\tau_1 > 4\rho^{-1} r N^{2/3}\}$. Define a coordinate $\widetilde{m}$ and bound it as below:

$$\begin{aligned}\widetilde{m} = \lfloor n\lambda^{-2}(1-\lambda)^2 \rfloor &\geqslant N\rho^2\lambda^{-2}(1-\lambda)^2 - a_0\lambda^{-2}(1-\lambda)^2 N^{2/3} - 1 \\ &\geqslant N\rho^2\lambda^{-2}(1-\lambda)^2 - a_0\rho^{-2}(1-\rho)^2 N^{2/3} - 1.\end{aligned}$$

This choice is made so that $(\widetilde{m}, n)$ points in the characteristic direction of λ, up to a constant error from integer parts. We have the bound

$$\begin{aligned}0 < m - \widetilde{m} &\leqslant N\big((1-\rho)^2 - \rho^2\lambda^{-2}(1-\lambda)^2\big) + a_0\big(1 + \rho^{-2}(1-\rho)^2\big)N^{2/3} + 1 \\ &\leqslant N\frac{\lambda + \rho - 2\lambda\rho}{\lambda^2}(\lambda - \rho) + a_0\rho^{-2}N^{2/3} + 1 \\ &\leqslant 2\rho^{-1} r N^{2/3} + a_0\rho^{-2}N^{2/3}\end{aligned}$$

for $N \geqslant N_0(\rho, r)$, for a suitably chosen $N_0(\rho, r)$. By Lemma A.2 in the appendix, and then by the upper bound (5.30), taking into consideration (5.46),

$$\begin{aligned}\mathbb{P}^{\lambda}_{0,(m,n)}\{\tau_1 > 4\rho^{-1} r N^{2/3}\} &= \mathbb{P}^{\lambda}_{0,(\widetilde{m},n)}\{\tau_1 > 4\rho^{-1} r N^{2/3} - (m - \widetilde{m})\} \\ &\leqslant \mathbb{P}^{\lambda}_{0,(\widetilde{m},n)}\{\tau_1 > \rho^{-1} r N^{2/3}\} \leqslant \frac{c_4}{r^3}\end{aligned}$$

where $c_4 = c_4(\rho)$ contains c_3 from (5.30).

Next we derive an upper bound for $\mathbb{P}^{\lambda}_{0,(m,n)}\{\tau_1 < (1-\rho) r N^{2/3}\}$. Let

$$(\bar{m}, \bar{n}) = \big(\lfloor N(1-\lambda)^2 \rfloor, \lfloor N\lambda^2 \rfloor\big) \tag{5.48}$$

point in the characteristic direction λ. Bound these differences:

$$\begin{aligned}m - \bar{m} &\geqslant N((1-\rho)^2 - (1-\lambda)^2) - a_0 N^{2/3} - 1 \\ &= N(\lambda - \rho)(2 - \rho - \lambda) - a_0 N^{2/3} - 1 \\ &= (2 - \rho - \lambda) r N^{2/3} - a_0 N^{2/3} - 1 \geqslant (1-\rho) r N^{2/3}\end{aligned}$$

and

$$\bar{n} - n \geqslant N(\lambda^2 - \rho^2) - a_0 N^{2/3} - 1 = N(\lambda - \rho)(\rho + \lambda) - a_0 N^{2/3} - 1 \geqslant \rho r N^{2/3},$$

again for large enough N relative to (ρ, r). By Lemma A.3 in the appendix, and then by the upper bound (5.30),

$$\begin{aligned}\mathbb{P}^{\lambda}_{0,(m,n)}\big\{\tau_1 < (1-\rho) r N^{2/3}\big\} &\leqslant \mathbb{P}^{\lambda}_{0,(m,n)}\big\{\tau_1 < m - \bar{m}\big\} \\ &= \mathbb{P}^{\lambda}_{0,(\bar{m},\bar{n})}\big\{\tau_2 > \bar{n} - n\big\} \\ &\leqslant \mathbb{P}^{\lambda}_{0,(\bar{m},\bar{n})}\big\{\tau_2 > \rho r N^{2/3}\big\} \leqslant \frac{c_5}{r^3}\end{aligned} \tag{5.49}$$

where $c_5 = c_5(\rho)$ contains c_3 from (5.30).

Combine the bounds:

$$\begin{aligned}\mathbb{P}^{\lambda}_{0,(m,n)}&\big\{(1-\rho)rN^{2/3} \leqslant \tau_1 \leqslant 4\rho^{-1}rN^{2/3}\big\}\\&= 1 - \mathbb{P}^{\lambda}_{0,(m,n)}\{\tau_1 > 4\rho^{-1}rN^{2/3}\} - \mathbb{P}^{\lambda}_{0,(m,n)}\{\tau_1 < (1-\rho)rN^{2/3}\}\\&\geqslant 1 - C_1 r^{-3}.\end{aligned}$$

∎

This completes the proof.

For the subsequent parts of the proof we could have defined $A_{N,r} = \{1 \leqslant \tau_1 \leqslant 4\rho^{-1}rN^{2/3}\}$ instead of with the tighter lower bound in (5.45). But we still would have had to prove an estimate such as (5.49).

Following the first four lines of the computation in (5.38),

$$\begin{aligned}\mathbb{E}^{\lambda}[G_{0,(m,n)}] - \mathbb{E}^{\rho}[G_{0,(m,n)}] &= m\Big(\frac{1}{1-\lambda} - \frac{1}{1-\rho}\Big) + n\Big(\frac{1}{\lambda} - \frac{1}{\rho}\Big)\\&= \frac{N}{\lambda(1-\lambda)}(\lambda-\rho)^2 + \left\{ \frac{\kappa_N^1}{(1-\lambda)(1-\rho)} - \frac{\kappa_N^2}{\lambda\rho} \right\}(\lambda-\rho)\\&\geqslant \frac{r^2N^{1/3}}{\lambda(1-\lambda)} - \frac{a_0 rN^{1/3}}{C(\rho)} \geqslant c_6 r^2 N^{1/3}\end{aligned} \tag{5.50}$$

where $c_6 = c_6(\rho) > 0$ is a constant chosen small enough to satisfy the inequality above for all $N \geqslant N_0$ and $r \geqslant b(\rho)a_0$ where $b(\rho)$ is a constant that depends on ρ.

Let Λ_N denote the set of paths $x_\bullet \in \Pi_{0,(m,n)}$ that satisfy $x_1 = e_1$ and $x_k \cdot e_2 > 0$ for $k > \lfloor 4\rho^{-1}rN^{2/3}\rfloor$. In other words, the path stays on the e_1-axis for a while after leaving the origin, but not beyond the point $\lfloor 4\rho^{-1}rN^{2/3}\rfloor e_1$. For any given weights $\{\omega_x\}$ on the rectangle $\{0,\dots,m\} \times \{0,\dots,n\}$, let

$$G_{0,(m,n)}(\Lambda_N) = \max_{x_\bullet \in \Lambda_N} \sum_{k=1}^{m+n} \omega_{x_k} \tag{5.51}$$

denote the LPP value whose maximum is restricted to paths in Λ_N. Observe that $G_{0,(m,n)}(\Lambda_N) = G_{0,(m,n)}$ if event $A_{N,r}$ of (5.45) occurs for weights $\{\omega_x\}$. (This would be true even if the lower bound in $A_{N,r}$ would be relaxed to $\tau_1 \geqslant 1$ instead of $\tau_1 \geqslant (1-\rho)rN^{2/3}$.)

We derive our second probability bound. Define the event

$$B_{N,r} = \big\{\omega : G_{0,(m,n)}(\Lambda_N) \geqslant \mathbb{E}^{\rho}[G_{0,(m,n)}] + \tfrac{1}{2}c_6 r^2 N^{1/3}\big\} \tag{5.52}$$

where $c_6 = c_6(\rho)$ is the constant from (5.50).

LEMMA 5.12. *There exists a constant $C_2 = C_2(\rho)$ such that the bound below holds for $r \geqslant b(\rho)a_0 \vee 1$ and $N \geqslant N_0$:*

$$\mathbb{P}^{\lambda}_{0,(m,n)}(B_{N,r}) \geqslant 1 - C_2 r^{-3}. \tag{5.53}$$

PROOF. Since by (5.50) for $r \geqslant b(\rho)a_0$,

$$\mathbb{E}^{\rho}[G_{0,(m,n)}] + \tfrac{1}{2}c_6 r^2 N^{1/3} \leqslant \mathbb{E}^{\lambda}[G_{0,(m,n)}] - \tfrac{1}{2}c_6 r^2 N^{1/3},$$

we can bound the complementary probability as follows. Constant C changes from line to line. Below we use Lemma 5.7.

$$\begin{aligned}
\mathbb{P}^{\lambda}_{0,(m,n)}(B^c_{N,r}) &= \mathbb{P}^{\lambda}_{0,(m,n)}\{G_{0,(m,n)}(\Lambda_N) < \mathbb{E}^{\rho}[G_{0,(m,n)}] + \tfrac{1}{2}c_6 r^2 N^{1/3}\} \\
&\leqslant \mathbb{P}^{\lambda}_{0,(m,n)}\{G_{0,(m,n)}(\Lambda_N) < \mathbb{E}^{\lambda}[G_{0,(m,n)}] - \tfrac{1}{2}c_6 r^2 N^{1/3}\} \\
&\leqslant \mathbb{P}^{\lambda}_{0,(m,n)}\{G_{0,(m,n)} < \mathbb{E}^{\lambda}[G_{0,(m,n)}] - \tfrac{1}{2}c_6 r^2 N^{1/3}\} + \mathbb{P}^{\lambda}_{0,(m,n)}(A^c_{N,r}) \\
&\leqslant \frac{C}{r^4 N^{2/3}} \mathbb{V}\mathrm{ar}^{\lambda}[G_{0,(m,n)}] + \frac{C_1}{r^3} \\
&\leqslant \frac{C}{r^4 N^{2/3}} \big(\mathbb{V}\mathrm{ar}^{\rho}[G_{0,(m,n)}] + m(\lambda-\rho)\big) + \frac{C_1}{r^3} \\
&\leqslant \frac{C}{r^4} + \frac{C_1}{r^3} \leqslant \frac{C_2}{r^3}. \qquad \square
\end{aligned}$$

With the preliminary work done, we turn to the proof of Theorem 5.10.

PROOF OF THEOREM 5.10. We construct a coupling of three environments. Let ω^ρ and ω^λ denote environments as described in (5.43) with parameters ρ and λ. We assume that these environments are coupled so that in the bulk, for $x \in \mathbb{Z}^2_{>0}$, $\omega^\rho_x = \omega^\lambda_x = \omega_x$, while the boundary variables $\{\omega^\rho_{ie_1}, \omega^\rho_{je_2}, \omega^\lambda_{ie_1}, \omega^\lambda_{je_2} : i, j \in \mathbb{Z}_{>0}\}$ are mutually independent.

Construct a mixed environment $\widehat{\omega}$ as follows:

$$\begin{aligned}
&\widehat{\omega}_{ie_1} = \omega^\lambda_{ie_1} \quad \text{for} \quad 1 \leqslant i \leqslant \lfloor 4\rho^{-1} r N^{2/3} \rfloor \\
\text{and} \qquad &\widehat{\omega}_x = \omega^\rho_x \quad \text{for} \quad x \notin \{ie_1 : 1 \leqslant i \leqslant \lfloor 4\rho^{-1} r N^{2/3} \rfloor\}.
\end{aligned}$$

Thus in the bulk all weights agree and are i.i.d. Exp(1): for $x \in \mathbb{Z}^2_{>0}$, $\widehat{\omega}_x = \omega^\rho_x = \omega^\lambda_x = \omega_x$. On the boundary $\widehat{\omega}$ follows ω^λ on the segment that is relevant for the event $B_{N,r}$ and elsewhere $\widehat{\omega}$ follows ω^ρ. Note that $\omega^\lambda \in B_{N,r}$ iff $\widehat{\omega} \in B_{N,r}$.

Let the distributions of the three environments ω^ρ, ω^λ and $\widehat{\omega}$, restricted to rectangle $\{0, \ldots, m\} \times \{0, \ldots, n\}$, be denoted by $\mathbb{P}^\rho_{0,(m,n)}$, $\mathbb{P}^\lambda_{0,(m,n)}$ and $\widehat{\mathbb{P}}_{0,(m,n)}$, respectively. These are all probability measures on the product space $\mathbb{R}_+^{\{0,\ldots,m\}\times\{0,\ldots,n\}}$. The Radon-Nikodym derivative

$$f_N(\omega) = \frac{d\widehat{\mathbb{P}}_{0,(m,n)}}{d\mathbb{P}^\rho_{0,(m,n)}}(\omega) = \prod_{i=1}^{\lfloor 4\rho^{-1} r N^{2/3} \rfloor} \frac{\lambda}{\rho} e^{-(\lambda-\rho)\omega_{ie_1}}$$

is a product of the Radon-Nikodym derivatives of the exponential single weight marginal distributions on that segment of the boundary where ω^ρ and $\widehat{\omega}$ differ. Computation of the mean square gives

$$\begin{aligned}
\mathbb{E}^\rho_{0,(m,n)}[f_N^2] &= \left(\frac{\lambda^2}{\rho^2}\int_0^\infty e^{-2(\lambda-\rho)s} \rho e^{-\rho s}\, ds\right)^{\lfloor 4\rho^{-1} r N^{2/3} \rfloor} = \left(\frac{\lambda^2}{\rho(2\lambda-\rho)}\right)^{\lfloor 4\rho^{-1} r N^{2/3} \rfloor} \\
&= \exp\left\{\lfloor 4\rho^{-1} r N^{2/3} \rfloor \left[2\log\Big(1+\frac{r}{\rho N^{1/3}}\Big) - \log\Big(1+\frac{2r}{\rho N^{1/3}}\Big)\right]\right\} \\
&\leqslant e^{4r^3\rho^{-3}}.
\end{aligned}$$

The point is that the bound above is independent of N.

Consider $r \geqslant b(\rho)a_0 \vee 1$ and large enough relative to $C_2 = C_2(\rho)$ from (5.53) so that $C_2 r^{-3} < 1/2$.

$$\begin{aligned}\tfrac{1}{2} < 1 - C_2 r^{-3} &\leqslant \mathbb{P}^{\lambda}_{0,(m,n)}(B_{N,r}) = \widehat{\mathbb{P}}_{0,(m,n)}(B_{N,r}) = \mathbb{E}^{\rho}_{0,(m,n)}[\mathbb{1}_{E_{N,r}} f_N] \\ &\leqslant \{\mathbb{P}^{\rho}_{0,(m,n)}(B_{N,r})\}^{1/2}\{\mathbb{E}^{\rho}_{0,(m,n)}[f_N^2]\}^{1/2} \\ &\leqslant \{\mathbb{P}^{\rho}_{0,(m,n)}(B_{N,r})\}^{1/2} e^{2r^3\rho^{-3}}.\end{aligned}$$

Since $G_{0,(m,n)} \geqslant G_{0,(m,n)}(\Lambda_N)$, from this comes the lower bound

$$\begin{aligned}&\mathbb{P}^{\rho}_{0,(m,n)}\{\omega : G_{0,(m,n)} \geqslant \mathbb{E}^{\rho}[G_{0,(m,n)}] + \tfrac{1}{2}c_6 r^2 N^{1/3}\} \\ &\qquad \geqslant \mathbb{P}^{\rho}_{0,(m,n)}\{\omega : G_{0,(m,n)}(\Lambda_N) \geqslant \mathbb{E}^{\rho}[G_{0,(m,n)}] + \tfrac{1}{2}c_6 r^2 N^{1/3}\} \\ &\qquad = \mathbb{P}^{\rho}_{0,(m,n)}(B_{N,r}) \geqslant \tfrac{1}{4}e^{-4r^3\rho^{-3}}\end{aligned} \tag{5.54}$$

To complete the proof of inequality (5.41) of Theorem 5.10, set $s = \frac{1}{2}c_6 r^2$ and let $a_1(\rho), a_2(\rho)$ be suitable functions of ρ, c_6 and C_2.

To prove the second inequality (5.42), abbreviate temporarily $X = G_{0,(m,n)} - \mathbb{E}^{\rho}[G_{0,(m,n)}]$ and first derive this estimate from inequality (5.41):

$$0 = \mathbb{E}^{\rho}[X] = \mathbb{E}^{\rho}[X^+] - \mathbb{E}^{\rho}[X^-] \geqslant sN^{1/3}e^{-a_1(\rho)s^{3/2}} - \mathbb{E}^{\rho}[X^-] = 2tN^{1/3} - \mathbb{E}^{\rho}[X^-]$$

where we set

$$t = \tfrac{1}{2}s\,e^{-a_1(\rho)s^{3/2}}$$

and then fix $s \geqslant a_2(\rho)(1 + a_0^2)$ to maximize t. Next,

$$\begin{aligned}2tN^{1/3} \leqslant \mathbb{E}^{\rho}[X^-] &= \mathbb{E}^{\rho}[X^-,\, X^- < tN^{1/3}] + \mathbb{E}^{\rho}[X^-,\, X^- \geqslant tN^{1/3}] \\ &\leqslant tN^{1/3} + \big(\mathbb{E}^{\rho}[(X^-)^2]\big)^{1/2}\big(\mathbb{P}^{\rho}\{X^- \geqslant tN^{1/3}\}\big)^{1/2} \\ &\leqslant tN^{1/3} + \big(\mathbb{V}\mathrm{ar}^{\rho}[G_{0,(m,n)}]\big)^{1/2}\big(\mathbb{P}^{\rho}\{X^- \geqslant tN^{1/3}\}\big)^{1/2}.\end{aligned}$$

The above and the upper variance bound from Theorem 5.1 give, for a positive constant $a_3(\rho, a_0)$,

$$\mathbb{P}^{\rho}\{X^- \geqslant tN^{1/3}\} \geqslant a_3(\rho, a_0)t^2.$$

This inequality is the same as (5.42). This completes the proof of Theorem 5.10. □

5.6. Bounds for path fluctuations. In this section we prove Theorem 5.3 for fluctuations of the maximizing path $\pi_{\bullet}^{0,(m,n)}$ of the last-passage value $G^{\rho}_{0,(m,n)}$. The parameter $0 < \rho < 1$ is fixed and $(m,n) = (\lfloor N(1-\rho)^2 \rfloor, \lfloor N\rho^2 \rfloor)$.

Upper bound proof. Consider $t \in (0,1)$ fixed for a while. Let

$$v_0 = (m_0, n_0) = (\lfloor tm \rfloor, \lfloor tn \rfloor).$$

This is an integer point that lies immediately to the left and below the rectangle $A_{rN^{2/3}}(tm, tn)$, and coincides with the lower left corner (tm, tn) of $A_{rN^{2/3}}(tm, tn)$ if this point has integer coordinates. See Figure 5.1.

The proof utilizes the coupling described in Appendix A. As in Lemma A.1, let $G^{(0)}_{v_0,x}$ denote last-passage values for lattice points $x \in v_0 + \mathbb{Z}^2_{\geqslant 0}$ defined as in (3.13)–(3.14), using boundary weights $I_{v_0+ie_1}$ and $J_{v_0+je_2}$ for $i, j \geqslant 1$ and bulk weights ω_x

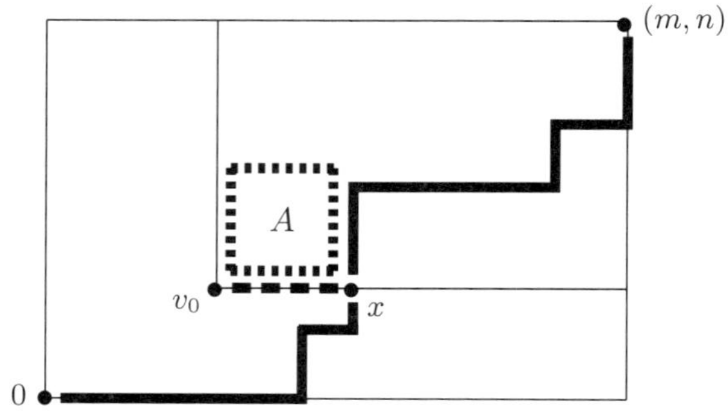

FIGURE 5.1. Illustration of the upper bound proof of Theorem 5.3. The maximizing path (solid thickset line) for $G^{\rho}_{0,(m,n)}$ goes through x and avoids the square $A = A_{rN^{2/3}}(tm, tn)$. By Lemma A.1, the maximizing path for $G^{(0)}_{v_0,(m,n)}$ goes from v_0 to (m,n) through x. The distance from v_0 to x equals $\tau_1^{(v_0)}$.

for $x \in v_0 + \mathbb{Z}^2_{>0}$. The bulk weights are the same as the ones used by $G^{\rho}_{0,(m,n)}$. The boundary edge weights are determined by the process $G^{\rho}_{0,\bullet}$ as in (3.19):

$$\begin{aligned} I_{v_0+ie_1} &= G^{\rho}_{0,\,v_0+ie_1} - G^{\rho}_{0,\,v_0+(i-1)e_1} \\ J_{v_0+je_2} &= G^{\rho}_{0,\,v_0+je_2} - G^{\rho}_{0,\,v_0+(j-1)e_2} \qquad \text{for } i,j \geqslant 1. \end{aligned}$$

Let $x_{\bullet}^{(v_0)*} \in \Pi_{v_0,x}$ denote the maximizing path for $G^{(0)}_{v_0,x}$. Analogously with (5.17), let $\tau_r^{(v_0)}$ denote the number of steps the maximizing path $x_{\bullet}^{(v_0)*}$ takes on the e_r-axis, relative to the origin v_0:

$$\tau_r^{(v_0)} = \max\{k \geqslant 0 : v_0 + ke_r \in x_{\bullet}^{(v_0)*}\}, \qquad r \in \{1,2\}.$$

Let $(m_1, n_1) = (m,n) - (m_0, n_0)$. This point satisfies

$$|\, m_1 - (1-t)N(1-\rho)^2 \,| \vee |\, n_1 - (1-t)N\rho^2 \,| \leqslant 1. \tag{5.55}$$

Due to the stationarity of the LPP process, we have the distributional equalities

$$\mathbb{P}^{\rho}_{0,(m,n)}\{\tau_r^{(v_0)} \geqslant j\} = \mathbb{P}^{\rho}_{0,(m_1,n_1)}\{\tau_r \geqslant j\}. \tag{5.56}$$

We are ready to prove the upper bound (5.14). If $\pi_{\bullet}^{0,(m,n)}$ does not intersect $A_{rN^{2/3}}(tm,tn)$, then the edge along which $\pi_{\bullet}^{0,(m,n)}$ enters the quadrant $(m_0,n_0) + \mathbb{Z}^2_{>0}$ is either $((k,n_0),(k,n_0+1))$ for some $k \geqslant m_0 + rN^{2/3}$ or $((m_0,\ell),(m_0+1,\ell))$ for some $\ell \geqslant n_0 + rN^{2/3}$. Lemma A.1 forces the occurrence of

$$\tau_1^{(v_0)} \geqslant rN^{2/3} \quad \text{or} \quad \tau_2^{(v_0)} \geqslant rN^{2/3}. \tag{5.57}$$

We apply the upper bound (5.30) of Proposition 5.9 to the right-hand side of (5.56). By (5.55) the scaling parameter "N" of Proposition 5.9 is now $N' = (1-t)N$. To ensure that $N' \geqslant 1$ we assume $N \geqslant (1-t)^{-1}$. Then we can take $N_0' = 1$, $a_0 = 1$ and $a_1 = 2$ in Proposition 5.9. By (5.30), for $r \geqslant 1$ the event in (5.57) has probability at most $C(\rho)r^{-3}$. This bound is uniform over $t \in (0,1)$, as long as $N \geqslant (1-t)^{-1}$. We have verified the upper bound (5.14).

Lower bound proof. Fix $t \in (0,1)$. Take N large enough so that $(m,n) \in \mathbb{Z}^2_{>0}$. This time put

$$v_0 = (m_0, n_0) = (\lceil tm \rceil - 1, \lceil tn \rceil - 1) \in \mathbb{Z}^2_{\geqslant 0}$$

so that v_0 is strictly below and to the left of the rectangle $A_{\delta N^{2/3}}(tm, tn)$. Let again $(m_1, n_1) = (m,n) - (m_0, n_0)$. By Lemma 5.8, variance identity (5.18), and the lower variance bound in (5.12), with a constant $C = C(\rho)$ that may vary,

$$\begin{aligned} \mathbb{E}^{\rho}_{0,(m,n)}[\tau_1^{(v_0)}] &= \mathbb{E}^{\rho}_{0,(m_1,n_1)}[\tau_1] \geqslant C^{-1}\mathbb{E}^{\rho}_{0,(m_1,n_1)}[S_{1,\tau_1}] - 1 \\ &\geqslant C^{-1}\mathbb{V}\mathrm{ar}[G^{\rho}_{0,(m_1,n_1)}] - C \\ &\geqslant C^{-1}N^{2/3} - C \geqslant C^{-1}N^{2/3}. \end{aligned}$$

In the last step we take N large enough relative to $C(\rho)$. Next, combine the lower bound from above with the upper bound of (5.31) for $q=2$. Then for $\delta_1 > 0$, again with $C = C(\rho)$ and for $N \geqslant N_0(\rho)$,

$$\begin{aligned} C^{-1}N^{2/3} &\leqslant \mathbb{E}^{\rho}_{0,(m,n)}[\tau_1^{(v_0)}] = \mathbb{E}^{\rho}_{0,(m_1,n_1)}[\tau_1] \\ &= \mathbb{E}^{\rho}_{0,(m_1,n_1)}[\tau_1,\, \tau_1 \leqslant \delta_1 N^{2/3}] + \mathbb{E}^{\rho}_{0,(m_1,n_1)}[\tau_1,\, \tau_1 > \delta_1 N^{2/3}] \\ &\leqslant \delta_1 N^{2/3} + \big(\mathbb{E}^{\rho}_{0,(m_1,n_1)}[\tau_1^2]\big)^{1/2}\big(\mathbb{P}^{\rho}_{0,(m_1,n_1)}\{\tau_1 > \delta_1 N^{2/3}\}\big)^{1/2} \\ &\leqslant \delta_1 N^{2/3} + CN^{2/3}\big(\mathbb{P}^{\rho}_{0,(m_1,n_1)}\{\tau_1 > \delta_1 N^{2/3}\}\big)^{1/2}. \end{aligned}$$

From this we deduce that, for small enough δ_1 relative to the constant $C(\rho)$, there exists $\varepsilon_1 > 0$ such that

$$\mathbb{P}^{\rho}_{0,(m,n)}\{\tau_1^{(v_0)} > \delta_1 N^{2/3}\} \geqslant \varepsilon_1 \qquad \text{for all } N \geqslant N_0(\rho, t).$$

Note that N has to be taken large enough relative to t also so that $\delta_1 N^{2/3} < m_1$, for otherwise the event above is empty. Hence the dependence of $N_0(\rho,t)$ on t. A similar estimate works for $\tau_2^{(v_0)}$.

Now the proof of the lower bound (5.15). If $\pi_{\bullet}^{0,(m,n)}$ intersects $A_{\delta N^{2/3}}(tm,tn)$, then the path $\pi_{\bullet}^{0,(m,n)}$ enters $A_{\delta N^{2/3}}(tm,tn)$ either through the south side or the west side, and hence either $1 \leqslant \tau_1^{(v_0)} \leqslant \delta N^{2/3}$ or $1 \leqslant \tau_2^{(v_0)} \leqslant \delta N^{2/3}$. (Note that if v_0 were the lower left corner of $A_{\delta N^{2/3}}(tm,tn)$, we could not make the last assertion because the path could run along either boundary $v_0 + e_r\mathbb{Z}_{>0}$ and give $\tau_r^{(v_0)}$ a larger value.) Take $\delta < \delta_1$ and utilize the fact that exactly one of $\tau_1^{(v_0)}$ and $\tau_2^{(v_0)}$ is positive.

$$\begin{aligned} &\mathbb{P}^{\rho}_{0,(m,n)}\{\pi_{\bullet}^{0,(m,n)} \text{ intersects } A_{\delta N^{2/3}}(tm,tn)\} \\ &\quad\leqslant \mathbb{P}^{\rho}_{0,(m,n)}\{1 \leqslant \tau_1^{(v_0)} \leqslant \delta N^{2/3}\} + \mathbb{P}^{\rho}_{0,(m,n)}\{1 \leqslant \tau_2^{(v_0)} \leqslant \delta N^{2/3}\} \\ &\quad= 1 - \mathbb{P}^{\rho}_{0,(m,n)}\{\tau_1^{(v_0)} > \delta N^{2/3}\} - \mathbb{P}^{\rho}_{0,(m,n)}\{\tau_2^{(v_0)} > \delta N^{2/3}\} \leqslant 1 - 2\varepsilon_1. \end{aligned}$$

This completes the proof of Theorem 5.3.

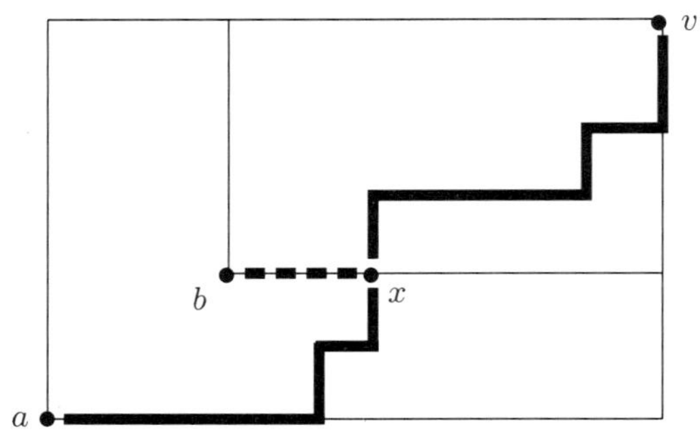

FIGURE A.1. Illustration of Lemma A.1. Path a-x-v is maximal for $G_{a,v}$ and path b-x-v is maximal for $G^{(a)}_{b,v}$.

Appendix A. Coupling the corner growth model

In this section we develop the couplings used in the proofs, beginning with a lemma for arbitrary deterministic weights. Fix a point $a \in \mathbb{Z}^2$. Suppose boundary weights $\{\omega_{a+ke_r} : k \in \mathbb{Z}_{>0}, r \in \{1,2\}\}$ on the south and west boundaries of the quadrant $a+\mathbb{Z}^2_{\geqslant 0}$ and bulk weights $\{\omega_x\}_{x\in a+\mathbb{Z}^2_{>0}}$ are given. Put an irrelevant weight $\omega_a = 0$ in the corner a. Let $G_{a,x}$ denote the LPP value for points $x \in a + \mathbb{Z}^2_{\geqslant 0}$, defined as in (3.1).

Let $b \geqslant a$ on $\mathbb{Z}^2$. On the lattice $b+\mathbb{Z}^2_{\geqslant 0}$, put a corner weight $\eta_b = 0$ and define boundary weights

$$\eta_{b+ke_r} = G_{a,\,b+ke_r} - G_{a,\,b+(k-1)e_r} \qquad \text{for } k \in \mathbb{Z}_{>0} \text{ and } r \in \{1,2\}. \tag{A.1}$$

In the bulk use $\eta_x = \omega_x$ for $x \in b + \mathbb{Z}^2_{>0}$. Denote the LPP process in $b+\mathbb{Z}^2_{\geqslant 0}$ that uses weights $\{\eta_x\}_{x\in b+\mathbb{Z}^2_{\geqslant 0}}$ by

$$G^{(a)}_{b,x} = \max_{x_\bullet \in \Pi_{b,x}} \sum_{i=0}^{|x-b|_1} \eta_{x_i}, \qquad x \in b + \mathbb{Z}^2_{\geqslant 0}. \tag{A.2}$$

The superscript (a) reminds us that the LPP process $G^{(a)}$ uses boundary weights determined by the LPP process $G_{a,\bullet}$ with lower left base point a. See Figure A.1 for an illustration of the next lemma.

LEMMA A.1. *Let $a \leqslant b \leqslant v$ in $\mathbb{Z}^2$. Then $G_{a,v} = G_{a,b} + G^{(a)}_{b,v}$. The restriction of any maximizing path for $G_{a,v}$ to $b+\mathbb{Z}^2_{\geqslant 0}$ is part of a maximizing path for $G^{(a)}_{b,v}$. The edges with one endpoint in $b+\mathbb{Z}^2_{>0}$ that belong to any particular maximizing path for $G^{(a)}_{b,v}$ extend to a maximizing path for $G_{a,b}$.*

PROOF. If $v = b + ke_r$ (that is, v is on the boundary of $b+\mathbb{Z}^2_{\geqslant 0}$) the situation is straightforward and we omit the details.

Suppose $v > b$ coordinatewise. Suppose a maximal path for $G_{a,v}$ enters the quadrant $b + \mathbb{Z}^2_{>0}$ through the edge $(x, y) = (b + ke_r, b + ke_r + e_{3-r})$. Suppose a maximal path for $G^{(a)}_{b,v}$ enters $b + \mathbb{Z}^2_{>0}$ through the edge $(\widetilde{x}, \widetilde{y}) = (b + \ell e_s, b + \ell e_s + e_{3-s})$. Then

$$\begin{aligned} G_{a,v} &= G_{a,x} + G_{y,v} = G_{a,b} + \sum_{i=1}^{k} \eta_{b+ie_r} + G_{y,v} \\ &\leqslant G_{a,b} + G^{(a)}_{b,v} = G_{a,b} + \sum_{i=1}^{\ell} \eta_{b+ie_s} + G_{\widetilde{y},v} \\ &= G_{a,\widetilde{x}} + G_{\widetilde{y},v} \leqslant G_{a,v}. \end{aligned}$$

Thus the inequalities above are in fact equalities. The claims of the lemma follow from this. □

We return to the exponential CGM. For lattice points $0 \leqslant v$, write $\mathbb{P}_{0,v}$ for the probability measure of the LPP process in the rectangle $[0, v]$ with boundary and bulk weights (5.8).

LEMMA A.2. *Let $1 \leqslant k < k + \ell \leqslant m$. Then*

$$\mathbb{P}_{0,(m,n)}(\tau_1 \geqslant k + \ell) = \mathbb{P}_{0,(m-k,n)}(\tau_1 \geqslant \ell).$$

PROOF. Take $a = 0$, $b = (k, 0)$ and $v = (m, n)$ in Lemma A.1. Under $\mathbb{P}_{0,(m,n)}$, the LPP process $G^{(a)}_{b,x}$ in $[b, v]$ has the same distribution, modulo the translation of the origin to b, as an LPP process under $\mathbb{P}_{0,(m-k,n)}$. By Lemma A.1 the maximizing paths from a and b to v agree in their portions inside $[k + 1, m] \times [0, n]$. □

LEMMA A.3. *Let $1 \leqslant \bar{m} < m$ and $1 \leqslant n < \bar{n}$. Then $\mathbb{P}_{0,(m,n)}(\tau_1 < m - \bar{m}) = \mathbb{P}_{0,(\bar{m},\bar{n})}(\tau_2 > \bar{n} - n)$.*

PROOF. We couple these LPP processes as follows. Let

$$a = (\bar{m} - m, 0), \quad a' = (0, n - \bar{n}) \quad \text{and} \quad v = (\bar{m}, n).$$

The origin 0 takes the role of b in Lemma A.1.

Let i.i.d. Exp(1) weights $\{\omega_x\}_{x \in \mathbb{Z}^2}$ be given. Then place independent boundary edge weights with distributions dictated by (5.8) on the south and west boundaries of the lattice region $(a + \mathbb{Z}^2_{\geqslant 0}) \cup (a' + \mathbb{Z}^2_{\geqslant 0})$:

(a) On horizontal boundary edges put Exp$(1 - \rho)$ weights $\sigma_{(i-1)e_1, ie_1}$ for $\bar{m} - m + 1 \leqslant i \leqslant 0$ and $\sigma_{a'+(i-1)e_1, a'+ie_1}$ for $i \in \mathbb{Z}_{>0}$.

(b) On vertical boundary edges put Exp(ρ) weights $\sigma_{(j-1)e_2, je_2}$ for $n - \bar{n} + 1 \leqslant j \leqslant 0$ and $\sigma_{a+(j-1)e_2, a+je_2}$ for $j \in \mathbb{Z}_{>0}$.

Next consider two LPP processes that emanate from a and a' and use the boundary weights described above in (a) and (b): $G_{a,\ell e_2}$ for points ℓe_2, $\ell \geqslant 1$, on the y-axis, and G_{a',ke_1} for points ke_1, $k \geqslant 1$, on the x-axis. Let these processes define boundary weights on $\mathbb{Z}^2_{\geqslant 0}$: $\eta_{(i-1)e_1, ie_1} = G_{a', ie_1} - G_{a',(i-1)e_1}$ and $\eta_{(j-1)e_2, je_2} = G_{a, je_2} - G_{a,(j-1)e_2}$ for $i, j \in \mathbb{Z}_{>0}$. See Figure A.2 for an illustration of the setting described here.

Now consider three LPP processes with lower left corners a, 0 and a':

(i) $\widetilde{G}_{a,x}$ uses boundary weights $\sigma_{(i-1)e_1, ie_1}$ for $\bar{m} - m + 1 \leqslant i \leqslant 0$ and $\eta_{(i-1)e_1, ie_1}$ for $i \in \mathbb{Z}_{>0}$ on the horizontal axis emanating from a, and boundary weights $\sigma_{a+(j-1)e_2, a+je_2}$ for $j \in \mathbb{Z}_{>0}$ on the vertical axis emanating from a.

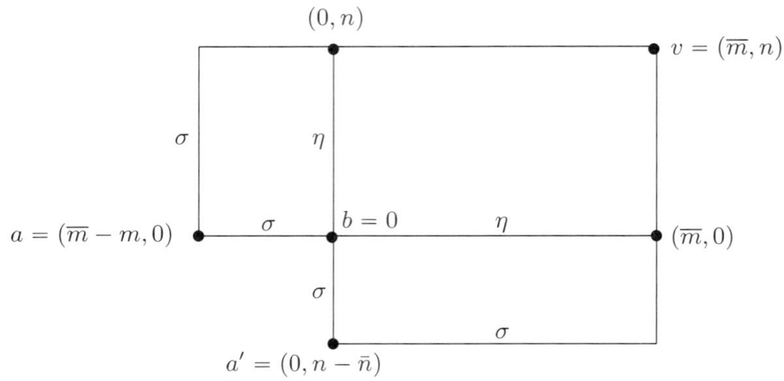

FIGURE A.2. Illustration of the proof of Lemma A.3. The η and σ labels on the south and west boundaries of rectangles indicate which boundary edge weights are placed on that side.

(ii) $\widetilde{G}_{0,x}$ uses boundary weights $\eta_{(i-1)e_1,\,ie_1}$ and $\eta_{(j-1)e_2,\,je_2}$ on the standard axes emanating from 0.
(iii) $\widetilde{G}_{a',x}$ uses boundary weights $\sigma_{a'+(i-1)e_1,\,a'+ie_1}$ for $i \in \mathbb{Z}_{>0}$ on the horizontal axis emanating from a', and boundary weights $\sigma_{(j-1)e_2,\,je_2}$ for $n-\bar{n}+1 \leqslant j \leqslant 0$ and $\eta_{(j-1)e_2,\,je_2}$ for $j \in \mathbb{Z}_{>0}$ on the vertical axis emanating from a'.

Let $\widetilde{\mathbb{P}}$ denote the probability measure under which this coupling has been constructed, that is, the probability measure of the independent weights ω_x and $\sigma_{x,\,x+e_k}$.

Let A be the event that the (a.s. unique) maximal path for $\widetilde{G}_{a,v}$ does not go through the origin. Let B the event that the (a.s. unique) maximal path for $\widetilde{G}_{a',v}$ goes through the point e_2. Lemma A.1 applies to the pair $\widetilde{G}_{a,v}$ and $\widetilde{G}_{0,v}$, and also to the pair $\widetilde{G}_{a',v}$ and $\widetilde{G}_{0,v}$. Thus the maximizing paths for $\widetilde{G}_{a,v}$ and $\widetilde{G}_{a',v}$ agree from that edge onwards through which they enter the positive first quadrant $\mathbb{Z}^2_{>0}$. Both A and B are equivalent to the statement that this edge emanates from some point ℓe_2 for $\ell \geqslant 1$. Hence $A = B$.

On the other hand, LPP processes $\{\widetilde{G}_{a,\,a+x}\}_{x\in\mathbb{Z}^2_{\geqslant 0}}$ and $\{\widetilde{G}_{a',\,a'+x}\}_{x\in\mathbb{Z}^2_{\geqslant 0}}$ both have the same distribution as the LPP process $\{G^{\rho}_{0,x}\}_{x\in\mathbb{Z}^2_{\geqslant 0}}$ with stationary increments. Event A is equivalent to the condition that the maximizing path for $\widetilde{G}_{a,v}$ takes at most $m-\bar{m}-1$ consecutive e_1-steps from a, which is the same as $\tau_1 < m-\bar{m}$ for $G^{\rho}_{0,(m,n)}$. Similarly, event B says that the maximizing path for $\widetilde{G}_{a',v}$ takes at least $\bar{n}-n+1$ consecutive e_2-steps from a', which for $G^{\rho}_{0,(\bar{m},\bar{n})}$ is the same as $\tau_2 > \bar{n}-n$. Thus

$$\mathbb{P}_{0,(m,n)}(\tau_1 < m-\bar{m}) = \widetilde{\mathbb{P}}(A) = \widetilde{\mathbb{P}}(B) = \mathbb{P}_{0,(\bar{m},\bar{n})}(\tau_2 > \bar{n}-n). \qquad \square$$

References

[1] Christopher Hoffman, *Geodesics in first passage percolation*, Ann. Appl. Probab. **18** (2008), no. 5, 1944–1969, DOI 10.1214/07-AAP510. MR2462555

[2] D. Aldous and P. Diaconis, *Hammersley's interacting particle process and longest increasing subsequences*, Probab. Theory Related Fields **103** (1995), no. 2, 199–213, DOI 10.1007/BF01204214. MR1355056

[3] Sven Erick Alm, *A note on a problem by Welsh in first-passage percolation*, Combin. Probab. Comput. **7** (1998), no. 1, 11–15, DOI 10.1017/S0963548397003301. MR1611098

[4] Sven Erick Alm and John C. Wierman, *Inequalities for means of restricted first-passage times in percolation theory*, Combin. Probab. Comput. **8** (1999), no. 4, 307–315, DOI 10.1017/S0963548399003843. Random graphs and combinatorial structures (Oberwolfach, 1997). MR1723645

[5] Antonio Auffinger and Michael Damron, *The scaling relation $\chi = 2\xi - 1$ for directed polymers in a random environment*, ALEA Lat. Am. J. Probab. Math. Stat. **10** (2013), no. 2, 857–880. MR3141825

[6] Antonio Auffinger and Michael Damron, *A simplified proof of the relation between scaling exponents in first-passage percolation*, Ann. Probab. **42** (2014), no. 3, 1197–1211, DOI 10.1214/13-AOP854. MR3189069

[7] Jinho Baik, Percy Deift, and Kurt Johansson, *On the distribution of the length of the longest increasing subsequence of random permutations*, J. Amer. Math. Soc. **12** (1999), no. 4, 1119–1178, DOI 10.1090/S0894-0347-99-00307-0. MR1682248

[8] Yuri Bakhtin, Eric Cator, and Konstantin Khanin, *Space-time stationary solutions for the Burgers equation*, J. Amer. Math. Soc. **27** (2014), no. 1, 193–238, DOI 10.1090/S0894-0347-2013-00773-0. MR3110798

[9] M. Balázs, E. Cator, and T. Seppäläinen, *Cube root fluctuations for the corner growth model associated to the exclusion process*, Electron. J. Probab. **11** (2006), no. 42, 1094–1132, DOI 10.1214/EJP.v11-366. MR2268539

[10] Márton Balázs, Júlia Komjáthy, and Timo Seppäläinen, *Microscopic concavity and fluctuation bounds in a class of deposition processes* (English, with English and French summaries), Ann. Inst. Henri Poincaré Probab. Stat. **48** (2012), no. 1, 151–187, DOI 10.1214/11-AIHP415. MR2919202

[11] Márton Balázs and Timo Seppäläinen, *Order of current variance and diffusivity in the asymmetric simple exclusion process*, Ann. of Math. (2) **171** (2010), no. 2, 1237–1265, DOI 10.4007/annals.2010.171.1237. MR2630064

[12] Alexei Borodin and Vadim Gorin, *Lectures on integrable probability*, Probability and statistical physics in St. Petersburg, Proc. Sympos. Pure Math., vol. 91, Amer. Math. Soc., Providence, RI, 2016, pp. 155–214. MR3526828

[13] Eric Cator and Piet Groeneboom, *Second class particles and cube root asymptotics for Hammersley's process*, Ann. Probab. **34** (2006), no. 4, 1273–1295, DOI 10.1214/009117906000000089. MR2257647

[14] Eric Cator and Leandro P. R. Pimentel, *Busemann functions and the speed of a second class particle in the rarefaction fan*, Ann. Probab. **41** (2013), no. 4, 2401–2425, DOI 10.1214/11-AOP709. MR3112921

[15] Sourav Chatterjee, *The universal relation between scaling exponents in first-passage percolation*, Ann. of Math. (2) **177** (2013), no. 2, 663–697, DOI 10.4007/annals.2013.177.2.7. MR3010809

[16] Henry Cohn, Noam Elkies, and James Propp, *Local statistics for random domino tilings of the Aztec diamond*, Duke Math. J. **85** (1996), no. 1, 117–166, DOI 10.1215/S0012-7094-96-08506-3. MR1412441

[17] Francis Comets, *Directed polymers in random environments*, Lecture Notes in Mathematics, vol. 2175, Springer, Cham, 2017. Lecture notes from the 46th Probability Summer School held in Saint-Flour, 2016. MR3444835

[18] Ivan Corwin, *The Kardar-Parisi-Zhang equation and universality class*, Random Matrices Theory Appl. **1** (2012), no. 1, 1130001, 76, DOI 10.1142/S2010326311300014. MR2930377

[19] I. Corwin, *Kardar-Parisi-Zhang universality*, Notices Amer. Math. Soc. **63** (2016), no. 3, 230–239, DOI 10.1090/noti1334. MR3445162

[20] Michael Damron and Jack Hanson, *Bigeodesics in first-passage percolation*, Comm. Math. Phys. **349** (2017), no. 2, 753–776, DOI 10.1007/s00220-016-2743-3. MR3595369

[21] Michael Damron, Firas Rassoul-Agha, and Timo Seppäläinen, *Random growth models*, Notices Amer. Math. Soc. **63** (2016), no. 9, 1004–1008, DOI 10.1090/noti1400. MR3525713

[22] Frank den Hollander, *Random polymers*, Lecture Notes in Mathematics, vol. 1974, Springer-Verlag, Berlin, 2009. Lectures from the 37th Probability Summer School held in Saint-Flour, 2007. MR2504175

[23] Jean-Dominique Deuschel and Ofer Zeitouni, *Limiting curves for i.i.d. records*, Ann. Probab. **23** (1995), no. 2, 852–878. MR1334175

[24] Jean-Dominique Deuschel and Ofer Zeitouni, *On increasing subsequences of I.I.D. samples*, Combin. Probab. Comput. **8** (1999), no. 3, 247–263, DOI 10.1017/S0963548399003776. MR1702546

[25] P. A. Ferrari and L. R. G. Fontes, *Current fluctuations for the asymmetric simple exclusion process*, Ann. Probab. **22** (1994), no. 2, 820–832. MR1288133

[26] Pablo A. Ferrari, James B. Martin, and Leandro P. R. Pimentel, *A phase transition for competition interfaces*, Ann. Appl. Probab. **19** (2009), no. 1, 281–317, DOI 10.1214/08-AAP542. MR2498679

[27] Pablo A. Ferrari and Leandro P. R. Pimentel, *Competition interfaces and second class particles*, Ann. Probab. **33** (2005), no. 4, 1235–1254, DOI 10.1214/009117905000000080. MR2150188

[28] Hans-Otto Georgii, *Gibbs measures and phase transitions*, De Gruyter Studies in Mathematics, vol. 9, Walter de Gruyter & Co., Berlin, 1988. MR956646

[29] Nicos Georgiou, Firas Rassoul-Agha, and Timo Seppäläinen, *Variational formulas and cocycle solutions for directed polymer and percolation models*, Comm. Math. Phys. **346** (2016), no. 2, 741–779, DOI 10.1007/s00220-016-2613-z. MR3535900

[30] Nicos Georgiou, Firas Rassoul-Agha, and Timo Seppäläinen, *Stationary cocycles and Busemann functions for the corner growth model*, Probab. Theory Related Fields **169** (2017), no. 1-2, 177–222, DOI 10.1007/s00440-016-0729-x. MR3704768

[31] Nicos Georgiou, Firas Rassoul-Agha, Timo Seppäläinen, and Atilla Yilmaz, *Ratios of partition functions for the log-gamma polymer*, Ann. Probab. **43** (2015), no. 5, 2282–2331, DOI 10.1214/14-AOP933. MR3395462

[32] J. M. Hammersley, *A few seedlings of research*, Proceedings of the Sixth Berkeley Symposium on Mathematical Statistics and Probability (Univ. California, Berkeley, Calif., 1970/1971), Univ. California Press, Berkeley, Calif., 1972, pp. 345–394. MR0405665

[33] Christopher Hoffman, *Coexistence for Richardson type competing spatial growth models*, Ann. Appl. Probab. **15** (2005), no. 1B, 739–747, DOI 10.1214/105051604000000729. MR2114988

[34] William Jockusch, James Propp, and Peter Shor, *Random domino tilings and the arctic circle theorem*, `arXiv:math/9801068`.

[35] Kurt Johansson, *Shape fluctuations and random matrices*, Comm. Math. Phys. **209** (2000), no. 2, 437–476, DOI 10.1007/s002200050027. MR1737991

[36] ———, *Random matrices and determinantal processes*, Les Houches lecture notes, 2005, `arXiv:math-ph/0510038`.

[37] Jeong Han Kim, *On increasing subsequences of random permutations*, J. Combin. Theory Ser. A **76** (1996), no. 1, 148–155, DOI 10.1006/jcta.1996.0095. MR1405997

[38] Elena Kosygina, *Homogenization of stochastic Hamilton-Jacobi equations: brief review of methods and applications*, Stochastic analysis and partial differential equations, Contemp. Math., vol. 429, Amer. Math. Soc., Providence, RI, 2007, pp. 189–204, DOI 10.1090/conm/429/08237. MR2391536

[39] Arjun Krishnan, *Variational formula for the time constant of first-passage percolation*, Comm. Pure Appl. Math. **69** (2016), no. 10, 1984–2012, DOI 10.1002/cpa.21648. MR3541856

[40] B. F. Logan and L. A. Shepp, *A variational problem for random Young tableaux*, Advances in Math. **26** (1977), no. 2, 206–222, DOI 10.1016/0001-8708(77)90030-5. MR1417317

[41] James B. Martin, *Limiting shape for directed percolation models*, Ann. Probab. **32** (2004), no. 4, 2908–2937, DOI 10.1214/009117904000000838. MR2094434

[42] Gregorio R. Moreno Flores, Timo Seppäläinen, and Benedek Valkó, *Fluctuation exponents for directed polymers in the intermediate disorder regime*, Electron. J. Probab. **19** (2014), no. 89, 28, DOI 10.1214/EJP.v19-3307. MR3263646

[43] Charles M. Newman, *A surface view of first-passage percolation*, Proceedings of the International Congress of Mathematicians, Vol. 1, 2 (Zürich, 1994), Birkhäuser, Basel, 1995, pp. 1017–1023. MR1404001

[44] Firas Rassoul-Agha and Timo Seppäläinen, *Quenched point-to-point free energy for random walks in random potentials*, Probab. Theory Related Fields **158** (2014), no. 3-4, 711–750, DOI 10.1007/s00440-013-0494-z. MR3176363

[45] Firas Rassoul-Agha, Timo Seppäläinen, and Atilla Yilmaz, *Variational formulas and disorder regimes of random walks in random potentials*, Bernoulli **23** (2017), no. 1, 405–431, DOI 10.3150/15-BEJ747. MR3556777

[46] Jeffrey M. Rosenbluth, *Quenched large deviation for multidimensional random walk in random environment: A variational formula*, ProQuest LLC, Ann Arbor, MI, 2006. Thesis (Ph.D.)–New York University. MR2708406

[47] H. Rost, *Nonequilibrium behaviour of a many particle process: density profile and local equilibria*, Z. Wahrsch. Verw. Gebiete **58** (1981), no. 1, 41–53, DOI 10.1007/BF00536194. MR635270

[48] Timo Seppäläinen, *A microscopic model for the Burgers equation and longest increasing subsequences*, Electron. J. Probab. **1** (1996), no. 5, approx. 51 pp. (electronic). MR97d:60162

[49] T. Seppäläinen, *Coupling the totally asymmetric simple exclusion process with a moving interface*, Markov Process. Related Fields **4** (1998), no. 4, 593–628. I Brazilian School in Probability (Rio de Janeiro, 1997). MR1677061

[50] T. Seppäläinen, *Hydrodynamic scaling, convex duality and asymptotic shapes of growth models*, Markov Process. Related Fields **4** (1998), no. 1, 1–26. MR1625007

[51] Timo Seppäläinen, *Large deviations for increasing sequences on the plane*, Probab. Theory Related Fields **112** (1998), no. 2, 221–244, DOI 10.1007/s004400050188. MR1653841

[52] Timo Seppäläinen, *Current fluctuations for stochastic particle systems with drift in one spatial dimension*, Ensaios Matemáticos [Mathematical Surveys], vol. 18, Sociedade Brasileira de Matemática, Rio de Janeiro, 2010. MR2724800

[53] Timo Seppäläinen, *Scaling for a one-dimensional directed polymer with boundary conditions*, Ann. Probab. **40** (2012), no. 1, 19–73, DOI 10.1214/10-AOP617. MR2917766

[54] A. M. Veršik and S. V. Kerov, *Asymptotic behavior of the Plancherel measure of the symmetric group and the limit form of Young tableaux* (Russian), Dokl. Akad. Nauk SSSR **233** (1977), no. 6, 1024–1027. MR0480398

Department of Mathematics, 480 Lincoln Drive, University of Wisconsin–Madison, Madison, Wisconsin 53711

Email address: seppalai@math.wisc.edu

Proceedings of Symposia in Applied Mathematics
Volume **75**, 2018
https://doi.org/10.1090/psapm/075/00661

Exactly solving the KPZ equation

Ivan Corwin

Abstract. We present a complete proof of the exact formula for the one-point distribution for the narrow-wedge Hopf-Cole solution to the Kardar-Parisi-Zhang (KPZ) equation. This presentation is intended to be self-contained so no previous knowledge about stochastic PDEs, or exactly solvable systems is presumed.

Contents

1. Introduction

1.1. Preface. The study of the Kardar-Parisi-Zhang (KPZ) equation and its associated universality class has spurred and paralleled the development of an immense amount of deep and fascinating mathematics and physics. The core study of the KPZ equation and universality class warrants a detailed treatment, and so do many of the related fields developed in parallel (e.g. the study of non-linear stochastic PDES, or of integrable probabilistic systems). This article purports to do much less – the only goal being to present the main steps of a proof of the exact formula for the distribution of a particular solution to the KPZ equation. In so doing, we will touch on a number of the aforementioned related developments. Exercising great self-control, we will forego diversions into these areas.

2010 *Mathematics Subject Classification.* 60K35, 60K37, 82B43.

Before delving into methods, let us quickly (and informally) answer two questions – what is the KPZ equation and what does it mean to exactly solve it?

The KPZ equation was introduced in 1986 by Kardar, Parisi and Zhang [**25**] as a model for random interface growth. In one-spatial dimension (sometimes also called $(1+1)$-dimension to emphasize that time is one dimension too) it describes the evolution of a function $h(t,x)$ recording the height of an interface at time $t \geqslant 0$ above position $x \in \mathbb{R}$. The equation for $h(t,x)$ is the following non-linear stochastic partial differential equation (SPDE)

$$\partial_t h(t,x) = \tfrac{1}{2}\partial_x^2 h(t,x) + \tfrac{1}{2}\big(\partial_x h(t,x)\big)^2 + \xi(t,x),$$

which is driven by a space-time white noise ξ. The KPZ equation is supposed to be a representative model within a large class (called the KPZ universality class) in the sense that its long-time, large-scale behavior should be shared by many other growth models. The key characteristic of this KPZ universality class is supposed to be the presence of a local growth mechanism which involves a non-linear dependence on the local slope (here, the term $\frac{1}{2}\big(\partial_x h(t,x)\big)^2$), and the presence of space-time random forcing (here, the term $\xi(t,x)$). The Laplacian provides a smoothing mechanism to compete with the roughening by the noise – this mechanism is not believed to be particularly necessary for models to lie in the KPZ universality class.

Putting aside, for the moment, the interesting and important mathematical question of what it actually means to be a solution to this equation[1], a physically important question regarding the KPZ equation is to understand what its solutions look like, from a statistical perspective. Since the white noise ξ is random, the solution $h(t,x)$ will likewise be random. So, we seek to understand the distribution of this solution as well as its dependence on the initial data $h(0,x)$. The holy-grail is to have an exact formula for the distribution of the solution (or for certain characterizing functions, or marginals of the solution). Given such exact formulas, we might be able to exactly describe the long-time, large-scape behavior of the model (and hence predict the behavior of the entire KPZ universality class). For instance, is there a limiting deterministic growth velocity v so that

$$\lim_{t\to\infty} \frac{h(t,x)}{t} = v? \tag{1.1}$$

If so, after centering by that velocity, what is the scale χ of the remaining fluctuations and what is their distribution $F(\cdot)$ so that for all $s \in \mathbb{R}$,

$$\lim_{t\to\infty} \mathbb{P}\left(\frac{h(t,x) - vt}{t^{\chi}} \leqslant s\right) = F(s)? \tag{1.2}$$

It is certainly not assured (and generally not true) that an SPDE will have tractable formulas for the distribution of its solution (for instance, for the one-point marginal distribution at a fixed time and space location). However, the KPZ equation happens to be an exception as it is closely related to certain structures from integrable systems and representation theory.

[1]Sometimes, see e.g. [**20**], people use the term "solve the KPZ" to mean constructing solutions. Here, we use the phrase "exactly solve" in place of "solve" to emphasize that we are interested not just in the existence of solutions but the exact formulas for their distribution.

1.2. In what way is the KPZ equation exactly solvable? The first hint that there is a nice structure behind the KPZ equation comes from looking at the moments of its exponential (or Cole-Hopf transform)

$$z(t,x) = e^{h(t,x)}.$$

Formally, this transformation turns the KPZ equation into the stochastic heat equation (SHE) with multiplicative noise:

$$\partial_t z(t,x) = \tfrac{1}{2}\partial_{xx} z(t,x) + z(t,x)\xi(t,x).$$

Writing $\mathbb{E}$ to denote the expected value of a random variable with respect to the randomness induced by the noise ξ, notice that since ξ is a centered Gaussian process, $\bar{z}(t,x) := \mathbb{E}\big[z(t,x)\big]$ actually solves the deterministic heat equation $\partial_t \bar{z}(t,x) = \frac{1}{2}\partial_{xx}\bar{z}(t,x)$. Amazingly, higher moments

$$\bar{z}(t,\vec{x}) := \mathbb{E}\Big[\prod_{i=1}^{k} z(t,x_i)\Big],$$

with $\vec{x} = (x_1,\dots,x_k)$, also solve simple, deterministic evolution equations. This observation goes back to Kardar [**24**] and independently Molchanov [**30**] wherein they argued that for any k,

$$\partial_t \bar{z}(t,\vec{x}) = \big(H\bar{z}\big)(t,x),$$

where H acts in the $\vec{x}$-variables as

$$H = \tfrac{1}{2}\sum_{i=1}^{k}\partial_{x_i^2} + \sum_{1\leqslant i<j\leqslant k}\delta(x_i - x_j = 0).$$

Here $\delta(x=0)$ is the delta functions at 0 (so $\int f(x)\delta(x=0) = f(0)$ for a suitable class of f). H is called the attractive, imaginary time delta Bose gas or Lieb-Liniger Hamiltonian, and it has been known since work of Lieb and Liniger from 1963 [**27**] that this Hamiltonian can be diagonalized explicitly using a tool known as the Bethe ansatz. Using those tools, one can show (see for instance [**7**]) that for $x_1 \leqslant \cdots \leqslant x_k$,

$$\bar{z}(t;\vec{x}) = \int_{\alpha_1 - \mathrm{i}\infty}^{\alpha_1+\mathrm{i}\infty}\frac{dz_1}{2\pi\,\mathrm{i}}\cdots\int_{\alpha_k - \mathrm{i}\infty}^{\alpha_k+\mathrm{i}\infty}\frac{dz_k}{2\pi\,\mathrm{i}}\prod_{1\leqslant A<B\leqslant k}\frac{z_A - z_B}{z_A - z_B - 1}\prod_{j=1}^{k} e^{\frac{t}{2}z_j^2 + x_j z_j},$$

where we assume that $\alpha_1 > \alpha_2 + 1 > \alpha_3 + 2 > \cdots > \alpha_k + (k-1)$ (note that z_i are complex variables of integration in the above formula on the right-hand side).

Armed with such a precise formula, one might expect that it is easy to recover the desired answers to questions (1.1) and (1.2). There is, however, a fundamental problem – the moments of the SHE do not characterize the distribution of the solution! Indeed, from the above formula one can show that the kth moment grows like exponential of k^3, which is far to fast to lead to a well-posed moment problem. Indeed, this is because the upper tail of z is sufficiently heavy so that the moments really only tease out the nature of that tail. This phenomena is related to the concept of intermittency.

The key idea which will resolve this roadblock of an ill-posed moment problem is that we should "be wise and discretize". In other words, we will look for discrete regularizations of the KPZ equation which preserves the above mentioned algebraic properties, yet no longer suffers from this moment problem. As we will see in this article, q-TASEP provides one such option (there are others).

1.3. Outline. This article is split in two parts – the first three sections are devoted to defining the KPZ equation and proving convergence of some discrete models to it. The remaining four sections then turn to the exact solvability, in particular focusing on the discrete model q-TASEP, as well as the semi-discrete directed polymer which arises as an intermediate limit of q-TASEP to the KPZ equation. In more detail, Section 2 defines what it means to solve the stochastic heat equation and constructs the delta initial data solution via chaos series. Section 3 turns to the KPZ equation and provides an explanation for why it arises as a continuum limit for many systems. Section 4 works out such a convergence result for the semi-discrete directed polymer in detail. Section 5 introduces the interacting particle system q-TASEP and gives its Markov duality. Section 6 uses this to give explicit contour integral formulas for certain q-moments of q-TASEP, and then describes the relation of duality and these formulas to the polymer replica method. Section 7 studies the intermittency of these moments (from the perspective of the parabolic Anderson model) and describes how to expand the contour integral formulas. Section 8 uses this expansion to gather all q-moments as a q-Laplace transform and performs the asymptotic analysis necessary to arrive at the KPZ equation statistics advertised above in the introduction. Section 9 explains the origins of the integral formuls in Section 6 in terms of the coordinate Bethe ansatz and the spectral theory for the q-Boson process. The appendices provide background on stochastic integration with respect to white noise, the Tracy-Widom distribution, and steepest descent asymptotic analysis.

1.4. Acknowledgments. This article is based on the author's lectures at a 2014 MSRI summer school on stochastic PDEs (and then revised for the 2017 AMS MAA Joint Meeting Short Course. The author thanks Jeffrey Kuan for writing up the handwritten notes, as well as for providing some valuable background materials, and Xuan Wu for making a close reading of the file and adding indexing to it; as well as Yuri Bahktin and James Nolin for co-organizing the school with him. The author was partially supported by the NSF through DMS-1208998, DMS-1664650, the Clay Mathematics Institute through a Clay Research Fellowship, the Institute Henri Poincare through the Poincare Chair, and the Packard Foundation through a Packard Fellowship for Science and Engineering.

2. Mild solution to the stochastic heat equation

The stochastic heat equation (SHE) with multiplicative noise looks (in differential form) like

$$\begin{cases} \partial_t z &= \frac{1}{2}\partial_{xx} z + z\xi \\ z(0,x) &= z_0(x) \end{cases}$$

where $z : \mathbb{R}_+ \times \mathbb{R} \to \mathbb{R}$ and z_0 is (possibly random) initial data which is independent of the white noise ξ. Recall that formally ξ has covariance

$$\mathbb{E}\big[\xi(t,x)\xi(s,y)\big] \text{``}=\text{''} \delta_{t=s}\delta_{x=y},$$

though this is only true in a weak, or integrated sense. See Appendix A for background on ξ. The noise ξ is constructed on a probability space $L^2(\Omega, \mathcal{F}, \mathbb{P})$.

DEFINITION 2.1. A *mild solution* to the SHE satisfies for all $t > 0, x \in \mathbb{R}$ the Duhamel form equation

$$z(t,x) = \int_{\mathbb{R}} p(t, x-y) z_0(y) dy + \int_0^t \int_{\mathbb{R}} p(t-s, x-y) z(s,y) \xi(s,y) dy ds,$$

where $p(t,x) = \frac{1}{\sqrt{2\pi t}} e^{-x^2/2t}$ is the heat kernel or fundamental solution to $\partial_t p = \partial_{xx} p$, with $p(0,x) = \delta_{x=0}$. In this, we must have that $\int_0^t \int_{\mathbb{R}} p^2(t-s, x-y) \mathbb{E}\big[z^2(s,y)\big] dy ds < \infty$ for the Itô integrals to make sense and be finite.

In this article we will be concerned only with the special delta initial data solution to the SHE. We will use a chaos series (see also Appendix A) to construct the solution to the SHE with $z_0(x) = \delta_{x=0}$.

THEOREM 2.2. *Consider the class of solutions $z(t,x)$ to the mild formulation of SHE which satisfy (i) for all test functions ψ, $\int_{\mathbb{R}} \psi(x) z(t,x) dx \to \int_{\mathbb{R}} \psi(x) \delta_{x=0} dx = \psi(0)$ as $t \searrow 0$ with $L^2(\Omega, \mathcal{F}, \mathbb{P})$ convergence; (ii) for all $T > 0$ there exists $C = C(T) > 0$ such that for all $0 < t \leqslant T$, and $x \in \mathbb{R}$, the bound $\mathbb{E}\big[z^2(t,x)\big] \leqslant C p^2(t,x)$ holds.*

Within the above class of functions there exists a unique solution to SHE with $\delta_{x=0}$ initial data, and that solution takes the following form as a convergent (in $L^2(\Omega, \mathcal{F}, \mathbb{P})$) chaos series:

$$z(t,x) = \sum_{k=0}^{\infty} \int_{\Delta_k(t)} \int_{\mathbb{R}^k} P_{k;t,x}(\vec{s}, \vec{y}) d\xi^{\otimes k}(\vec{s}, \vec{y}) =: \sum_{k=0}^{\infty} I_k(t,x),$$

where $\Delta_k(t) = \{0 \leqslant s_1 < \cdots < s_k \leqslant t\}$, and

$$P_{k;t,x}(\vec{s}, \vec{y}) = p(t - s_k, x - y_k) p(s_k - s_{k-1}, y_k - y_{k-1}) \cdots p(s_2 - s_1, y_2 - y_1) p(s_1, y_1),$$

or in other words, represents the transition densities of a one-dimensional Brownian motion started at $(0,0)$ to go through the (time,space) points $(s_1, y_1), \ldots, (s_k, y_k)$, (t,x). Also, recall from Appendix A that $d\xi^{\otimes k}(\vec{s}, \vec{y})$ is the multiple stochastic integral.

PROOF. We proceed according to the following steps:

(1) Show that the chaos series is convergent.
(2) Prove that $\mathbb{E}\big[z^2(t,x)\big] \leqslant C p^2(t,x)$.
(3) Show that the series solves SHE with $\delta_{x=0}$ initial data.
(4) Argue uniqueness.

Step (1): Consider a random variable $X \in L^2(\Omega, \mathcal{F}, \mathbb{P})$ such that

$$X = \sum_{k=0}^{\infty} \int_{\Delta_k(\infty)} \int_{\mathbb{R}^k} f_k(\vec{t}, \vec{x}) d\xi^{\otimes k}(\vec{t}, \vec{x}).$$

for functions $f_k \in L^2(\Delta_k(\infty) \times \mathbb{R}^k), k = 0, 1, 2, \ldots$. (From Appendix A we know that all X have a unique decomposition into this form.) By the multiple stochastic integral covariance isometry

$$\mathbb{E}\left[\int_{\Delta_k(\infty)} \int_{\mathbb{R}^k} f(\vec{t}, \vec{x}) d\xi^{\otimes k}(\vec{t}, \vec{x}) \int_{\Delta_j(\infty)} \int_{\mathbb{R}^k} g(\vec{t}, \vec{x}) d\xi^{\otimes j}(\vec{t}, \vec{x})\right] = \mathbf{1}_{j=k} \langle f, g \rangle_{L^2(\Delta_k \times \mathbb{R}_k)}.$$

Thus,

$$\mathbb{E}\big[X^2\big] = \sum_{k=0}^{\infty} \|f_k\|^2_{L^2(\Delta_k\times\mathbb{R}^k)}.$$

This can be applied to the series for $z(t,x)$.

EXERCISE 2.3. Prove that

$$p^2(t,y) = \frac{1}{\sqrt{2\pi t}}p(t,\sqrt{2}y) = \frac{1}{\sqrt{4\pi t}}p(t/2,y)$$

and

$$\int_{\mathbb{R}} p^2(t-u,x-z)p^2(u-s,z-y)dz = \sqrt{\frac{t-s}{4\pi(t-u)(u-s)}}p^2(t-s,x-y).$$

LEMMA 2.4.

$$\mathbb{E}\big[I_k^2(t,x)\big] = \frac{t^{k/2}}{(4\pi)^{k/2}}\frac{\Gamma(\frac{1}{2})^k}{\Gamma(\frac{k}{2})}p^2(t,x).$$

PROOF. The multiple stochastic integral covariance isometry implies that

$$\mathbb{E}\big[I_k^2(t,x)\big] = \int_{\Delta_k(t)}\int_{\mathbb{R}^k} P^2_{k;t,x}(\vec{s},\vec{y})d\vec{y}d\vec{s},$$

or more explicitly,

$$\mathbb{E}\big[I_k^2(t,x)\big] = \int_{\Delta_k(t)} d\vec{s}\int_{\mathbb{R}^k} d\vec{y}\, p^2(s_1,y_1)p^2(s_2-s_1,y_2-y_1)\cdots$$
$$\cdots p^2(s_k-s_{k-1},y_k-y_{k-1})p^2(t-s_k,x-y_k).$$

Taking the integrals starting with y_k down to y_1 and using Exercise 2.3,

$$\mathbb{E}\big[I_k^2(t,x)\big] = \int_{\Delta_k(t)} d\vec{s}\, p^2(t,x)\sqrt{\frac{t}{4\pi(t-s_1)s_1}}\sqrt{\frac{t-s_1}{4\pi(t-s_2)(s_2-s_1)}}\cdots$$
$$\cdots\sqrt{\frac{t-s_{k-1}}{4\pi(t-s_k)(s_k-s_{k-1})}}$$
$$= \frac{t^{1/2}p^2(t,x)}{(4\pi)^{k/2}}\int_{\Delta_k(t)} d\vec{s}\frac{1}{\sqrt{s_1}}\frac{1}{\sqrt{s_2-s_1}}\cdots\frac{1}{\sqrt{s_k-s_{k-1}}}\frac{1}{\sqrt{t-s_k}}.$$

Changing variables to factor out the t's yields

$$\mathbb{E}\big[I_k^2(t,x)\big] = \frac{t^{k/2}}{(4\pi)^{k/2}}p^2(t,x)\int_{\Delta_k(1)} d\vec{s}\frac{1}{\sqrt{s_1}}\frac{1}{\sqrt{s_2-s_1}}\cdots\frac{1}{\sqrt{s_k-s_{k-1}}}\frac{1}{\sqrt{1-s_k}}.$$

EXERCISE 2.5. For $\alpha_i > 0$, $1 \leqslant i \leqslant k$ show that

$$\int_{\substack{0\leqslant x_1,\cdots,x_k\leqslant 1\\ \sum x_i=1}} d\vec{x}\prod_{i=1}^{k} x_i^{\alpha_i-1} = \frac{\prod_{i=1}^k\Gamma(\alpha_i)}{\Gamma(\sum_{i=1}^k\alpha_i)}.$$

This is called the Dirichlet(α) distribution.

Using the exercise we conclude that

$$\mathbb{E}\big[I_k^2(t,x)\big] = \frac{t^{k/2}}{(4\pi)^{k/2}}\frac{\Gamma(1/2)^k}{\Gamma(k/2)}p^2(t,x). \qquad \square$$

Since $\Gamma(k/2) \sim (k/2)!$ this implies that the chaos series converges in $L^2(\Omega, \mathcal{F}, \mathbb{P})$ thus completing step 1.

Step (2): By Lemma 2.4 and the Itô isometry,

$$\mathbb{E}\big[z^2(t,x)\big] = \sum_{k=0}^{\infty} \mathbb{E}\big[I_k^2(t,x)\big] = p^2(t,x) \sum_{k=0}^{\infty} \frac{t^{k/2}}{(4\pi)^{k/2}} \frac{\Gamma(1/2)^k}{\Gamma(k/2)} \leqslant Cp^2(t,x),$$

where the constant $C = C(T)$ can be chosen fixed as t varies in $[0,T]$.

Steps (3) and (4): Assume $z(t,x)$ solves SHE with delta initial data. We will show that $z(t,x)$ equals the chaos series. The mild form of SHE implies that

$$z(t,x) = p(t,x) + \int_0^t \int_{\mathbb{R}} p(t-s, x-y) z(s,y) \xi(s,y) dy ds.$$

Iterate this (like Picard iteration) to obtain

$$z(t,x) = \sum_{k=0}^{n} I_k(t,x) + \int_{\Delta_{n+1}(t)} \int_{\mathbb{R}^{n+1}} p(t-s_{n+1}, x-y_{n+1}) \cdots$$
$$\cdots p(s_2 - s_1, y_2 - y_1) z(s_1, y_1) \xi^{\otimes k}(\vec{s}, \vec{y}) d\vec{s} d\vec{y}.$$

Comptue the $L^2(\Omega, \mathcal{F}, \mathbb{P})$ norm of the remainder as

$$\int_{\Delta_{n+1}(t)} \int_{\mathbb{R}^{n+1}} p^2(t-s_{n+1}, x-y_{n+1}) \cdots p^2(s_2 - s_1, y_2 - y_1) \cdot \mathbb{E}\big[z^2(s_1, y_1)\big] d\vec{y} d\vec{s}.$$

Use bound $\mathbb{E}[z^2(s_1,y_1)] \leqslant cp^2(s_1,y_1)$ and notice that the above expression is then bounded as

$$\leqslant c\,\mathbb{E}[I_{n+1}(t,x)] = c \frac{t^{(n+1)/2}}{(4\pi)^{n/2}} \frac{\Gamma(1/2)^{n+1}}{\Gamma(n/2)} p^2(t,x),$$

which goes to zero as $n \to \infty$ hence showing that z is the chaos series and that the chaos series solves the mild form of SHE. □

We close by demonstrating one use of the chaos series to probe the short time behavior of $z(t,x)$.

COROLLARY 2.6. *The following one-point*[2] *weak convergence holds:*

$$\lim_{\epsilon \to 0} \epsilon^{-1/4} \big[z(\epsilon t, \epsilon^{1/2} x) - p(\epsilon t, \epsilon^{1/2} x)\big] = \int_0^t \int_{\mathbb{R}} p(t-s, x-y) \xi(s,y) p(s,y) dy ds.$$

PROOF. This comes from the first term in the chaos series whereas our estimate on $\mathbb{E}[I_k(t,x)^2]$ enables us to discard all subsequent terms. Recall

$$I_1(t,x) = \int_0^t \int_{\mathbb{R}} p(t-s, x-y) \xi(s,y) p(s,y) dy ds.$$

Apply scalings to study now $\epsilon^{-1/4} I_1(\epsilon t, \epsilon^{1/2} x)$.

EXERCISE 2.7. Show that in distribution,

$$\xi(t,x) = \epsilon^{(z+1)/2} \xi(\epsilon^z t, \epsilon x).$$

Use it to show that $\epsilon^{-1/4} I_1(\epsilon t, \epsilon^{1/2} x) = I_1(t,x)$ in distribution whereas we have $\epsilon^{-1/4} I_k(\epsilon t, \epsilon^{1/2} x) = \epsilon^{a_k} I_k(t,x)$. Compute $a_k > 0$.

From this exercise, it is now easy to conclude the corollary. □

---

[2]Extending to space-time process convergence is not much harder.

Unfortunately, the chaos series is not very useful in studying large t behaviors (e.g. for $\log z(t,x)$) and also the chaos series do not behave very well under applications of functions. However, they can be useful in proving approximation as we will soon see.

3. Scaling to the KPZ equation

The heat equation with deterministic potential $v(t,x)$ is written in differential form as

$$\partial_t z = \frac{1}{2}\partial_{xx} z + vz.$$

EXERCISE 3.1. Show that if $h := \log z$ and $u := \partial_x h$ then

$$\partial_t h = \frac{1}{2}\partial_{xx} h + \frac{1}{2}(\partial_x h)^2 + v,$$
$$\partial_t u = \frac{1}{2}\partial_{xx} u + \frac{1}{2}\partial_x (u^2) + \partial_x v.$$

The first equation is a Hamilton Jacobi equation with forcing and the second equation is a viscous Burgers equation with conservative forcing.

Define the mollifier $p_\kappa(x) = p(\kappa, x)$ (i.e. the heat kernel at time κ). Note that as $\kappa \to 0$, this approaches a delta function as its maximal height grows like $\kappa^{-1/2}$ and standard deviation shrinks like $\kappa^{1/2}$. We could just as well work with any similar mollifier sequence. Define mollified white-noise via

$$\xi_\kappa(t,x) = \int_{\mathbb{R}} p_\kappa(x-y)\xi(t,y)dy = (p_\kappa * \xi)(t,x).$$

By the Itô isometry,

$$\mathbb{E}\big[\xi_\kappa(t,x)\xi_\kappa(s,y)\big] = \delta_{t=s} c_k(x-y), \quad c_\kappa(x) = (p_\kappa * p_\kappa)(x).$$

Note that as $\kappa \to 0$, we have $c_\kappa(x) \to \infty$.

EXERCISE 3.2. Define z_κ to be the solution (same definition of solution as before) to

$$\partial_t z_\kappa = \frac{1}{2}\partial_{xx} z_\kappa + \xi_\kappa z_\kappa,$$

and $h_\kappa := \log z_\kappa$. Show that

$$\partial_t h_\kappa = \frac{1}{2}\partial_{xx} h_\kappa + \big[\frac{1}{2}(\partial_x h_\kappa)^2 - c_\kappa(0)\big] + \xi_\kappa.$$

Also show that $z_\kappa \to z$ as $\kappa \searrow 0$. (Suggestion: use the chaos series.)

The exercise shows that in some sense, we can think of $h = \log z$ as the solution to

$$\partial_t h = \frac{1}{2}\partial_{xx} h + \big[\frac{1}{2}(\partial_x h_k)^2 - \infty\big] + \xi.$$

Since when studying interface fluctuations we generally need to subtract off overall height shifts, this $-\infty$ should not concern us so much. This suggests the definition

DEFINITION 3.3. The Hopf–Cole solution $h : \mathbb{R}_+ \times \mathbb{R} \to \mathbb{R}$ to the KPZ equation with $h_0 : \mathbb{R} \to \mathbb{R}$ initial data is defined as $h = \log z$ where z is the solution to the SHE with initial data $z_0(x) = e^{h_0(x)}$.

For some initial conditions such as $\delta_{x=0}$, it does not make sense to take its logarithm. However, for all positive t and $x \in \mathbb{R}$, we know that $z(t,x)$ is strictly positive (see [**29, 31**]), hence the logarithm is well defined. The solution to KPZ corresponding to this delta initial data is called the narrow wedge solution (or KPZ with narrow wedge initial data). The idea is that in short time the log of $p(t,x)$ looks like $-x^2/2t$ which is a steep parabola. Hence the initial data corresponds to growth out of a wedge type environment. We will see a justification of this later.

Now let us ask: what kind of discrete models converge to the KPZ equation? To motivate this, let us do some rescaling of the KPZ equation.

Define the scaled KPZ solution (we temporarily use z here as the "dynamic scaling exponent" in accordance with other literature)

$$h^\epsilon(t,x) = \epsilon^b h(\epsilon^{-z}t, \epsilon^{-1}x).$$

We find that under this scaling,

$$\begin{aligned}
\partial_t h &= \epsilon^{z-b}\partial_t h^\epsilon \\
\partial_x h &= \epsilon^{1-b}\partial_x h^\epsilon, \qquad (\partial_x h)^2 = \epsilon^{2-2b}(\partial_x h^\epsilon)^2 \\
\xi(t,x) &\overset{(d)}{=} \epsilon^{-(z+1)/2}\xi(\epsilon^{-z}t, \epsilon^{-1}x).
\end{aligned}$$

Note that above, when writing $\partial_t h$ and $\partial_x h$ we mean to take the derivatives and then evaluate them at $\epsilon^{-z}t$ and $\epsilon^{-1}x$ respectively. Thus

$$\partial_t h^\epsilon = \frac{1}{2}\epsilon^{2-z}\partial_{xx}h^\epsilon + \frac{1}{2}\epsilon^{2-z-b}(\partial_x h^\epsilon)^2 + \epsilon^{b-z/2+1/2}\xi.$$

Note that the noise on the right-hand side is not the same for different ϵ, but in terms of its distribution it is.

We now pose a natural question: are there any scalings of the KPZ equation under which it is invariant? If so, then we can hope to scale growth processes in the same way to arrive at the KPZ equation. However, one can check that there is no way to do this! On the other hand, there are certain weak scalings which fix the KPZ equation. By weak, we mean that simultaneously as we scaling time, space and fluctuations, we also can put tuning parameters in front of certain terms in the KPZ equation and scale them with ϵ. In other words, we simultaneously scale time, space and fluctuations, as well as the model. Let us consider two such weak scalings.

Weak non–linearity scaling: Take $b = 1/2, z = 2$. The first and third terms stay fixed, but the middle term blows up. Thus, insert the constant λ_ϵ in front of the nonlinear term $(\partial x h)^2$ and set $\lambda_\epsilon = \epsilon^{1/2}$. Under this scaling, the KPZ equation is mapped to itself.

Weak noise scaling: Take $b = 0, z = 2$. Under this scaling, the linear $\partial_{xx}h$ and nonlinear $(\partial_x h)^2$ terms stay fixed, but now the noise blows up. So insert β_ϵ in front of the noise term and set $\beta_\epsilon = \epsilon^{1/2}$, and again the KPZ equation stays invariant.

One can hope that these rescalings are attractive. In other words, if you take models with a parameter (nonlinearity or noise) that can be tuned, that these models will all converge to the same limiting object. We will see that weak non–linearity scaling can be applied to particle growth processes and weak noise scaling can be applied to directed polymers.

We would be remise at this point if we did not mention the broader notion of KPZ universality (see [**15, 37**] for surveys on this subject). It was predicted by [**18, 25**] is that under $b = 1/2$ and $z = 3/2$ scaling (called KPZ scaling) the KPZ equation should have non-trivial limiting behavior. The exact nature of this limiting behavior (called the KPZ fixed point in [**17**]) is still somewhat nebulous as we will briefly observe. The fixed point should be the limit of particle growth processes and directed polymers. To understand more about the KPZ fixed point, we will need to develop exact solvability, which will be done in later sections.

Here is a fact (which can be shown, for instance from the approximation of the KPZ equation from ASEP). If we start KPZ with two–sided Brownian motion, then later it will be a two–sided Brownian motion plus a height shift. One can see this by showing that particle systems preserve random walks. This suggests $b = 1/2$ (the Brownian scaling). In this case, the only way to avoid blowup is to take $z = 3/2$. So this suggests that KPZ scaling is, indeed, a natural candidate. But in this case, it looks like the noise and the linear term disappears. What would be left would be the inviscid Burgers equation. But this is not true, since, for example, the inviscid Burgers equation is deterministic and also since the KPZ equation must preserve Brownian motion, but the inviscid Burgers equation does not.

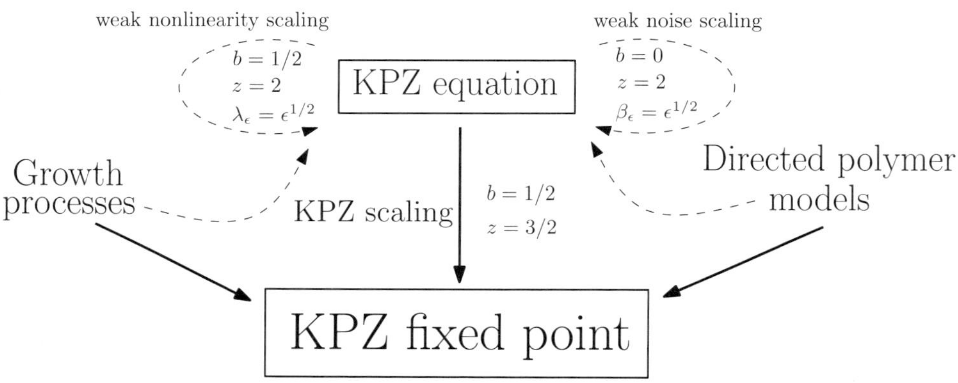

FIGURE 3.1. Three types of scalings for the KPZ equation. Weak noise and weak non–linearity scaling fix the equation whereas under KPZ scaling, the KPZ equation should go to the (nebulous) KPZ fixed point. It is believed (and in some cases shown) that these ideas extend to a variety of growth processes and directed polymer models.

Because we presently lack a stochastic description of the KPZ fixed point, most of what we know about it comes from asymptotic analysis of a few exactly solvable models. Moreover, most information comes in the form of knowledge of certain marginal distributions of the space-time KPZ fixed point process.

In order to illustrate the ideas of weak universality of the KPZ equation, lets consider the corner growth model, which is equivalent to ASEP. Growth of local minima into local maxima occurs after independent exponentially distributed waiting times of rate p and the opposite (local maxima shrinking into local minima) occurs with rate q, with $p + q = 1$ and $p - q =: \gamma > 0$. The step initial data corresponds to setting $h^\gamma(0,x) = |x|$. Here we use $h^\gamma(t,x)$ to record the height function at time t above position x, for the model with asymmetry γ.

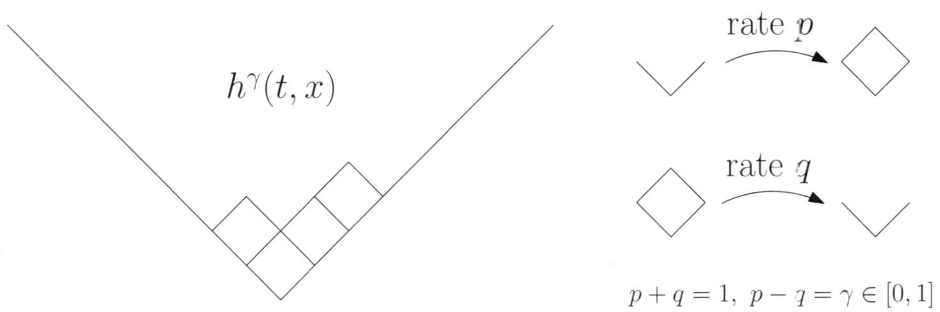

FIGURE 3.2. Corner growth model, or equivalently asymmetric simple exclusion process (ASEP) started from step initial data.

The following theorem is an example of how the KPZ equation arises under weak non–linearity scaling.

THEOREM 3.4 ([**3**]). *For the step initial data corner growth model,*

$$z_\epsilon(t,x) = \frac{1}{2}\epsilon^{-1/2}\exp\left(-\epsilon^{1/2}h^{\epsilon^{1/2}}(\epsilon^{-2}t,\epsilon^{-1}x) + \epsilon^{-3/2}\frac{t}{2}\right)$$

converges to the solution to the SHE with $\delta_{x=0}$ initial distribution. Convergence here means as probability measures on $C([\tilde{\delta},T],\mathbb{C}(\mathbb{R}))$ (continuous functions of time to the space of continuous spatial trajectories) for any $\tilde{\delta},T>0$.

What about weak noise scaling? One example of how this leads to the KPZ equation is through directed polymers, a class of models introduced in [**22**]. To motivate this, lets return to the heat equation with deterministic potential:

$$\partial_t z = \frac{1}{2}\partial_{xx}z + vz.$$

Recall the Feynman-Kac representation. Consider a Brownian motion run backwards in time, fixed so that $b(t)=x$, and let $\mathcal{E}_{b(t)=x}$ represent the expectation with respect to this Brownian motion. Then

$$z(t,x) = \mathcal{E}_{b(t)=x}\left[z_0(b(0))\exp\left(\int_0^t v(s,b(s))ds\right)\right].$$

EXERCISE 3.5. Prove this. Hint: This can be proved via a chaos series in v.

Even if z_0 is not a function, this can still make sense. For example, if $z_0(x) = \delta_{x=0}$ then the entire expression is killed if you don't end up at 0. Thus we can replaced Brownian motion with a Brownian bridge at the cost of a factor of $p(t,x)$.

What if $v = \xi_\kappa = (p_\kappa * \xi)(t,x)$? Then the solution, z_κ, to the mollified equation

$$\partial_t z_\kappa = \frac{1}{2}\partial_{xx}z_\kappa + \xi_\kappa z_\kappa$$

can be represented via the following path integral

$$z_\kappa(t,x) = \mathcal{E}_{b(t)=x}\left[z_0(b(0))\exp\left(\int_0^t \xi_\kappa(s,b(s))ds - \frac{c_\kappa(0)t}{2}\right)\right],$$

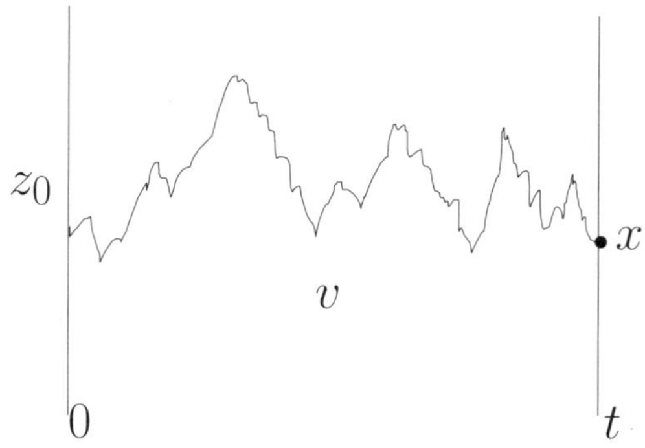

FIGURE 3.3. Path integral solution to heat equation with potential v.

where $c_\kappa(x) = (p_\kappa * p_\kappa)(x)$. See [**5**] for more on this.

Let us discretize this whole picture. We will discretize the white noise by i.i.d. random variables and the Brownian motion by a simple symmetric random walk. Consider up/right paths π on $(\mathbb{Z}_+)^2$. Let w_{ij} be i.i.d. random variables and $\beta > 0$ be inverse temperature. Let the partition function be

$$Z^\beta(M,N) = \sum_{\pi:(1,1)\to(M,N)} e^{\beta \sum_{(i,j)\in\pi} w_{ij}},$$

where the summation is over all paths π (as in Figure 3.4) which start at $(1,1)$, and take unit steps in the $(1,0)$ or $(0,1)$ directions and terminate at (M,N).

This satisfies a discrete SHE

$$Z^\beta(M,N) = e^{\beta w_{M,N}}\big(Z^\beta(M-1,N) + Z^\beta(M,N-1)\big).$$

The similarity to the SHE becomes more apparent if you make the change of variables $Z^\beta(M,N) = 2^{M+N}\tilde{Z}^\beta(M,N)$, $e^{\beta w_{i,j}} = 1 + \xi_{i,j}$ and iterate the recursion once more. The resulting equation can then be organized as

$$\begin{aligned}&\tilde{Z}^\beta(M,N) - \tilde{Z}^\beta(M-1,N-1)\\ &= \frac{1}{2}\Big(\tilde{Z}^\beta(M-2,N) - 2\tilde{Z}^\beta(M-1,N-1) + \tilde{Z}^\beta(M,N-2)\Big) + \text{multiplicative noise},\end{aligned}$$

where multiplicative noise is the sum of products of ξ's multiplied by single $\tilde{Z}^\beta$ terms. If $M+N$ is interpreted as time, and $M-N$ as space, this looks like a discrete SHE.

THEOREM 3.6 ([**1**]). *Let $w_{i,j}$ have finite eighth moment. Then there exist explicit constants $C(t,x,N)$ (depending on the distribution of $w_{i,j}$) such that*

$$Z_N(t,x) := C(t,x,N) Z^{2^{-1/2}N^{-1/4}}(tN + x\sqrt{N}, tN - x\sqrt{N})$$

converges as $N \to \infty$ to $z(2t,2x)$ where z is the solution of the SHE with $\delta_{x=0}$ initial data.

Here N is playing the role of ϵ^{-2} from before, and this is readily recognized as a weak noise scaling limit.

In the next section we will consider a semi-discrete polymer in which M has been taken large and the discrete sums of $w_{i,j}$ become Brownian increments. For a

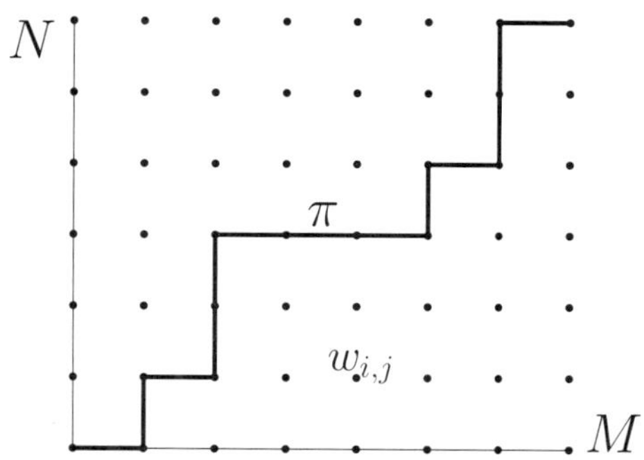

FIGURE 3.4. Discrete polymer path π moving through random potential of $w_{i,j}$.

modification of this semi-discrete polymer model we will give a proof of an analogous weak noise scaling limit.

4. Weak noise convergence of polymers to SHE

The O'Connell–Yor partition function (also called the semi–discrete polymer or semi-discrete SHE) at inverse temperature $\beta > 0$ is defined by

$$Z^\beta(T, N) := \int\limits_{0 \leqslant s_1 < s_2 < \cdots < s_{N-1} < T} d\vec{s} e^{\beta\left(B_1^N(0,s_1) + B_2^N(s_1,s_2) + \ldots B_N^N(s_{N-1},T)\right)},$$

where $B_i^N(a,b) = B_i^N(b) - B_i^N(a)$ with $\{B_i^N\}_{i=1}^N$ independent Brownian motions.

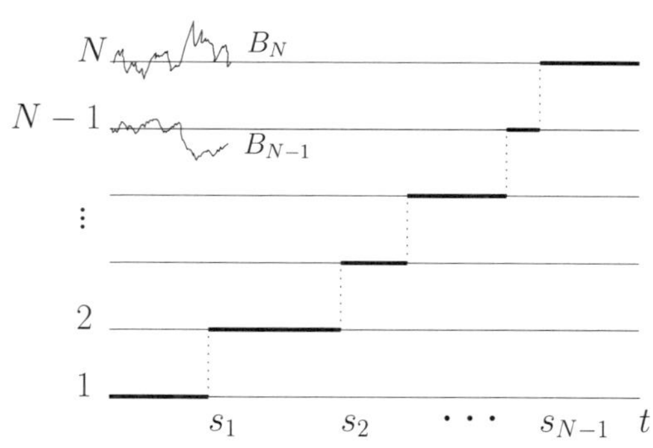

FIGURE 4.1. O'Connell–Yor polymer partition function.

THEOREM 4.1 ([**32**] (see also [**34**]). *There exist explicit constants $C(t,x,N)$ such that*

$$Z_N(t,x) := C(t,x,N) Z^{N^{-1/4}}(tN + x\sqrt{N}, tN)$$

converges as $N \to \infty$ to $z(t,x)/p(t,x)$ where z is the solution of the SHE with $\delta_{x=0}$ initial data and $p(t,x)$ is the heat kernel. Moreover, there exists a coupling of the Brownian motions B_j^N with ξ so the convergence occurs in $L^2(\Omega, \mathcal{F}, \mathbb{P})$ as well.

This theorem is proved by developing $Z^{\beta}(T,N)$ into a semi-discrete chaos series. Let us, therefore, recall the continuous chaos series for the solution to the SHE with delta initial data:

$$\frac{z(1,0)}{p(1,0)} = \sum_{k=0}^{\infty} \tilde{I}_k(1,0)$$

where, due to the division by $p(1,0)$ we have slightly modified chaos terms

$$\tilde{I}_k(1,0) = \int_{\Delta_k} d\vec{s} \int_{\mathbb{R}^k} d\vec{x} \rho_k(\vec{s},\vec{x}) \xi^{\otimes k}(\vec{s},\vec{x}).$$

Here we have again used the notation

$$\Delta_k(t) = \{0 \leqslant s_1 < \cdots < s_k < t\}, \qquad \text{with} \quad \Delta_k(1) =: \Delta_k$$

and

$$\rho_k(\vec{s},\vec{x}) = \frac{p_{k;1,0}(\vec{s},\vec{x})}{p(1,0)} = \frac{p(1-s_k,x_k)p(s_k-s_{k-1},x_k-x_{k-1})\cdots p(s_1,x_1)}{p(1,0)}.$$

Note that $\rho_k(\vec{s},\vec{x})$ represents the transition density of a Brownian bridge from $(0,0)$ to $(1,0)$ going through the points $\vec{x}$ at times $\vec{s}$. We will not prove Theorem 4.1, but instead will prove a result about a slightly modified polymer partition function, which is itself a key step in proving Theorem 4.1.

DEFINITION 4.2. The modified semi-discrete polymer partition function is defined as

$$\tilde{Z}^{\beta}(T,N) = |\Delta_{N-1}(T)|^{-1} \int_{\Delta_{N-1}(T)} d\vec{s} \prod_{j=1}^{N} \big(1+\beta B_j(s_{j-1},s_j)\big),$$

where by convention $s_0 = 0$ and $s_N = T$.

Note that for a measurable set $A \in \mathbb{R}^n$, we use $|A|$ represent the Lebesgue measure of the set. Also, observe that for small β we have the first order expansion $e^{\beta B_j(s_{j-1},s_j)} \approx 1+\beta B_j(s_{j-1},s_j)$.

THEOREM 4.3 ([**32**]). *As $N \to \infty$, we have the following weak convergence*

$$\tilde{Z}^{N^{-1/4}}(N,N) \to \frac{z(1,0)}{p(1,0)}.$$

Moreover, there exists a coupling of the B_j^N with ξ so that this convergence occurs in $L^2(\Omega,\mathcal{F},\mathbb{P})$.

Note that we have included superscript N on the B_j^N since, though for each N $\{B_j^N\}_{j=1}^N$ are marginally independent Brownian motions, as N varies these collections of curves are not independent and, in fact, are built from the same probability space on which ξ is defined. The $L^2(\Omega,\mathcal{F},\mathbb{P})$ convergence certainly implies the weak convergence.

PROOF OF THEOREM 4.3. We proceed in three steps. In Step (1) we describe the coupling of the B_j^N with ξ. In Step (2) we express $\tilde{Z}^{\beta}(N,T)$ as a semi–discrete chaos series and then using the coupling to express this as a continuous chaos series in ξ. Finally, in Step (3) we use convergence of the Poisson jump process to Brownian bridge to conclude termwise convergence of the chaos series, and the convergence of the entire chaos series too.

Step (1): Define the function

$$\varphi^N(s,x) = \left(\frac{s}{N}, \frac{x-s}{\sqrt{N}}\right).$$

This maps the square $[0,N]^2$ to the rhombus $\{(t,y) : 0 \leqslant t \leqslant 1, -t\sqrt{N} \leqslant y \leqslant (1-t)\sqrt{N}\}$. Call

$$I_j^N(t) = \left[\frac{j-1}{\sqrt{N}} - \sqrt{N}t, \frac{j}{\sqrt{N}} - \sqrt{N}t\right].$$

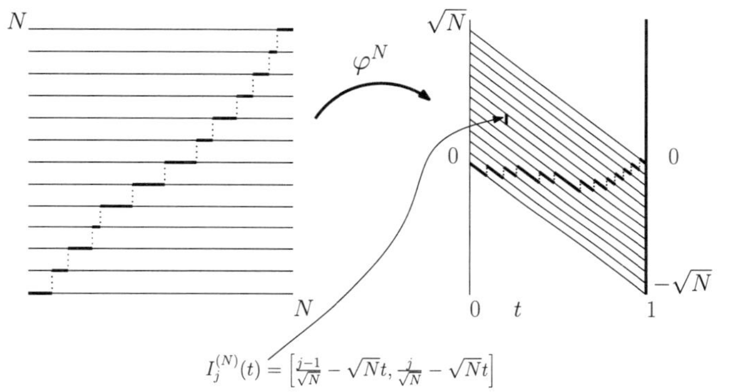

FIGURE 4.2. The mapping φ^N used to build the coupling of the B_j^N with ξ.

EXERCISE 4.4. Show that since φ^N has Jacobian $N^{-3/2}$, if we define

$$B_j^N(s_1,s_2) = N^{3/4}\int_{s_1/N}^{s_2/N} ds \int_{I_j^N(s)} dx\xi(s,x),$$

then $\{B_j^N(0,s)\}_{j\in[N],s\in[0,N]}$ are independent standard Brownian motions.

We will use this coupling.

Step (2): Let us rewrite $\tilde{Z}^\beta(T,N)$ via an expectation. Let $X_\bullet$ be distributed as the trajectory on $[0,T]$ of a Poisson jump process with $X_0 = 1$ (i.e. starting at 1 and increasing by 1 after independent exponential waiting times).

EXERCISE 4.5. Show that conditioning $X_\bullet$ on the event that $X_T = N$, the jump times of $X_\bullet$ (labeled $s_1, \ldots, s_{N-1}$) are uniform over the simplex $0 < s_1 < \ldots < s_{N-1} < T$ with measure

$$\frac{d\vec{s}}{|\Delta_{N-1}(T)|}\mathbf{1}_{\vec{x}\in\Delta_{N-1}(T)}.$$

We may rewrite

$$\tilde{Z}^{\beta}(T,N) = \mathbb{E}\left[\prod_{j=1}^{N}\left(1+\beta\int_0^T \mathbf{1}_{X_s=j}dB_j(s)\right)\Big|X_T=N\right],$$

$$= \sum_{k=0}^{N}\beta^k\tilde{J}_k(T,N)$$

$$\tilde{J}_k(T,N) = \mathbb{E}\left[\int_{[0,T]^k}\sum_{\vec{i}\in D_k^N}\prod_{j=1}^{k}\mathbf{1}_{X_{s_j}=i_j}dB_{i_j}(s_j)|X_T=N\right],$$

where

$$D_k^N = \{i\in\mathbb{Z}_+^k : 1\leqslant i_1 < i_2 < \cdots < i_k \leqslant N\}.$$

and (for use later)

$$\partial D_k^N = \{i\in\mathbb{Z}_+^k : 1\leqslant i_1 \leqslant i_2 \leqslant \cdots \leqslant i_k \leqslant N \text{ with at least one equality}\}.$$

The expectation can be taken inside yielding

$$\tilde{J}_k(T,N) = \int_{[0,T]^k}\sum_{\vec{i}\in D_k^N}\mathbb{P}(X_{s_1}=i_1,\ldots,X_{s_k}=i_k|X_T=N)dB_{i_1}(s_1)\cdots dB_{i_k}(s_k).$$

We would like to use the coupling now, so it makes sense to change our probabilities into the variables of the image of φ^N. Call

$$\rho^N(s,t,x,y) = \mathbb{P}\left(X_{N(t-s)} = \lceil\sqrt{N}y+Nt\rceil - \lceil\sqrt{N}x+Ns\rceil\right)$$

and

$$\rho_k^N(\vec{s},\vec{x}) = \frac{N^{k/2}\rho^N(s_k,1;x_k,0)\prod_{j=1}^k\rho^N(s_{j-1},s_j,x_{j-1},x_j)}{\rho^N(0,1;0,0)},$$

where $\vec{s}\in\Delta_k$ and $\vec{x}\in\mathbb{R}^k$. This is the semi–discrete analog of $\rho_k(\vec{s},\vec{x})$. The scaling $N^{k/2}$ comes from local limit theorems. For $f\in L^2(\Delta_k\times\mathbb{R}^k)$ define

$$\|f\|_2^2 = \int_{\Delta_k}\int_{\mathbb{R}^k} f(\vec{s},\vec{x})^2 d\vec{s}d\vec{x}.$$

Using our coupling and the formula for ρ_k^N we can rewrite

$$\tilde{J}_k(N,N) = N^{k/4}\int_{[0,1]^k}d\vec{s}\sum_{\vec{i}\in D_k^N}\prod_{j=1}^{k}\int_{I_{i_j}^N(s_j)}dx_j\rho_k^N(\vec{s},\vec{x})\xi^{\otimes k}(\vec{s},\vec{x}).$$

The $N^{1/4}$ is the product of $N^{-1/2}$ from the definition of ρ_k^N and $N^{3/4}$ from the coupling.

Step (3): Let us start by stating two lemma (we will not prove them here, though their proofs are not so hard).

LEMMA 4.6. *There exists a constant $c\in(0,\infty)$ such that for all $k,N\geqslant 1$, $\|\rho_k^N\|_2^2\leqslant c^k\|\rho_k\|_2^2$.*

LEMMA 4.7. *There exists a constant $c\in(0,\infty)$ such that for all $N\geqslant 1$ we have $\sum_{k=1}^N\|\rho_k^N\|_2^2\leqslant C$ and for each $k\geqslant 1$, $\lim_{N\to\infty}\|\rho_k^N-\rho_k\|_2^2 = 0$.*

Given these lemmas, we deduce the following (which we prove).

LEMMA 4.8. *The following convergence occurs in* $L^2(\Omega, \mathcal{F}, \mathbb{P})$*:*

$$\lim_{N\to\infty} N^{-k/4}\tilde{J}_k(N,N) = \tilde{I}_k(1,0).$$

PROOF. Observe that

$$N^{-k/4}\tilde{J}_k(N,N) = \int_{[0,1]^k} \sum_{\vec{i}\in[N]^k} \prod_{j=1}^{k} \int_{I_{i_j}^N(s_j)} dx_j \rho_k^N(\vec{s},\vec{x})\xi^{\otimes k}(\vec{s},\vec{x})$$
$$- \int_{[0,1]^k} \sum_{\vec{i}\in\partial D_k^N} \prod_{j=1}^{k} \int_{I_{i_j}^N(s_j)} dx_j \rho_k^N(\vec{s},\vec{x})\xi^{\otimes k}(\vec{s},\vec{x})$$

In the first term we can replace $[0,1]^k$ by $\Delta_k(1)$ and the $\sum\prod\int$ can be replaced by $\int_{\mathbb{R}^k}$ (all since ρ_k^N vanishes otherwise). By Lemma 4.7, we can conclude that the first term limits to $\tilde{I}_k(1,0)$. Call E_k^N the second term. By Lemma 4.6

$$\mathbb{E}\big[(E_k^N)^2\big] \leqslant c^k \int_{\Delta_k} d\vec{s} \sum_{i\in\partial D_k^N} \prod_{j=1}^{k} \int_{I_{i_j}^N(s_j)} dx_j (\rho_k(\vec{s},\vec{x}))^2 = \mathcal{O}(c^k/N). \tag{4.1}$$

The $1/N$ factor can be understood as the cost of staying in the same level for two samples of times. □

We can now complete the proof of Theorem 4.3. From Lemma 4.8 we know that for each $M \geqslant 1$,

$$\lim_{N\to\infty} \sum_{k=0}^{M} N^{-k/4}\tilde{J}_k(N,N) = \sum_{k=0}^{M} \tilde{I}_k(1,0)$$

with convergence in $L^2(\Omega,\mathcal{F},\mathbb{P})$. From Section 3 we know that for all $\epsilon > 0$ there exists $M \geqslant 1$ such that

$$\sum_{k=M}^{\infty} \tilde{I}_k(1,0) < \epsilon.$$

Thus, it remains to show that for all $\epsilon > 0$ there exists $M \geqslant 1$ such that

$$\sum_{k=M}^{\infty} N^{-k/4}\tilde{J}_k(N,N) < \epsilon,$$

once $N > M$. By Lemma 4.6 we can bound each term on the left-hand side of the above expression by a constant times

$$\|\rho_k^N(\vec{t},\vec{x})\|_2^2 + \mathbb{E}\big[(E_k^N)^2\big].$$

By Lemma 4.6 and equation (4.1) these are both bounded by constants times $\|\rho_k(\vec{t},\vec{x})\|_2^2$. This shows that

$$\sum_{k=M}^{\infty} N^{-k/4}\tilde{J}_k(N,N) \leqslant c \sum_{k=M}^{\infty} \tilde{I}_k(1,0)$$

and hence taking M large enough this can be bounded by ϵ as desired. □

5. Moments of q–TASEP via duality and Bethe ansatz

The past few sections have focused on making sense of the KPZ equation (via the SHE) and then deriving various approximations for it. We turn to one such approximation which enjoys certain exact solvability. This exact solvability goes through in the limit to the KPZ equation / SHE. In that context, the exact solvability amounts to the ability to write down simple and exact formulas for the expectation (over the white noise ξ) of the solution to the SHE with delta initial data. Unfortunately, the moment problem for the SHE is not well-posed, meaning the knowledge of these moments does not determine its distribution. Thus, despite the many formulas, this does not provide a direct rigorous route to exactly solving the equation. By going to the below discretization, we preserve this exact solvability while also resolving the issue of the moment problem well-posedness.

The interacting particle system q–TASEP will be the main subject of the remaining sections (for more on this model, see [**8**, **13**, **15**, **16**]). It is a model for traffic on a one–lane road where cars slow down as they approach the next one. The scale on which this occurs is controlled by $q \in (0, 1)$.

Particles are labeled right to left as $x_1(t) > x_2(t) > \ldots$, where $x_i(t)$ records the location of particle i at time t. In continuous time each particle can jump right by one according to independent exponential clocks of rate $1 - q^{\text{gap}}$, where gap is the number of empty sites before the next right particle. In other words, for particle i the gap is $x_{i-1} - x_i - 1$.

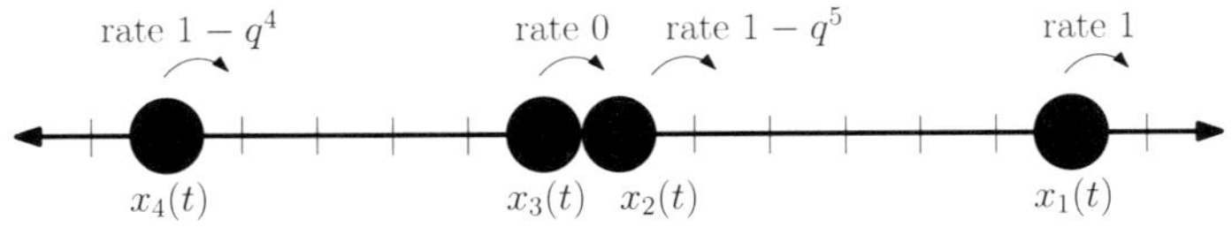

FIGURE 5.1. The q–TASEP with particles labeled $x_i(t)$ and jump rates written in.

Since the jumping of particle i only depends on $i - 1$, it suffices to consider an N–particle restriction. In this case, the state space is given by

$$X^N := \{\vec{x} = (x_0, \ldots, x_N) : \infty = x_0 > x_1 > \ldots > x_N\}.$$

The role of fixing $x_0 = \infty$ is to have a virtual particle to make notation nicer. This system can be encoded as a continuous time Markov process on X^N.

Recall the following facts about Markov processes on X (see [**28**] for more background on Markov processes and interacting particle systems). A Markov process is defined via a semigroup $\{S_t\}_{t\geqslant 0}$ with $S_{t_1}S_{t_2} = S_{t_1+t_2}$ and $S_0 = \text{Id}$, acting on a suitable domain of functions $f : X \to \mathbb{R}$. For $f : X \to \mathbb{R}$ in the domain of S, define $\mathbb{E}^x\big[f(x(t))\big] = S_t f(x)$, where X is the state space and $\mathbb{E}^x$ is expectation with respect to starting state $x(0) = x$. The generator of a Markov process is defined as

$$L = \lim_{t\to 0} \frac{S_t - \text{Id}}{t}$$

and captures the process since

$$S_t = e^{tL} = \sum_{k\geqslant 0} \frac{1}{k!}(tL)^k.$$

It follows then that $\frac{d}{dt}S_t = S_t L = L S_t$, so for $f : X \to \mathbb{R}$

$$\frac{d}{dt}\mathbb{E}^x\big[f(x(t))\big] = \mathbb{E}^x\big[Lf(x(t))\big] = L\mathbb{E}^x\big[f(x(t))\big].$$

The generator of q–TASEP acts on $f : X^N \to \mathbb{R}$ as

$$(L^{q-TASEP} f)(\vec{x}) := \sum_{i=1}^{N}(1 - q^{x_{i-1}-x_i-1})\left(f(\vec{x_i}^{\,+}) - f(\vec{x})\right),$$

where $\vec{x_i}^{\,+} = (x_0, x_1, \ldots, x_i + 1, \ldots, x_N)$ represents the movement of particle i by 1 to the right. Note that we will not worry here about specifying the domain of the q–TASEP generator.

Consider the functions $f_1, \ldots, f_N$ defined by $f_n(\vec{x}) = q^{x_n}$. How can $q^{x_n(t)}$ change in an instant of time?

$$\begin{aligned} dq^{x_n}(t) &= (1 - q^{x_{n-1}-x_n-1})(q^{x_n+1} - q^{x_n})dt + \text{ noise (i.e. martingale)} \\ &= (1-q)(q^{x_{n-1}-1} - q^{x_n})dt + \text{ noise }. \end{aligned}$$

Now taking expectations gives

$$\frac{d}{dt}\mathbb{E}[q^{x_n(t)}] = (1-q)\left(\mathbb{E}[q^{x_{n-1}(t)-1}] - \mathbb{E}[q^{x_n(t)}]\right).$$

which shows that as a function of $n \in \{1, \ldots, N\}$ the expectations $\mathbb{E}[q^{x_n(t)}]$ solve a triangular, closed system of linear ordinary differential equations.

In order to generalize this system, and then ultimately solve it, we will use a Markov duality.

DEFINITION 5.1. Suppose $x(\cdot), y(\cdot)$ are independent Markov processes with state spaces X, Y and generators L^X, L^Y. Let $H : X \times Y \to \mathbb{R}$ be bounded and measurable. Then we say $x(\cdot)$ and $y(\cdot)$ are dual with respect to H if for all $x \in X, y \in Y$, $L^X H(x,y) = L^Y H(x,y)$.

EXERCISE 5.2. Show that duality implies $\mathbb{E}^x[H(x(t), y)] = \mathbb{E}^y[H(x, y(t))]$ for all t and hence

$$\frac{d}{dt}\mathbb{E}^x\big[H(x(t), y)\big] = L^X\mathbb{E}^x\big[H(x(t), y)\big] = L^Y\mathbb{E}^x\big[H(x(t), y)\big].$$

It turns out that q–TASEP enjoys a simple duality with a Markov process called the q–Boson process [**13**, **38**] (which is a special totally asymmetric zero range process) with state space

$$Y^N = \big\{\vec{y} = (y_0, \ldots, y_N) | y_i \in \mathbb{Z}_{\geqslant 0}\big\},$$

in which there can be y_i particles above site i and in continuous time. According to independent exponential clocks, one particle moves from i to $i-1$ at rate $1 - q^{y_i}$, nothing enters site N and nothing exits site 0 (so the total number of particles is conserved over time).

The generator of the q–Boson process acts on functions $h : Y^N \to \mathbb{R}$ as

$$(L^{\text{q-Boson}} h)(\vec{y}) := \sum_{i=1}^{N}(1 - q^{y_i})\big(h(\vec{y}^{\,i,i-1}) - h(\vec{y})\big),$$

where $\vec{y}^{\,i,i-1} = (y_0, y_1, \ldots, y_{i-1}+1, y_i - 1, \ldots, y_N)$ represents moving a particle from i to $i-1$. Note that q–TASEP gaps evolve according to the q–Boson jump rates. So, in a sense, the duality we now present is actually a self-duality.

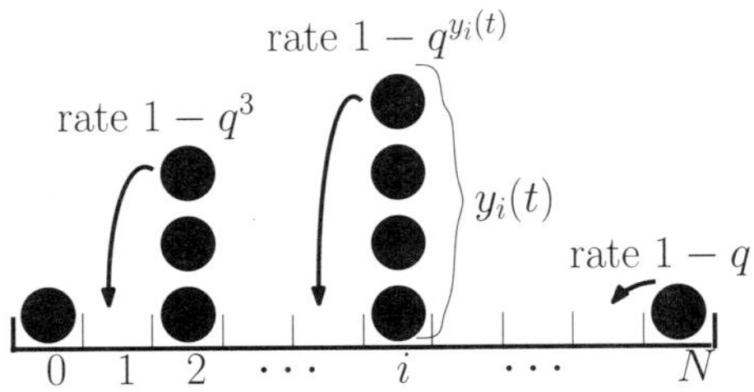

FIGURE 5.2. The q–Boson process.

THEOREM 5.3 ([**13**]). *As Markov processes, q–TASEP $\vec{x}(t) \in X^N$ and q–Boson process $\vec{y}(t) \in Y^N$ are dual with respect to*

$$H(\vec{x},\vec{y}) = \prod_{i=0}^{N} q^{(x_i+i)y_i}$$

with the convention that $H=0$ if $y_0>0$ and otherwise the product starts at $i=1$.

PROOF. For all $\vec{x},\vec{y}$,

$$\begin{aligned}(L^{\text{q-TASEP}}H)(\vec{x},\vec{y}) &= \sum_{i=1}^{N}(1-q^{x_{i-1}-x_i-1})(H(\vec{x}_i^{+},\vec{y})-H(\vec{x},\vec{y}))\\ &= \sum_{i=1}^{N}(1-q^{x_{i-1}-x_i-1})(q^{y_i}-1)\prod_{j=0}^{N}q^{(x_j+j)y_j}\\ &= \sum_{i=1}^{N}(1-q^{y_i})(H(\vec{x},\vec{y}^{i,i-1})-H(\vec{x},\vec{y}))\\ &= (L^{\text{q-Boson}}H)(\vec{x},\vec{y})\end{aligned}$$

□

We will use this theorem to solve for joint moments of $q^{x_n(t)+n}$ as n varies in $\{1,\ldots,N\}$. In particular if we treat $\vec{x}$ as fixed then the duality implies

$$\frac{d}{dt}\mathbb{E}^{\vec{x}}\left[\prod_{i=0}^{N}q^{(x_i(t)+i)y_i}\right] = L^{\text{q-Boson}}\mathbb{E}^{\vec{x}}\left[\prod_{i=0}^{N}q^{(x_i(t)+i)y_i}\right], \tag{5.1}$$

where $L^{\text{q-Boson}}$ acts in the $\vec{y}$ variables.

PROPOSITION 5.4. *Fix q–TASEP initial data $\vec{x} \in X^N$. If $h:\mathbb{R}_{\geq 0}\times Y^N \to \mathbb{R}$ solves*

(1) For all $\vec{y}\in Y^N$ and $t \in \mathbb{R}_{\geq 0}$

$$\frac{d}{dt}h(t;\vec{y}) = L^{q\text{-}Boson}h(t;\vec{y});$$

(2) For all $\vec{y}\in Y^N$,

$$h(0;\vec{y}) = h_0(\vec{y}) := H(\vec{x},\vec{y});$$

Then for all $\vec{y}\in Y^N$ and $t\in\mathbb{R}_{\geq 0}$, $\mathbb{E}^{\vec{x}}\big[H(\vec{x}(t),\vec{y})\big] = h(t;\vec{y})$.

PROOF. By (5.1), Proposition 5.4(1) must hold for $\mathbb{E}^{\vec{x}}[H(\vec{x}(t), \vec{y})]$ and Proposition 5.4(2) follows by definition. Uniqueness follows because $L^{\text{q-Boson}}$ preserves the number of particles and restricts to a triangular system of coupled ODEs on each k–particle subspace. Then, standard ODE uniqueness results [**14**] imply uniqueness. In other words, the system is closed due to particle conservation of the q–Boson process. □

Since the q–Boson process acts on k–particle subspaces, it preserves the state spaces

$$Y_k^N = \{\vec{y} \in Y^N : \sum y_i = k\}.$$

For $\vec{y} \in Y_k^N$ we can associate a vector of weakly ordered particle locations in

$$W_{\geqslant 0}^k = \{\vec{n} = (n_1 \geqslant \ldots \geqslant n_k \geqslant 0)\}.$$

Write $\vec{n}(\vec{y})$ and $\vec{y}(\vec{n})$ for this association. For example, if $N = 3$ and $k = 4$ and $\vec{y} = (y_0 = 0, y_1 = 3, y_2 = 0, y_3 = 1)$ then $\vec{n} = (3, 1, 1, 1)$. We will abuse notation and write $h(t, \vec{n}) := h(t, \vec{y}(n))$.

In order to solve the system of equations for h, its more convenient to work in these n coordinates. Let us consider how the system looks for different k.

For $k = 1$, $\vec{n} = (n)$ and the evolution equation becomes

$$\tfrac{d}{dt} h(t; \vec{n}) = (1 - q)\nabla h(t; \vec{n}),$$

where $(\nabla f)(n) = f(n - 1) - f(n)$. This is just the generator of a single q–Boson particle moving.

For $k = 2$, $\vec{n} = (n_1 \geqslant n_2)$ and we must consider two cases

- If $n_1 > n_2$, then we have for Proposition 5.4(1)

$$\frac{d}{dt} h(t; n_1, n_2) = \sum_{i=1}^{2} (1 - q)\nabla_i h(t; n_1, n_2).$$

- If $n_1 = n_2 = n$ then

$$\frac{d}{dt} h(t; n, n) = (1 - q^2)\nabla_2 h(t; n, n),$$

  where we have chosen ∇_2 in order to preserve the order of $\vec{n}$ so as to stay in $W_{\geqslant 0}^k$. Unfortunately this is not constant coefficient or separable so it is not a priori clear how to solve it. Moreover, as k grows the number of boundary cases may grow quite rapidly like 2^{k-1}.

To resolve this, we use an idea of Bethe [**6**] from 1931. We try to rewrite in terms of solution to k particle free evolution equation subject to $k - 1$ two–body boundary conditions. Usually this is not possible and one has many body boundary conditions, but if it is possible then we say the system is coordinate Bethe ansatz solvable. Let us see this idea in motion.

For $k = 2$ consider $u : \mathbb{R}_{\geqslant 0} \times \mathbb{Z}_{\geqslant 0}^2 \to \mathbb{R}$ which satisfies

$$\frac{d}{dt} u(t; \vec{n}) = \sum_{i=1}^{2} (1 - q)\nabla_i u(t; \vec{n}).$$

Then, when $n_1 > n_2$ this right–hand–side exactly matches that of the true evolution equation. However, for $n_1 = n_2 = n$ the two right hand sides differ by

$$\sum_{i=1}^{2}(1-q)\nabla_i u(t;n,n) - (1-q^2)\nabla_2 u(t;n,n). \tag{5.2}$$

If we could find a u such that (5.2) $\equiv 0$ then when we restrict u to $\{\vec{n} : n_1 \geqslant n_2\}$, it will actually solve the true evolution equation: $u|_{\vec{n}\in W^2_{\geqslant 0}} = h$. The boundary condition is equivalent to

$$(\nabla_1 - q\nabla_2)u|_{n_1=n_2} \equiv 0.$$

The only way we might hope to find such a u is to tinker with the initial data outside the set $W^2_{\geqslant 0}$. In a sense, this is like an advanced version of the reflection principle.

For $k = 3$, we might need more than two body boundary conditions (e.g. $n_1 = n_2 = n_3$), but amazingly all higher order cases follow from the two body cases. Hence we have a coordinate Bethe ansatz solvable system. This is shown by the following result.

PROPOSITION 5.5. *If* $u : \mathbb{R}_{\geqslant 0} \times \mathbb{Z}^k_{\geqslant 0} \to \mathbb{R}$ *solves*
(1) For all $\vec{n} \in \mathbb{Z}^k_{\geqslant 0}$ *and* $t \in \mathbb{R}_{\geqslant 0}$,

$$\frac{d}{dt}u(t;\vec{n}) = \sum_{i=1}^{k}(1-q)\nabla_i u(t;\vec{n});$$

(2) For all $\vec{n} \in \mathbb{Z}^k_{\geqslant 0}$ *such that* $n_i = n_{i+1}$ *for some* $i \in \{1,\dots,k-1\}$ *and all* $t \in \mathbb{R}_{\geqslant 0}$,

$$(\nabla_i - q\nabla_{i+1})u(t;\vec{n}) = 0;$$

(3) For all $\vec{n} \in W^k_{\geqslant 0}$, $u(0;\vec{n}) = h_0(\vec{n})$;
Then for all $\vec{n} \in W^k_{\geqslant 0}$, $h(t;\vec{n}) = u(t;\vec{n})$.

EXERCISE 5.6. Prove the proposition. Hint: if $n_1 = \cdots = n_c > n_{c+1}$ then use the boundary condition to replace $\sum_{i=1}^{c}(1-q)\nabla_i$ by $(1-q^c)\nabla_c$, and use this to rewrite the free evolution equation into $L^{\text{q-Boson}}$ (when written in terms of the $\vec{n}$ variables).

It is possible to solve this for general h, cf. [**11**]. We will focus only on the particular case $h_0(\vec{n}) = \prod_{i=1}^{k}\mathbf{1}_{n_i>0}$. This initial data corresponds (via Proposition 5.4) with studying step initial data q–TASEP where $x_i(0) = -i$ for $i = 1,\dots,N$. Then $H\big(\vec{x},\vec{y}(\vec{n})\big) = \prod_{i=1}^{k} q^{x_{n_i}(t)+n_i} = 1$ as long as all $n_i \geqslant 0$. Thus, taking $h_0(\vec{n}) = \prod_{i=1}^{k}\mathbf{1}_{n_i>0}$ and solving the system in Proposition 5.5 we find that for $\vec{n} \in W^k_{\geqslant 0}$,

$$\mathbb{E}^{\text{step}}\left[\prod_{i=1}^{k} q^{x_{n_i}(t)+n_i}\right] = u(t;\vec{n}).$$

The idea for solving this (in fact for general initial data) also traces back to Bethe [**6**]. First, solve the one particle free evolution equation (fundamental solution). Then, use linearity to take superpositions of fundamental solutions in order to try to satisfy boundary and initial conditions. It is not a priori obvious that this will work ... but it does.

6. Exact formulas and the replica method

The goal of the first part of this section is to explicitly solve the true evolution equation with initial data $h_0(\vec{n}) = \prod_{i=1}^{k} \mathbf{1}_{n_i>0}$. This will yield a formula for $\mathbb{E}^{\text{step}}[\prod_{i=1}^{k} q^{x_{n_i}(t)+n_i}]$. We will utilize Proposition 5.5 which reduces this to solving the free evolution equation subject to two–body boundary conditions. Let us start with the case $k = 1$ (for which there are no boundary conditions). For $z \in \mathbb{C}\backslash\{1\}$ define

$$u_z(t;n) := \frac{e^{(q-1)tz}}{(1-z)^n}.$$

It is immediate to check that Proposition 5.5(1) is satisfied:

$$\frac{d}{dt}u_z(t;n) = (1-q)\nabla u_z(t;n).$$

Since $k = 1$, Proposition 5.5(2) is not present so it remains to check 3). Note that linear combinations of $u_z(t;n)$ over z's still solve Proposition 5.5(1). Consider then

$$u(t;n) = \frac{-1}{2\pi i}\oint u_z(t;n)\frac{dz}{z},$$

where the contour contains 1 but not 0. When $t = 0$ and $n \geqslant 1$ expand contours to infinity. We have at least z^{-2} decay at infinity hence no residue (recall from [**2**]) at infinity. However, we crossed a first order pole at $z = 0$ which gives $u(0;n) \equiv 1$. When $n \leqslant 0$, there is, on the other hand, no pole at at $z = 1$, hence the integral is necessarily zero. This shows Proposition 5.5(3).

Hence we have proved that

$$\mathbb{E}^{\text{step}}\big[q^{x_n(t)+n}\big] = u(t;n) = \frac{-1}{2\pi i}\oint \frac{e^{(q-1)tz}}{(1-z)^n}\frac{dz}{z}.$$

We will now state the general k version of this result. This may appear as something of a rabbit pulled out of a hat. The origins of this formula are demystified in Section 9.

THEOREM 6.1. *Proposition* 5.5 *with* $h_0(\vec{n}) = \prod_{i=1}^{k} \mathbf{1}_{n_i>0}$ *is solved by*

$$u(t;\vec{n}) = \frac{(-1)^k q^{k(k-1)/2}}{(2\pi i)^k}\oint\cdots\oint \prod_{1\leqslant A<B\leqslant k}\frac{z_A - z_B}{z_A - qz_B}\prod_{j=1}^{k} u_{z_j}(t;n_j)\frac{dz_j}{z_j},$$

with contours such that the z_A *contour contains* $\{qz_B\}_{B>A}$ *and* 1 *but not* 0.

COROLLARY 6.2. *For all* $\vec{n} \in W^k_{\geqslant 0}$,

$$\begin{aligned}\mathbb{E}^{step}\left[\prod_{i=1}^{k} q^{x_{n_i}(t)+n_i}\right] &= u(t;\vec{n})\\ &= \frac{(-1)^k q^{k(k-1)/2}}{(2\pi i)^k}\oint\cdots\oint \prod_{1\leqslant A<B\leqslant k}\frac{z_A - z_B}{z_A - qz_B}\prod_{j=1}^{k}\frac{e^{(q-1)tz_j}}{(1-z_j)^{n_j}}\frac{dz_j}{z_j}.\end{aligned}$$

Since $q \in (0,1)$ and $x_n(t)+n \geqslant 0$, the observable $q^{x_n(t)+n} \in (0,1]$. Hence, knowledge of the joint moments of $\{q^{x_n(t)+n}\}_{n\in\{1,\dots,N\}}$ for fixed t completely characterizes their distribution, and hence also the distribution of $\{x_n(t)\}_{n\in\{1,\dots,N\}}$! Later, we will see how to extract some useful distributional formulas from these moments.

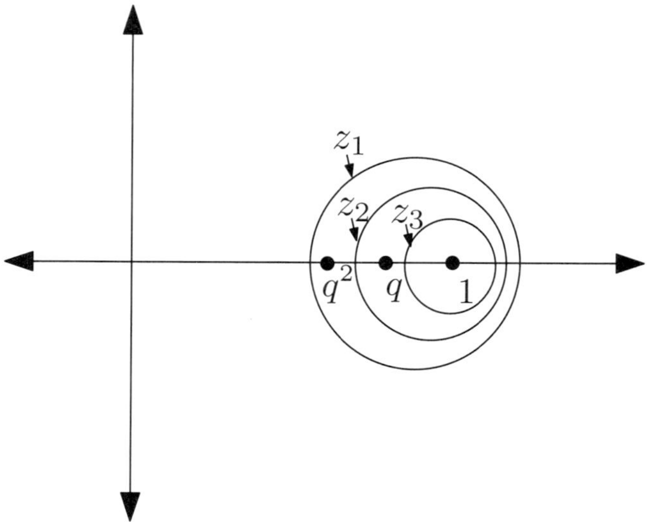

FIGURE 6.1. Nested contours for Theorem 6.1.

EXERCISE 6.3. Prove Theorem 6.1. Parts (1) and (3) clearly generalize the $k = 1$ case. For (2) study the effect of applying $\nabla_i - q\nabla_{i+1}$ to integrand, and show integral of result is 0.

In the remaining part of this section I will study how q–TASEP and its duality and moment formulas behave as $q \nearrow 1$. Doing so, we will encounter a semi–discrete stochastic heat equation or equivalently the O'Connell–Yor semi–discrete directed polymer model. In our next section we will return to q–TASEP and study some distributional properties.

From duality we know that the q–TASEP dynamics satisfy (for some Martingale $M_n(t)$)

$$\begin{aligned} dq^{x_n(t)+n} &= (1-q)\nabla q^{x_n(t)+n}dt + q^{x_n(t)+n}dM_n(t) \\ q^{x_n(0)+n} &\equiv \mathbf{1}_{n\geqslant 1} \text{ (for step initial data).} \end{aligned}$$

This suggests that as $q \nearrow 1$ the (properly scaled) observable $q^{x_n(t)+n}$ might converge to the semi–discrete SHE.

DEFINITION 6.4. A function $z : \mathbb{R}_{\geqslant 0} \times \mathbb{Z}_{\geqslant 0} \to \mathbb{R}_{\geqslant 0}$ solves the semi–discrete SHE with initial data $z_0 : \mathbb{Z}_{\geqslant 0} \to \mathbb{R}_{\geqslant 0}$ if:

$$\begin{aligned} dz(\tau;n) &= \nabla z(\tau;n)d\tau + z(\tau;n)dB_n(\tau) \\ z(0;n) &= z_0(n). \end{aligned}$$

THEOREM 6.5 ([**8**]). *For q–TASEP with step initial data, set*

$$q = e^{-\epsilon}, \qquad t = \epsilon^{-2}\tau, \qquad x_n(t) = \epsilon^{-2}\tau - (n-1)\epsilon^{-1}\log\epsilon^{-1} - \epsilon^{-1}\mathcal{F}_\epsilon(\tau;n)$$

(this third equation above defines $\mathcal{F}_\epsilon$) and call $z_\epsilon(\tau;n) = \exp\{-3\tau/2 + \mathcal{F}_\epsilon(\tau;n)\}$. Then as a space time process $z_\epsilon(\tau;n)$ converges weakly to $z(\tau;n)$ with $z_0(n) = \mathbf{1}_{n=1}$ initial data.

HEURISTIC PROOF SKETCH. The initial data is clear since $z_\epsilon(0;n) = \epsilon^{n-1}e^{\epsilon n} \to \mathbf{1}_{n=1}$.

As for the dynamics, observe that for $d\tau$ small,

$$\begin{aligned} d\mathcal{F}_\epsilon(\tau;n) &\approx \mathcal{F}_\epsilon(\tau;n) - \mathcal{F}_\epsilon(\tau - d\tau;n) \\ &= \epsilon^{-1}d\tau - \epsilon\big(x_n(\epsilon^{-2}\tau) - x_n(\epsilon^{-2}\tau - \epsilon^{-2}d\tau)\big) \end{aligned}$$

Under scaling, the q–TASEP jump rates are given by

$$1 - q^{x_{n-1}(t)-x_n(t)-1} = 1 - \epsilon e^{\mathcal{F}_\epsilon(\tau;n-1)-\mathcal{F}_\epsilon(\tau;n)} + O(\epsilon^2).$$

So in time $\epsilon^{-2}d\tau$, by convergence of Poisson point process to Brownian motion

$$\begin{aligned} &\epsilon\big(x_n(\epsilon^{-2}\tau) - x_n(\epsilon^{-2}\tau - \epsilon^{-2}d\tau)\big) \\ &\qquad \approx \epsilon^{-1}d\tau - e^{\mathcal{F}_\epsilon(\tau;n-1)-\mathcal{F}_\epsilon(\tau;n)}d\tau - (B_n(\tau) - B_n(\tau - d\tau)). \end{aligned}$$

Thus we see that (using little oh notation)

$$d\mathcal{F}_\epsilon(\tau;n) \approx e^{\mathcal{F}_\epsilon(\tau;n-1)-\mathcal{F}_\epsilon(\tau;n)}d\tau + dB_n(\tau) + o(1)$$

Exponentiating and applying Itô's lemma gives

$$de^{\mathcal{F}_\epsilon(\tau;n)} = \left(\frac{1}{2}e^{\mathcal{F}_\epsilon(\tau;n)} + e^{\mathcal{F}_\epsilon(\tau;n-1)}\right)d\tau + e^{\mathcal{F}_\epsilon(\tau;n)}dB_n(\tau) + o(1)$$

or going to z_ϵ,

$$dz_\epsilon(\tau;n) = \nabla z_\epsilon(\tau;n)d\tau + z_\epsilon(\tau;n)dB_n(\tau) + o(1)$$

and as $\epsilon \searrow 0$ we recover the semi–discrete SHE. □

Theorem 6.5 implies that for $q = e^{-\epsilon}$,

$$q^{x_n(\epsilon^{-2}\tau)}e^{\epsilon^{-1}\tau}\epsilon^{n-1}e^{-3\tau/2} \longrightarrow z(\tau;n).$$

Though not shown in the proof, it is quite reasonable to imagine (and perhaps prove, without too much additional work) moments converge as well. We will check this (somewhat indirectly) in two steps. First, we will take limits of our known q–TASEP moments. Second, we will compute directly the semi–discrete SHE moments. Consider

$$\begin{aligned} &\mathbb{E}^{\text{step}}\left[\prod_{i=1}^{k} q^{x_n(\epsilon^{-2}\tau)}e^{\epsilon^{-1}\tau}\epsilon^{n_i-1}e^{-3\tau/2}\right] \qquad (6.1)\\ &= \frac{(-1)^k q^{k(k-1)/2}}{(2\pi i)^k}\oint\cdots\oint \prod_{1\leqslant A<B\leqslant k}\frac{z_A - z_B}{z_A - qz_B}\prod_{j=1}^{k}\frac{e^{(q-1)\epsilon^{-2}\tau z_j + \epsilon^{-1}\tau - 3\tau/2}}{(\epsilon^{-1}(1-z_j))^{r_j}}\frac{\epsilon^{-1}dz_j}{dz_j}. \end{aligned}$$

with $q = e^{-\epsilon}$. Change variables $z = e^{-\epsilon\tilde{z}}$, then as $\epsilon \searrow 0$

$$\begin{aligned} \frac{z_A - z_B}{z_A - qz_B} &\to \frac{\tilde{z}_A - \tilde{z}_B}{\tilde{z}_A - \tilde{z}_B - 1}, \\ \frac{1}{(\epsilon^{-1}(1-z_j))^{n_j}} &\to \frac{1}{(\tilde{z}_j)^{n_j}}, \\ \frac{\epsilon^{-1}dz_j}{z_j} &\to -d\tilde{z}_j, \\ (q-1)\epsilon^{-2}\tau z_j + \epsilon^{-1}\tau - 3\frac{\tau}{2} &\to \tau(\tilde{z}_j - 1) \end{aligned}$$

and hence

$$\lim_{\epsilon \searrow 0} (6.1) = \frac{1}{(2\pi i)^k} \oint \cdots \oint \prod_{1 \leqslant A < B < \leqslant k} \frac{z_A - z_B}{z_A - z_B - 1} \prod_{j=1}^{k} \frac{e^{\tau(z_j - 1)}}{z_j^{n_j}} dz_j,$$

with the z_A contour containing $\{z_B + 1\}_{B > A}$ and 0.

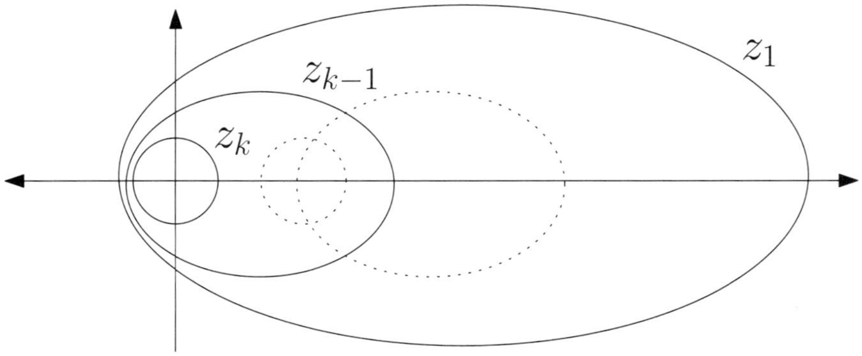

FIGURE 6.2. Nested contours for semi-discrete SHE moment formulas. The dashed contours represent the image under addition by 1 of the z_k and z_{k-1} contours. Notice that the z_{k-1} contour includes both the z_k contour and its image under addition by 1. This is the nature of the nesting.

We will now check that

$$\lim_{\epsilon \searrow 0} (6.1) = \mathbb{E}\left[\prod_{j=1} z(\tau; n_j) \right].$$

This is done through a limit version of the duality for q–TASEP. This now goes by the name of the polymer replica method. For this we should recall the Feymann–Kac representation, as briefly do now.

Consider a homogeneous Markov process generator L and deterministic potential v. We will provide a path integral or directed polymer interpretation for the solution to

$$\begin{aligned} \frac{d}{dt} z(t,x) &= (Lz)(t,x) + v(t,x) z(t,x) \\ z(0,x) &= z_0(x). \end{aligned}$$

Let $\varphi(\cdot)$ be the Markov process with generator L run backwards in time from t to 0 and let $\mathcal{E}^{t,x}$ be the associated expectation operator. For example, for $L = \nabla$, trajectories of φ look like in Figure 6.3.

Let $p(t,x) = \mathcal{E}^{t,x}\left[\mathbf{1}_{\varphi(0)=0}\right]$ be the heat kernel for L.

For $v \equiv 0$, by superposition / linearity of expectation we have

$$z(t,x) = \mathcal{E}^{t,x}\left[z_0(\varphi(0))\right].$$

When v is turned on, Duhamel's principle allows us to write

$$z(t,x) = \int_{\mathbb{R}} p(t, x-y) z_0(y) d\mu(y) + \int_0^t ds \int_{\mathbb{R}} d\mu(y) p(t-s, y-x) z(s,y) v(s,y),$$

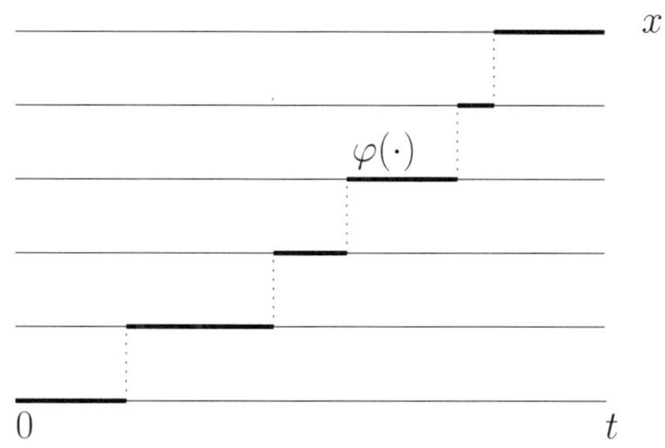

FIGURE 6.3. A possible trajectory in the Feynman-Kac representation.

where μ is measure on the state space (e.g. Lebesgue or counting measure). This identity can be applied repeatedly to yield an infinite series (which is convergent under mild hypothesis) and which can be identified with

$$z(t,x) = \mathcal{E}^{t,x}\left[z_0(\varphi(0)) \cdot \exp\left\{\int_0^t v(s,\varphi(s))d\mu(s)\right\}\right],$$

which is called the Feynman-Kac representation.

EXERCISE 6.6. Prove the Feynman-Kac representation using chaos series in v.

When v is random (Gaussian) care is needed, as we must deal with stochastic integrals. This leads to something called the Wick or Girsanov correction. For $z(\tau;n)$ with $v(\tau;n)d\tau = dB_n(\tau)$ and $L = \nabla$ this yields

$$z(\tau;n) = \mathcal{E}^{t,x}\left[z_0(\varphi(0)) \exp\left\{\int_0^\tau dB_{\varphi(s)}(s) - \frac{ds}{2}\right\}\right],$$

which is called the O'Connell Yor polymer partition function. The study of general directed polymer models is quite interesting but will be too far off–topic.

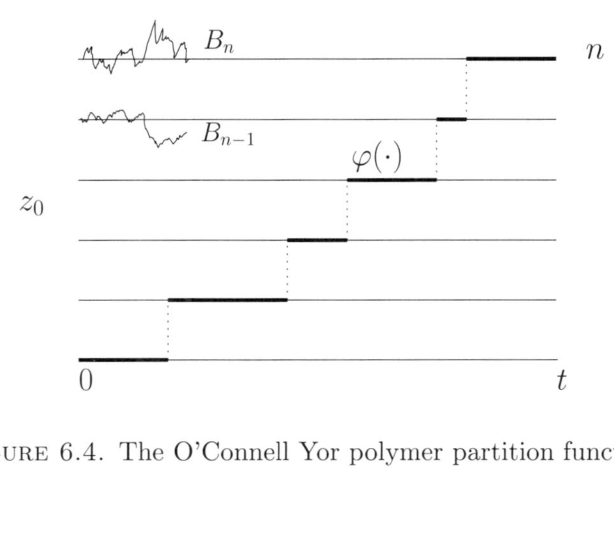

FIGURE 6.4. The O'Connell Yor polymer partition function.

Define $\bar{z}(\tau;\vec{n}) = \mathbb{E}\left[\prod_{i=1}^{k} z(\tau;n_i)\right]$ then applying the Feynman-Kac representation to each $z(\tau;n_i)$ with $\varphi_i(\cdot)$ and $\mathcal{E}_i$ associated to z_i, we find

$$\bar{z}(\tau;\vec{n}) = \mathbb{E}\left[\prod_{i=1}^{k} \mathcal{E}_i^{\tau,n_i}\left[z_0(\varphi_i(0))\cdot\exp\left\{\int_0^\tau dB_{\varphi_i(s)}(s) - \frac{ds}{2}\right\}\right]\right].$$

By interchanging the $\mathbb{E}$ and $\mathcal{E}_1^{\tau,n_1}\cdots\mathcal{E}_k^{\tau,n_k}$ we get

$$\bar{z}(\tau;\vec{n}) = \mathcal{E}_1^{\tau,n_1}\cdots\mathcal{E}_k^{\tau,n_k}\left[\prod_{i=1}^{k} z_0(\varphi_i(0))\cdot\mathbb{E}\left[\exp\left\{\int_0^t\sum_{i=1}^{k} dB_{\varphi_i(s)}(s) - \frac{ds}{2}\right\}\right]\right].$$

Using $\mathbb{E}[e^{kX}] = e^{k^2\sigma^2/2}$ for $X\sim\mathcal{N}(0,\sigma^2)$ we find:

EXERCISE 6.7.

$$\mathbb{E}\left[\exp\left\{\int_0^t\sum_{i=1}^{k} dB_{\varphi_i(s)}(s) - \frac{ds}{2}\right\}\right] = \exp\left\{\int_0^t\sum_{1\leqslant i<j\leqslant k}\mathbf{1}_{\varphi_i(s)=\varphi_j(s)}ds\right\}$$

or in other words, the exponential of the pair local overlap time.

Thus

$$\bar{z}(\tau;\vec{n}) = \mathcal{E}^{\tau;\vec{n}}\left[\prod_{i=1}^{k} z_0(\varphi_i(0))\exp\left\{\int_0^\tau\sum_{1\leqslant i<j\leqslant k}\mathbf{1}_{\varphi_i(s)=\varphi_j(s)}ds\right\}\right].$$

EXERCISE 6.8. Applying Feynman-Kac again shows that $z(\tau;\vec{n})$ with $\tau\in\mathbb{R}_{\geqslant 0}, \vec{n}\in\mathbb{Z}^k_{>0}$ is the unique solution to the semi–discrete delta Bose gas:

$$\frac{d}{dt}\bar{z}(\tau;\vec{n}) = \left(\sum_{i=1}^{k}\nabla_i + \sum_{1\leqslant i<j\leqslant k}\mathbf{1}_{n_i=n_j}\right)\bar{z}(\tau;\vec{n}),$$

$$\bar{z}(0;\vec{n}) = \prod_{i=1}^{k} z_0(n_i).$$

Notice that the above system involves functions that are symmetric in $n_1,\ldots,n_k$ whereas the nested contour integral formula we computed above for $\bar{z}(\tau;\vec{n})$ is only valid on $n_1\geqslant n_2\geqslant\ldots\geqslant n_k$, and clearly not symmetric. It is possible to check that extending the integral formula symmetrically we get a solution to the above system. Alternatively we can show how the above delta Bose gas can be rewritten in terms of a free evolution equation with boundary conditions.

PROPOSITION 6.9 ([**13**]). *If* $u:\mathbb{R}_{\geqslant 0}\times\mathbb{Z}^k_{\geqslant 0}\to\mathbb{R}_{\geqslant 0}$ *solves*
(1) For all $\vec{n}\in(\mathbb{Z}_{\geqslant 0})^k$ *and* $\tau\in\mathbb{R}_{\geqslant 0}$,

$$\frac{d}{dt}u(\tau;\vec{n}) = \sum_{i=1}^{k}\nabla_i u(\tau;\vec{n}).$$

(2) For all $\vec{n}\in\mathbb{Z}^k_{\geqslant 0}$ *such that for some* $i\in\{1,\ldots,k-1\}, n_i = n_{i+1}$,

$$(\nabla_i - \nabla_{i+1} - 1)u(\tau;\vec{n}) = 0.$$

(3) For all $\vec{n}\in W^k_{\geqslant 0}$, $u(0;\vec{n}) = \bar{z}_0(\vec{n})$.
Then for all $\vec{n}\in W^k_{\geqslant 0}$, $u(\tau;\vec{n}) = \bar{z}(\tau;\vec{n})$.

EXERCISE 6.10. Prove this proposition.

EXERCISE 6.11. Check our limiting integral formulas satisfy proposition, hence proving that their restriction to $\vec{n} \in W^k_{\geqslant 0}$ gives $\bar{z}(\tau;\vec{n})$.

EXERCISE 6.12. Reverse engineer the proof of the proposition in order to produce a delta Bose gas at the q–Boson level.

7. Unnesting the contours and moment asymptotics

In the last two sections we proved moment formulas for step initial data q–TASEP and $\mathbf{1}_{n=1}$ initial data semi–discrete SHE: For $n_1 \geqslant \ldots \geqslant n_k > 0$

$$\mathbb{E}\left[\prod_{j=1}^{k} q^{x_{n_j}(t)+n_j}\right] = \frac{(-1)^k q^{k(k-1)/2}}{(2\pi i)^k} \oint \cdots \oint \prod_{1\leqslant A<B\leqslant k} \frac{z_A - z_B}{z_A - q z_B} \prod_{j=1}^{k} \frac{e^{(q-1)t z_j}}{(1-z_j)^{n_j}} \frac{dz_j}{z_j},$$

$$\mathbb{E}\left[\prod_{j=1}^{k} z(\tau;n_j)\right] = \frac{1}{(2\pi i)^k} \oint \cdots \oint \prod_{1\leqslant A<B\leqslant k} \frac{z_A - z_B}{z_A - z_B - 1} \prod_{j=1}^{k} \frac{e^{\tau(z_j-1)}}{z_j^{n_j}} dz_j$$

with nested contours.

We will now look to study the growth, for instance, of $\mathbb{E}\big[z(\tau,n)^k\big]$ as τ and n go to infinity. This, it turns out, will require us to figure out how to unnest the contours so as to perform asymptotic analysis. But first, let us motivate this investigation with a story and a new concept – intermittency.

Parabolic Anderson Models are a class of models for mass transport or population dynamics in a disordered environment. We will be focusing on one spatial dimension, but as motivation consider algae in the sea. Algae can grow (or duplicate), die or move. The growth / death is very much a function of the local conditions (heat, sun light, nutrients, etc.) while the movement can be thought as due to steady currents.

We would like to understand certain intermittent behaviors, like the occurrence of large spikes in population for a little while in certain isolated areas.

Let us model the ocean as $\mathbb{Z}$ and think of algae as unit mass particles at the sites of $\mathbb{Z}$. There are no restrictions on particles per site, and in fact each particle evolves independent of others, only coupled through interaction with a common environment. In particular in continuous time, particles do three things independently:

- split into two unit mass particles at exponential rate $r_+(\tau,n)$;
- die at exponential rate $r_-(\tau,n)$;
- jump to the right by 1 at rate 1.

The functions r_+ and r_- represent an environment.

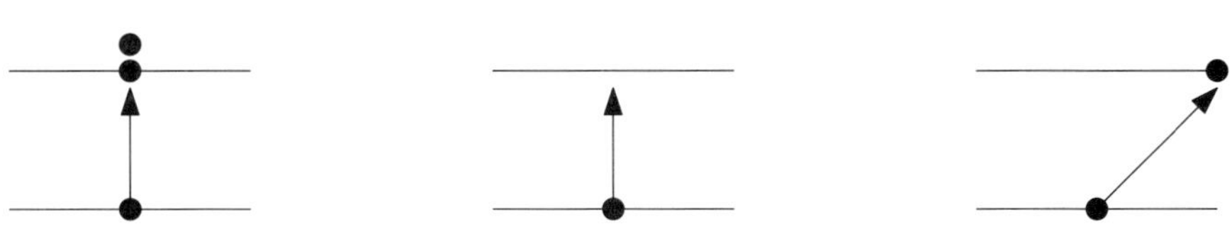

FIGURE 7.1. Particles choose amongst: (Left) duplication; (center) death; (right) jump right.

Call $m(\tau,n)$ the expected total mass (expectation over the random evolution). Then one sees that

$$\frac{d}{d\tau}m(\tau,n) = \nabla m(\tau,n) + \big(r_+(\tau,n) - r_-(\tau,n)\big)m(\tau,n).$$

If the individual masses are very small, yet very many then this actually describes the mass density evolution.

If the environment r_+, r_- is quickly and randomly evolving, then it may be appropriate to model $r_+(\tau,n) - r_-(\tau,n)$ by white noise in which case

$$dm(\tau,n) = \nabla m(\tau,n)d\tau + dB_n(\tau)m(\tau,n)$$

and we recover the semi–discrete SHE.

Initial data $m(0,n) = \mathbf{1}_{n=1}$ corresponds with starting a single cluster of particles at $n=1$ and letting them evolve.

It makes sense when studying this in large time, to scale τ and n proportionally like $n = \nu\tau$ ($\nu = 1$ making the most sense). We expect to see the overall population die out (though never reach 0) at a certain exponential rate. However, we would like to understand how often or likely it is to see spikes above this background level. One measure of this is through Lyapunov exponents.

DEFINITION 7.1. Assuming they exist, the almost sure Lyapunov exponent is

$$\tilde{\gamma}_1(\nu) := \lim_{\tau\to\infty} \frac{1}{\tau}\log z(\tau,\nu\tau)$$

and for $k \geqslant 1$, the k-th moment Lyapunov exponent is

$$\gamma_k(\nu) := \lim_{\tau\to\infty} \frac{1}{\tau}\log \mathbb{E}\big[z(\tau,\nu\tau)^k\big].$$

If these exponents are strictly ordered like

$$\tilde{\gamma}_1(\nu) < \gamma_1(\nu) < \frac{\gamma_2(\nu)}{2} < \ldots < \frac{\gamma_k(\nu)}{k} < \ldots$$

then the system is called intermittent.

Intermittency implies that moments are only determined by tail behavior, and higher moments probe higher into the tails.

EXERCISE 7.2. Use Jensen's inequality to prove that there is always weak inequality among the Lyapunov exponents.

EXERCISE 7.3. Prove that if z is intermittent then for any α such that $\frac{\gamma_k(\nu)}{k} < \alpha < \frac{\gamma_{k+1}(\nu)}{k+1}$,

$$\mathbb{P}\big(z(\tau,\nu\tau) > e^{\alpha\tau}\big) \leqslant e^{-(\alpha - \frac{\gamma_k(\nu)}{k})\tau}.$$

Unfortunately, intermittency of z will preclude deducing the distribution of z from its moments. We will return later to a way around this.

We will explicitly compute all of these exponents, starting with the γ_k's. For $\tilde{\gamma}_1$, this will come later in a slightly different way.

In order to perform asymptotics of $\mathbb{E}\big[z(\tau,\nu\tau)^k\big]$, we should deform all contours to ones on which we can use steepest descent analysis (see Appendix C for a simple worked out example). Such deformations may cross poles and hence pick up residues. We will see how this works.

Consider the case $k = 1$:

$$\mathbb{E}[z(\tau, \nu\tau)] = \frac{1}{2\pi i} \oint_{|z|=1} \frac{e^{\tau(z-1)}}{z^{\nu\tau}} dz = \frac{1}{2\pi i} \oint_{|z|=1} e^{\tau(z-1-\nu\log z)} dz.$$

Call $f(z) = (z-1) - \nu \log z$ and observe that $f'(z) = 1 - \nu/z$ and $f''(z) = \nu/z^2$. So f has a critical point at $z_c = \nu$ and $f''(z_c) = 1/\nu$. To study the growth of the integral we should look at $\mathrm{Re} f(z)$ and deform contours to go through the critical point.

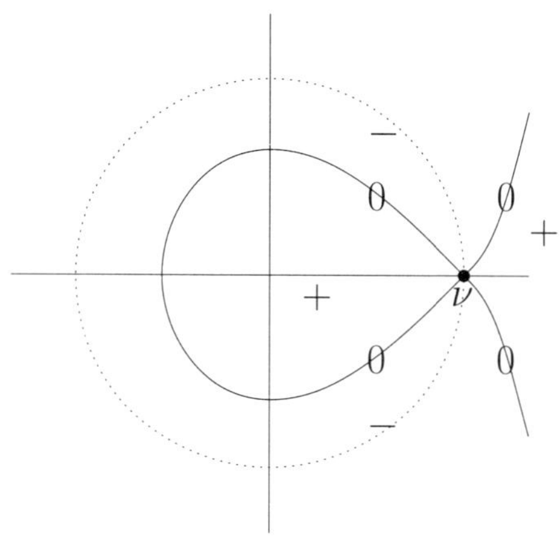

FIGURE 7.2. Contour plot of $\mathrm{Re} f(z) - f(z_c)$ so dotted line has maximum at ν and is monotonically decreasing away.

EXERCISE 7.4. Show that $\mathrm{Re} f(\nu e^{i\theta})$ is strictly decreasing on $\theta \in [0, \pi]$.

This implies that as $\tau \to \infty$, the integral localizes to its value at the critical point. Since $f(\nu) = \nu - 1 - \nu \log \nu$ this means

$$\gamma_1(\nu) = \nu - 1 - \nu \log \nu.$$

Consider the $k = 2$ case:

$$\mathbb{E}\big[z(\tau, \nu\tau)^2\big] = \frac{1}{(2\pi i)^2} \oint\oint \frac{z_1 - z_2}{z_1 - z_2 - 1} e^{f(z_1)+f(z_2)} dz_1 dz_2,$$

with nested contours for z_1 and z_2. In order to deform both contours to the steepest descent curve we must deform z_1 to z_2. Doing this we encounter a pole at $z_1 = z_2 + 1$. The residue theorem implies

$$\mathbb{E}\big[z(\tau, \nu\tau)^2\big] = \frac{1}{(2\pi i)^2} \oint \oint_{|z_1|=|z_2|=1} \frac{z_1 - z_2}{z_1 - z_2 - 1} e^{f(z_1)+f(z_2)} dz_1 dz_2 + \frac{1}{2\pi i} \oint e^{f(z_2+1)+f(z_2)} dz_2.$$

We must study growth of each term. The first term behaves like $e^{2f(\nu)\tau}$ while the second term requires closer consideration.

Set $\tilde{f}(z) = f(z) + f(z+1)$, then $\tilde{f}'(z) = 2 - \nu/z - \nu/(z+1)$ and so $\tilde{f}$ has a critical point at $z_c = \frac{1}{2}(\nu - 1 + \sqrt{\nu^2+1})$. The plot of $\mathrm{Re}\tilde{f}(z) - \tilde{f}(z_c)$ looks quite similar to those for f, and hence the steepest descent proceeds similarly.

So the second term is like $e^{\tilde{f}(z_c)\tau}$. Comparing $2f(\nu)$ to $\tilde{f}(\frac{1}{2}(\nu - 1 + \sqrt{\nu^2+1}))$ we find the second is larger, hence determines the growth of $\mathbb{E}\big[z(\tau, \nu\tau)^2\big]$.

THEOREM 7.5 ([**9**]). *The k-th Lyapunov exponent for the semi-discrete SHE are given by*

$$\gamma_k(\nu) = H_k(z_{c,k}),$$

where

$$H_k(z) = \frac{k(k-3)}{2} + kz - \nu \log\left(\prod_{i=0}^{k-1}(z+1)\right)$$

and $z_{c,k}$ is the unique solution to $H_k'(z) = 0, z \in (0, \infty)$.

PROOF SKETCH. For general k case there are many terms to consider in deforming contours. However, there is an argument (which I won't make here) that shows its the term coming from taking the residue at

$$z_1 = z_k + k - 1, \quad z_2 = z_k + k - 2, \ldots, z_{k-1} = z_k + 1$$

which matters. In terms of exponential terms this yields $e^{\tau H_k(z_k)}$ and steepest descent analysis gives the result. □

Let us record $\tilde{\gamma}_1$, to be proved later.

THEOREM 7.6 ([**33**]). *The almost sure Lyapunov exponent is given by*

$$\tilde{\gamma}_1(\nu) = -\frac{3}{2} + \inf_{s>0}\big(s - \nu\psi(s)\big)$$

where $\psi(s) = (\log\Gamma)'(s)$ is the digamma function.

By comparing these exponents we see (in very high resolution) the intermittency behavior of this model. We also see that the k–moments grow like e^{ck^2} so they do not determine the distribution of z.

We finish this section by showing how nested contour integrals (at the q–level) expand into residual subspaces as all contours are deformed to that of z_k. This will also be important in studying the q-moment generating function in the next section.

THEOREM 7.7. *Given nested contours $\gamma_1, \ldots, \gamma_k$ and a function $F(z_1, \ldots, z_k)$ which is analytic between the nested contours $\gamma_1, \ldots, \gamma_k$ we have*

$$\begin{aligned} &\oint_{\gamma_1}\frac{dz_1}{2\pi i}\cdots\oint_{\gamma_k}\frac{dz_k}{2\pi i}\prod_{1\leqslant A<B\leqslant k}\frac{z_A - z_B}{z_A - qz_B}F(z_1,\ldots,z_k) \\ &\qquad = \sum_{\lambda\vdash k}\underbrace{\oint_{\gamma_k}\cdots\oint_{\gamma_k}}_{\ell(\lambda)} d\mu_\lambda(\vec{w})E^q(\vec{w}\circ\lambda) \end{aligned} \tag{7.1}$$

where $\lambda \vdash k$ means λ partitions k, a partition $\lambda = (\lambda_1 \geqslant \lambda_2 \geqslant \ldots \geqslant \lambda_{\ell(\lambda)} > 0)$ with $\lambda_i \in \mathbb{Z}_{\geqslant 0}, \sum \lambda_i = k$, and $m_i = |\{j : \lambda_j = i\}|$, the (complex valued) measure

$$\begin{aligned} d\mu_\lambda(\vec{w}) = &\frac{(1-q)^k(-1)^k q^{-k(k-1)/2}}{m_1! m_2! \cdots}\det\left[\frac{1}{w_i q^{\lambda_i} - w_j}\right]_{i,j=1}^{\ell(\lambda)} \\ &\times \prod_{j=1}^{\ell(\lambda)} w_j^{\lambda_j} q^{\lambda_j(\lambda_j-1)/2}\frac{dw_j}{2\pi i}, \end{aligned} \tag{7.2}$$

and the function

$$E^q(z_1,\dots,z_k) = \sum_{\sigma\in S_k}\prod_{1\leqslant B<A\leqslant k}\frac{z_{\sigma(A)}-qz_{\sigma(B)}}{z_{\sigma(A)}-z_{\sigma(B)}}F(z_{\sigma(1)},\dots,z_{\sigma(k)})$$

with

$$\vec{w}\circ\lambda = (w_1, qw_1,\dots,q^{\lambda_1-1}w_1, w_2, qw_2,\dots,q^{\lambda_2-1}w_2,\dots,w_\ell, qw_\ell,\dots,q^{\lambda_l-1}w_\ell)$$

where $\ell=\ell(\lambda)$.

PROOF. Let us consider an example, $k=3$, to get an idea of residue expansion structure. The cross term (which gives poles) is

$$\frac{z_1-z_2}{z_1-qz_2}\frac{z_1-z_3}{z_1-qz_3}\frac{z_2-z_3}{z_2-qz_3}.$$

As γ_2 deforms to γ_3 we cross poles at $z_2=qz_3$. We can either take the residue or the integral.

- Residue: Cross term becomes $\frac{z_1-z_3}{z_1-qz_3}z_3(q-1)$ and only the z_1, z_2 integrals survive. As γ_1 deforms to γ_3 it crosses a pole at $z_1=q^2z_3$. Again we can either pick the residue or integral.
- Integral: Cross term remains the same. Now shrink γ_1 to γ_2. It crosses poles at $z_1=qz_2$ and $z_1=qz_3$. We can pick either of these residues or the integral.

In total, get five terms for various residual subspaces: $z_1=qz_2$ and $z_2=qz_3$; $z_1,z_2=qz_3$; $z_1=qz_2,z_3$; $z_1=qz_3,z_2$; z_1,z_2,z_3.

In general we index our residue expansion via partitions $\lambda\vdash k$ and strings

$$I = \begin{array}{c} i_1<i_2<\ldots<i_{\lambda_1},\\ j_1<j_2<\ldots<j_{\lambda_2},\\ \vdots \end{array}$$

where these indices are disjoint and union to $\{1,\dots,k\}$. Each row corresponds with taking residues at $z_{i_1}=qz_{i_2}, z_{i_2}=qz_{i_3},\dots,z_{j_1}=qz_{j_2},z_{j_2}=qz_{j_3},\dots,z_{j_{\lambda_2-1}}=qz_{j_{\lambda_2}},\dots$. This is written as $\mathrm{Res}^q_I f(z_1,\dots,z_k)$ and the output is a function of $z_{i_{\lambda_1}}, z_{j_{\lambda_2}},\dots$.

For example, for $k=3$ and $\lambda=(2,1)$ we can have $I\in\left\{\begin{smallmatrix}1<2\\3\end{smallmatrix},\begin{smallmatrix}1<3\\2\end{smallmatrix},\begin{smallmatrix}2<3\\1\end{smallmatrix}\right\}$.

Calling $S(\lambda)$ the set of all I's corresponding to λ, we thus see that

$$\mathrm{LHS}(7.1) = \sum_{\lambda\vdash k}\frac{1}{m_1!m_2!\dots}$$
$$\times\sum_{I\in S(\lambda)}\underbrace{\oint_{\gamma_k}\frac{dz_1}{2\pi i}\cdots\oint_{\gamma_k}\frac{dz_\ell}{2\pi i}}_{\ell(\lambda)}\mathrm{Res}^q_I\left(\prod_{1\leqslant A<B\leqslant k}\frac{z_A-z_B}{z_A-qz_B}F(z_1,\dots,z_k)\right)$$

where the $m_i!$'s come from symmetries of λ and multi–counting of residue subspaces arising from that. For each I we can relabel the z variables as

$$(z_{i_1},\dots,z_{i_{\lambda_1}})\mapsto(y_{\lambda_1},\dots,y_1)$$
$$(z_{j_1},\dots,z_{j_{\lambda_2}})\mapsto(y_{\lambda_1+\lambda_2},\dots,y_{\lambda_1+1})$$

and call $\sigma \in S_k$ the permutation taking $z_i \mapsto y_{\sigma(i)}$ and call the last y's $w_j = y_{\lambda_1+\ldots+\lambda_{j-1}+1}$. This change of variables puts all residual spaces with given λ into the form

$$\begin{aligned} y_{\lambda_1} &= qy_{\lambda_1-1}, \ldots, y_2 = qy_1 \\ y_{\lambda_1+\lambda_2} &= qy_{\lambda_1+\lambda_2-1}, \ldots, y_{\lambda_1+2} = qy_{\lambda_1+1} \\ &\vdots \end{aligned}$$

We denote the residue on this subspace Res^q_λ. Thus the left-hand side of (7.1) equals

$$\sum_{\lambda \vdash k} \frac{1}{m_1!m_2!\ldots} \times \sum_{\sigma\in S_k} \underbrace{\oint_{\gamma_k} \frac{dw_1}{2\pi i} \cdots \oint_{\gamma_k} \frac{dw_\ell}{2\pi i}}_{\ell(\lambda)} \mathrm{Res}^q_\lambda \left(\prod_{1\leqslant A<B\leqslant k} \frac{y_{\sigma(A)} - y_{\sigma(B)}}{y_{\sigma(A)} - qy_{\sigma(B)}} F(y_{\sigma(1)}, \ldots, y_{\sigma(k)}) \right).$$

Notice: not all $\sigma \in S_k$ arise from I's, but it is easy to see that those which do not have residue 0. Finally, rewrite

$$\prod_{1\leqslant A<B\leqslant k} \frac{y_{\sigma(A)} - y_{\sigma(B)}}{y_{\sigma(A)} - qy_{\sigma(B)}} = \prod_{A\neq B} \frac{y_A - y_B}{y_A - qy_B} \prod_{1\leqslant B<A\leqslant k} \frac{y_{\sigma(A)} - qy_{\sigma(B)}}{y_{\sigma(A)} - y_{\sigma(B)}}.$$

The S_k–symmetric part has no poles in the residue subspaces and can be factored, leaving

$$\mathrm{LHS}(7.1) = \sum_{\lambda\vdash k} \frac{1}{m_1!m_2!\cdots} \oint_{\gamma_k} \frac{dw_1}{2\pi i} \cdots \oint_{\gamma_k} \frac{dw_\ell}{2\pi i} \mathrm{Res}^q_\lambda \left(\prod_{A\neq B} \frac{y_A - y_B}{y_A - qy_B} \right) E^q(\vec{w} \circ \lambda).$$

Regarding Res^q_λ as a function of w's one checks that this yields $d\mu_\lambda(\vec{w})$. □

8. From moments to the q–Laplace transform Fredholm determinant

We will start by applying the residue expansion theorem to our nested contour integral formula for $\mathbb{E}^{\mathrm{step}}[q^{k(x_n(t)+n)}]$, i.e. taking all $n_i \equiv n$.

COROLLARY 8.1. *For step initial data q–TASEP, $k \geqslant 1$ and $n \geqslant 1$,*

$$\mathbb{E}^{step}\left[q^{k(x_n(t)+n)}\right] = k_q! \sum_{\lambda\vdash k} \frac{(1-q)^k}{m_1!m_2!\ldots} \oint_{\gamma} \frac{dw_1}{2\pi i} \cdots \oint_{\gamma} \frac{dw_{\ell(\lambda)}}{2\pi i} \det\left[\frac{1}{w_i q^{\lambda_i} - w_j}\right]_{i,j=1}^{\ell(\lambda)} \prod_{j=1}^{\ell(\lambda)} \frac{e^{t(q^{\lambda_j}-1)w_j}}{(w_j;q)^n_{\lambda_j}},$$

where the integral occurs over γ, a small contour containing 1. Here, $(a;q)_\ell = (1-a)(1-qa)\cdots(1-q^{\ell-1}a)$ is called the q–Pochhammer symbol, $k_q! = (q;q)_k/(1-q)^k$.

PROOF. Apply the residue expansion theorem with

$$F(z_1, \ldots, z_k) = (-1)q^{k(k-1)/2} \prod_{j=1}^{k} \frac{e^{t(q-1)z_j}}{(1-z_j)^n} \frac{1}{z_j}.$$

Observe that due to the symmetry of F in the z's

$$E^q(z_1,\ldots,z_k) = F(z_1,\ldots,z_k)\cdot \sum_{\sigma\in S_k}\prod_{1\leqslant B<A\leqslant k}\frac{z_{\sigma(A)}-qz_{\sigma(B)}}{z_{\sigma(A)}-z_{\sigma(B)}} = F(z_1,\ldots,z_k)\cdot k_q!$$

and the rest comes from simplifications and telescoping in the geometric substitutions. □

EXERCISE 8.2. Prove that

$$\sum_{\sigma\in S_k}\prod_{B<A}\frac{z_{\sigma(A)}-qz_{\sigma(B)}}{z_{\sigma(A)}-z_{\sigma(B)}} = k_q!$$

The formula for $\mathbb{E}\big[q^{k(x_n(t)+n)}\big]$ can be written as

$$\mathbb{E}\big[q^{k(x_n(t)+n)}\big] = k_q!\sum_{l=1}^{\infty}\frac{1}{l!}\sum_{\substack{\lambda_1,\ldots\lambda_\ell=1\\ \lambda_1+\cdots+\lambda_l=k}}^{\infty}\oint_\gamma\frac{dw_1}{2\pi i}\cdots\oint_\gamma\frac{dw_\ell}{2\pi i}\det\left[K_1(\lambda_i,w_i;\lambda_j,w_j)\right]_{i,j=1}^{l},$$

where

$$K_1(\lambda,w,\lambda',w') = \frac{(1-q)^\lambda e^{(q^\lambda-1)tw}(w;q)_\lambda^{-n}}{wq^\lambda-w'}.$$

Notice how the symmetry factor $\frac{1}{m_1!m_2!\ldots}$ was replaced by $\frac{1}{l!}$ due to the unordering of the $\lambda_1,\ldots,\lambda_k$.

In principle, since the moments of $q^{x_n(t)+n}$ determine its distribution, so too will any moment generating function. We will consider one well suited to the formulas. For $\zeta\in\mathbb{C}$ define (this is only consistent for small $|\zeta|$)

$$\begin{aligned}G(\zeta) &:= \sum_{k\geqslant 0}\mathbb{E}\big[q^{k(x_n(t)+n)}\big]\frac{\zeta^k}{k_q!}\\ &= 1+\sum_{l=1}^{\infty}\frac{1}{l!}\sum_{\lambda_1=1}^{\infty}\cdots\sum_{\lambda_l=1}^{\infty}\oint\frac{dw_1}{2\pi i}\cdots\oint\frac{dw_\ell}{2\pi i}\det\left[\tilde K_\zeta(w_i,\lambda_i;w_j,\lambda_j)\right]\\ &=: \det(1+\tilde K_\zeta)_{L^2(\mathbb{Z}_{\geqslant 0}\times\gamma)},\end{aligned}$$

where

$$\tilde K_\zeta(\lambda,w,\lambda',w') = \frac{\zeta^\lambda(1-q)^\lambda e^{(q^\lambda-1)tw}(w;q)_\lambda^{-n}}{wq^\lambda-w'}.$$

Notice that the kernel does not depend on λ', so the summations can be brought inside yielding

$$1+\sum_{l=1}^{\infty}\frac{1}{l!}\oint_\gamma\frac{dw_1}{2\pi i}\cdots\oint_\gamma\frac{dw_\ell}{2\pi i}\cdot\det\left[\tilde K_\zeta(w_i,w_j)\right]_{i,j=1}^{l}$$

with kernel

$$\tilde K_\zeta(w,w') = \sum_{\lambda=1}^{\infty} g(q^\lambda)\big((1-q)\zeta\big)^\lambda$$

and

$$g(q^\lambda) = \frac{e^{(q^\lambda-1)tw}\left(\frac{(q^\lambda w;q)_\infty}{(w;q)_\infty}\right)^n}{wq^\lambda-w'}.$$

The reason for rewriting in terms of g is that it is now seen to be an analytic function in λ (previously just in $\mathbb{Z}_{\geqslant 0}$) away from its poles. We could stop here, but we will do one more manipulation to $\tilde{K}_\zeta$ and replace the summation by an integral using a Mellin–Barnes representation. This is done with an eyes towards $q \nearrow 1$ asymptotics for which the sum of termwise limits in the sum defining $\tilde{K}_\zeta$ fails to be convergent, while the entire sum does have a nice limit. Going to contours away from $\mathbb{Z}_{\geqslant 0}$ allows us to show this and study the limit.

LEMMA 8.3. *For g satisfying the below conditions and ζ such that $|\zeta| < 1, \zeta \in \mathbb{C}\backslash\mathbb{R}_+$,*

$$\sum_{n=1}^{\infty} g(q^n)(\zeta)^n = \int_{C_\infty} \frac{\pi}{\sin(-\pi s)}(-\zeta)^s g(q^s)\frac{ds}{2\pi i}.$$

The function g and contour C_∞ must satisfy: The left-hand-side is convergent and the right-hand-side must be able to be approximated by integrals over C_k contours which enclose $\{1, \dots, k\}$ and no other singularities of the right-hand-side integrand and whose symmetric difference from C_∞ has integral going to zero as k goes to infinity.

EXERCISE 8.4. Prove this using residue calculus.

Applying the lemma brings us to:

THEOREM 8.5. *For $|\zeta|$ small enough,*

$$G(\zeta) = \sum_{k=0}^{\infty} \mathbb{E}\big[q^{k(x_n(t)+n)}\big]\frac{\zeta^k}{k_q!} = 1 + \sum_{l=1}^{\infty} \frac{1}{l!} \oint_\gamma \frac{dw_1}{2\pi i} \cdots \oint_\gamma \frac{dw_\ell}{2\pi i} \det\left[K_\zeta(w_i, w_j)\right]_{i,j=1}^{l}$$

with

$$K_\zeta(w, w') = \int_{-i\infty+1/2}^{i\infty+1/2} \frac{\pi}{\sin(\pi s)}\big(-(1-q)\zeta\big)^s \frac{e^{(q^s-1)tw}\left(\frac{(q^s w;q)_\infty}{(w;q)_\infty}\right)^n}{wq^s - w'} \frac{ds}{2\pi i}.$$

The important fact is that t, n, ζ come into this in a very simple way! So, $G(\zeta)$ should contain the distribution of $x_n(t) + n$, but how to get it out? We need an inverse transform of sorts. For this we will go a little deeper into the theory of q-series.

In 1949 Hahn [**19**] introduced two q-deformed exponential function:

$$e_q(x) := \frac{1}{((1-q)x;q)_\infty}, \qquad E_q(x) = (-(1-q)x;q)_\infty,$$

where recall that $(a;q)_\infty = (1-a)(1-qa)\cdots$.

EXERCISE 8.6. Show pointwise convergence of $e_q(x), E_q(x)$ to e^x as $q \nearrow 1$.

We will focus on $e_q(x)$. This has a Taylor series expansion for $|x|$ small

$$e_q(x) = \sum_{k=0}^{\infty} \frac{x^k}{k_q!},$$

which is a special case of the q-Binomial theorem [**4**].

EXERCISE 8.7. For $|x| < 1$, $|q| < 1$ prove the identity

$$\sum_{k=0}^{\infty} \frac{(a;q)_k}{(q;q)_k} x^k = \frac{(ax;q)_\infty}{(x;q)_\infty}.$$

This, along with the fact that $q^{x_n(t)+n} \leqslant 1$ implies that for $|\zeta|$ small enough,

$$\mathbb{E}\left[e_q(\zeta q^{x_n(t)+n})\right] = \sum_{k=0}^{\infty} \mathbb{E}\left[q^{k(x_n(t)+n)}\right] \frac{\zeta^k}{k_q!} = G(\zeta) = \det(I + K_\zeta)_{L^2(\gamma)}.$$

The interchange of expectation and summation is completely justified for small $|\zeta|$ and then the left–hand–side and right–hand–side are seen to be analytic in $\zeta \in \mathbb{C}\backslash\{\mathbb{R}_+\}$ thus allowing extension of the equality.

The left–hand–side is called the e_q–Laplace transform of $q^{x_n(t)+n}$ and the fact that it is given by a Fredholm determinant [**26**] is quite miraculous and as of yet, not so well understood.

THEOREM 8.8. *For $\zeta \in \mathbb{C}\backslash\{\mathbb{R}_+\}, q < 1, t > 0, n \geqslant 1$,*

$$\mathbb{E}\left[e_q(\zeta q^{x_n(t)+n})\right] = \det(I + K_\zeta)_{L^2(\gamma)} := 1 + \sum_{l=1}^{\infty} \frac{1}{l!} \oint_\gamma \cdots \oint_\gamma \det\left[K_\zeta(w_i, w_j)\right]_{i,j=1}^{l}$$

We close this section by demonstrating how to invert the e_q–Laplace transform.

DEFINITION 8.9. For a function $f \in \ell^1(\mathbb{Z}_{\geqslant 0})$ define for $\zeta \in \mathbb{C}\backslash\{q^{-m}\}_{m \geqslant 0}$

$$\hat{f}(\zeta) := \sum_{n=0}^{\infty} \frac{f(n)}{(\zeta q^n; q)_\infty} \quad \left(= \mathbb{E}^f\left[e_q\left(\frac{\zeta}{1-q} q^n\right)\right]\right).$$

with n distributed according to the measure f.

PROPOSITION 8.10. *The e_q-Laplace transform is inverted by*

$$f(n) = -q^n \int (q^{n+1}\zeta; q)_\infty \hat{f}^q(\zeta) \frac{d\zeta}{2\pi i}$$

with the ζ contour containing only $\zeta = q^{-m}, 0 \leqslant m \leqslant n$ poles.

EXERCISE 8.11. Prove this via residues.

So we have found a rather concise (and analyzable) formula for the probability distribution of $x_n(t) + n$, from which we can perform asymptotics. In fact, for our applications we can work with the e_q–Laplace transform instead of its inversion.

For instance, applying our knowledge from Theorem 6.5 of how q–TASEP goes to the semi–discrete SHE and performing some asymptotic analysis we prove:

THEOREM 8.12 ([**8, 10**]). *For semi–discrete SHE with $z_0(n) = \mathbf{1}_{n=1}$ and for* $\mathrm{Re}(u) \geqslant 0$,

$$\mathbb{E}\left[e^{-ue^{\frac{3\tau}{2}} z(\tau,n)}\right] = 1 + \sum_{l=1}^{\infty} \frac{1}{l!} \oint_\gamma \frac{dv_1}{2\pi i} \cdots \oint_\gamma \frac{dv_l}{2\pi i} \det\left[K_u(v_i, v_j)\right]_{i,j=1}^{l}$$

with contour γ a small circle around 0 and

$$K_u(v, v') = \int_{-i\infty+1/2}^{i\infty+1/2} \frac{\pi}{\sin(-\pi s)} \frac{g(v)}{g(v+s)} \frac{ds}{v+s-v'}$$

$$g(z) = \big(\Gamma(z)\big)^n u^{-z} e^{-\tau z^2/2}.$$

Note that had we tried to prove this theorem directly from our moment formulas we would have failed. The moment generating function is a divergent series. It is only by work at the q–level and explicitly summing the series that we could then take $q \nearrow 1$ and prove this result.

Here is one application:

THEOREM 8.13 ([**8**, **10**]). *For all $\nu > 0$*

$$\lim_{\tau\to\infty} \mathbb{P}\left(\frac{\log z(\tau,\nu\tau) - \tau\tilde{\gamma}_1(\nu)}{d(\nu)\tau^{1/3}} \leqslant s\right) = F_{GUE}(s),$$

where $d(\nu) = \big(-\nu\psi''(s(\nu))/2\big)^{1/3}$ with $s(\nu) = \arg\inf_{s>0}\big(s - \nu\psi(s)\big)$. Here, $\psi(s) = (\log\Gamma)'(s)$ is the digamma function and, as before,

$$\tilde{\gamma}_1(\nu) = -\frac{3}{2} + \inf_{s>0}(s - \nu\psi(s))$$

is the almost sure Lyapunov exponent.

This describes (in the parabolic Anderson model)the typical mass density fluctuations or the particle location fluctuations for a continuous space limit of q–TASEP and is consistent with the KPZ universality class belief.

Finally, utilizing Theorem 4.1, one can prove

THEOREM 8.14 ([**3**, **10**]). *Consider z, the solution to the SHE with $z_0(x) = \delta_{x=0}$. For all S with $Re(S) \geqslant 0$,*

$$\mathbb{E}\Big[e^{-Se^{t/24}z(t,0)}\Big] = \det(1 - K_S)_{L^2(\mathbb{R}_+)}$$

with kernel

$$K_S(\eta,\eta') = \int_{\mathbb{R}} dr \frac{S}{S + e^{-r(t/2)^{1/3}}} \operatorname{Ai}(r+\eta)\operatorname{Ai}(r+\eta').$$

Additionally,

$$\lim_{t\to\infty} \mathbb{P}\left(\frac{\log z(t,0) + t/24}{(t/2)^{1/3}} \leqslant r\right) = F_{\mathrm{GUE}}(r)$$

where F_{GUE} is the GUE Tracy-Widom distribution (see Appendix B*).*

9. q–Boson spectral theory

Recall that q–Boson particle system in $\vec{n} = (n_1 \geqslant \dots \geqslant n_k)$ coordinates has backwards generator

$$(L^{\text{q-Boson}} f)(\vec{n}) = \sum_{\text{cluster } i} (1 - q^{c_i})(f(\vec{n}^-_{c_1+\cdots c_i}) - f(\vec{n})).$$

Also recall the generator of k free (distant) particles is

$$(\mathcal{L}u)(\vec{n}) = \sum_{i=1}^{k} (\nabla_i u)(\vec{n}),$$

where $\nabla_i f(\vec{n}) = f(n-1) - f(n)$ in n_i variable.

We say that u satisfies the boundary conditions if

$$(\nabla_i - q\nabla_{i+1})u\big|_{n_i = n_{i+1}} = 0,$$

for $1 \leqslant i \leqslant k-1$.

The question we now confront is how to find the left and right eigenfunctions of $L^{\text{q-Boson}}$. We have already seen the below essentially contained in the proof of Proposition 5.5.

PROPOSITION 9.1. *If $u : \mathbb{Z}^k \to \mathbb{C}$ is an eigenfunction for $\mathcal{L}$ with eigenvalue λ, and u satisfies the boundary conditions, then u is an eigenfunction of $L^{q\text{-}Boson}$ with eigenvalue λ.*

In order to find such eigenfunctions we will use another idea going back to Bethe [**6**] in 1931 (see also [**27**]). This idea goes under the name of coordinate Bethe ansatz and takes the following general form.

PROPOSITION 9.2. *Consider a space X. Eigenfunctions for a sum of 1d operators acting on {functions on X}*

$$(\mathcal{L}\psi)(\vec{x}) = \sum_{i=1}^{k}(L_i\psi)(\vec{x}), \qquad \vec{x} = (x_1, \ldots, x_k) \in X^k$$

(with L_i acting in the i-th coordinate) that satisfy boundary conditions depending on B acting on {functions on X^2}

$$B_{i,i+1}\psi(\vec{x})|_{x_i = x_{i+1}} = 0, \ \text{for } 1 \leqslant i, \leqslant k-1$$

(with $B_{i,i+1}$ acting on the $i, i+1$ coordinate) can be found via:

1. *Diagonalizaing the 1d operators $L\psi_z = \lambda_z\psi_z$ where $\psi_z : X \to \mathbb{C}$ and z is a parameter (e.g. complex number)*
2. *Taking linear combinations*
$$\psi_{\vec{z}}(\vec{x}) = \sum_{\sigma \in S(k)} A_\sigma(\vec{z}) \prod_{j=1}^{k} \psi_{z_{\sigma(j)}}(x_j)$$
3. *Evaluating $A_\sigma(\vec{z})$ as*
$$A_\sigma(\vec{z}) = sgn(\sigma) \prod_{a>b} \frac{S(z_{\sigma(a)}, z_{\sigma(b)})}{S(z_a, z_b)}$$
where
$$S(z_1, z_2) = \frac{B(\psi_{z_1} \otimes \psi_{z_2})(x,x)}{\psi_{z_1}(x)\psi_{z_2}(x)}.$$

Then $\big(\mathcal{L}\psi_{\vec{z}})\vec{x}\big) = \big(\sum_{i=1}^{k} \lambda_{z_i}\big)\psi_{\vec{z}}(\vec{x})$.

PROOF. By Liebniz rule, it is clear that $\psi_{\vec{z}}(\vec{x})$ are eigenfunctions for the free generator. It remains to check that the choice of A_σ implies that the boundary conditions are satisfied. Let $\tau_i = (i, i+1)$ act on permutations by permuting i and $i+1$. It suffices to find A_σ (or show that our specific choice of A_σ) that satisfy that for all $\sigma \in S(k)$, and all $1 \leqslant i \leqslant k-1$, when $B_{i,i+1}$ is applied to

$$T_\sigma + T_{\tau_i\sigma}, \qquad \text{where} \quad T_\sigma = A_\sigma(\vec{z}) \prod_{j=1}^{k} \psi_{z_{\sigma(j)}}(x_j),$$

the result is zero whenever $x_i = x_{i+1}$. Indeed, if one sums the above left-hand side over all $\sigma \in S(k)$, the result is $2\big(B_{i,i+1}\psi_{\vec{z}}\big)(\vec{x})$ and if each summand can be made to equal zero (when $x_i = x_{i+1}$), so too with the entire sum by zero.

From definitions, one sees that

$$B_{i,i+1}T_\sigma\big|_{x_i=x_{i+1}} = S(z_{\sigma(i)}, z_{\sigma(i+1)})T_\sigma.$$

So

$$B_{i,i+1}\big(T_\sigma + T_{\tau_i\sigma}\big)\big|_{x_i=x_{i+1}} = S(z_{\sigma(i)}, z_{\sigma(i+1)})T_\sigma + S(z_{\sigma(i+1)}, z_{\sigma(i)})T_{\tau_i\sigma}.$$

Unwinding this, we find that in order for this sum to be zero, we must have

$$A_{\tau_i\sigma}(\vec{z}) = -\frac{S\big(z_{\sigma(i+1)}, z_{\sigma(i)}\big)}{S\big(z_{\sigma(i)}, z_{\sigma(i+1)}\big)}A_\sigma(\vec{z}).$$

The transpositions τ_i as $1 \leqslant i \leqslant k-1$ varies generates $S(k)$ and hence fixing $A_I(\vec{z}) \equiv 1$, the formula uniquely characterizes the $A_\sigma(\vec{z})$. This expression matches that claimed by the proposition. □

Let us apply this method when $L = \nabla$ and $(Bg)(x,y) = (\nabla_1 - q\nabla 2)g(x,y)$. We can write 1d eigenfunctions of L as $\psi_z(n) = (1-z)^{-n}$ (here n replaces x from the proposition). This choice of ψ_z has eigenvalue $(1-q)z$ and leads to $S(z_1, z_2) = -(z_1 - qz_2)$. Thus, Bethe ansatz produces left eigenfunctions for all $z_1, \ldots, z_k \in \mathbb{C}\backslash\{1\}$

$$\psi^\ell_{\vec{z}}(\vec{n}) := \sum_{\sigma\in S(k)} \prod_{1\leqslant B<A\leqslant k} \frac{z_{\sigma(A)} - qz_{\sigma(B)}}{z_{\sigma(A)} - z_{\sigma(B)}} \prod_{j=1}^{k} (1 - z_{\sigma(j)})^{-n_j}$$

such that when restricted to $\vec{n} = (n_1 \geqslant n_2 \geqslant \cdots \geqslant n_k)$,

$$(L^{\text{q-Boson}}\psi^\ell_{\vec{z}})(\vec{n}) = \sum_{j=1}^{k}(1-q)z_j\, \phi^\ell_{\vec{z}}(\vec{n}).$$

As a brief remark, notice that these are eigenfunctions for any choices of $z_j \in \mathbb{C}\backslash\{1\}$. If instead of working with the q-Boson process on $\mathbb{Z}$ we considered a periodic (or other boundary condition) portion of $\mathbb{Z}$, then in order to respect the boundary conditions of the finite lattice we would need to impose additional conditions on the $\vec{z}$. These restrictions are known as the Bethe ansatz equations and make things much more involved. It is only for solutions $\vec{z}$ of these algebraic equations that $\psi^\ell_{\vec{z}}(\vec{n})$ are eigenfunctions (on the finite lattice).

Having identified left eigenfunctions, we come now to the question of right eigenfunctions. Had our operator $L^{\text{q-Boson}}$ been self-adjoint, the left and right eigenfunctions would coincide and we could decompose and recompose functions with respect to this single set of functions. It is clear that $L^{\text{q-Boson}}$ is not self-adjoint, however it is not too far off, as it enjoys the property of being PT-invariant (in fact, this property is shared by all totally asymmetric zero range processes, provided their jump rate $g(k)$ is reasonable). PT-invariance means that the process generated by $L^{\text{q-Boson}}$ is invariant under joint space reflection and time inversion.

The product invariant measure μ for the q-Boson process in the $\vec{y}$ variables have one point marginal

$$\mu_\alpha(y_0) = \mathbf{1}_{y_0\geqslant 0}\frac{\alpha^{y_0}}{g(1)g(2)\cdots g(y_0)},$$

where $\alpha > 0$ controls the overall density of the invariant measure, and $g(k) = 1 - q^k$ is the q-Boson jump rate. Time reversal corresponds with taking the adjoint of $L^{\text{q-Boson}}$ in $L^2(\vec{y}, \mu_\alpha)$ (the choice of α does not, in fact, matter here). Then PT-invariance amounts to the fact that one readily shows [**11**] that

$$L^{\text{q-Boson}} = P\big(L^{\text{q-Boson}}\big)^* P^{-1}$$

where $\big(Pf\big)(\vec{y}) = f\big(\{y_{-i}\}_{i\in\mathbb{Z}}\big)$ is the space reflection operator (clearly $P = P^{-1}$).

PT-invariance can be written in terms of the $\vec{n}$ variables and using matrix notation as

$$L^{\text{q-Boson}} = \big(PC_q\big)\big(L^{\text{q-Boson}}\big)^{\text{transpose}}\big(PC_q\big)^{-1}$$

where now $\big(Pf\big)(n_1, \dots, n_k) = f(-n_k, \dots, -n_1)$ and C_q is the multiplication operator with

$$C_q(\vec{n}) = (-1)^k q^{-\frac{k(k-1)}{2}} \prod_{\text{cluster } i} \frac{(q;q)_{c_i}}{(1-q)^{c_i}}.$$

Note that we could have defined $C_q(\vec{n})$ as any function depending only on k, times the product over clusters of the $(q;q)_{c_i}$ terms. The choice above will be well suited to avoid messy constants in what follows.

Returning to the matter of right eigenfunctions, it is clear from PT-invariance that applying $\big(PC_q\big)^{-1}$ to left eigenfunction, produces right eigenfunctions. Thus, we define right eigenfunctions

$$\psi^r_{\vec{z}}(\vec{n}) := \sum_{\sigma\in S(k)} \prod_{1\leqslant B<A\leqslant k} \frac{z_{\sigma(A)} - q^{-1} z_{\sigma(B)}}{z_{\sigma(A)} - z_{\sigma(B)}} \prod_{j=1}^{k} \big(1 - z_{\sigma(j)}\big)^{n_j}$$

which satisfy

$$\Big(\big(L^{\text{q-Boson}}\big)^{\text{transpose}} \psi^r_{\vec{z}}\Big)(\vec{n}) = \sum_{j=1}^{k} (1-q) z_j \psi^r_{\vec{z}}(\vec{n}).$$

Note that $\psi^r_{\vec{z}}(\vec{n}) = q^{-\frac{k(k-1)}{2}} \big(PC_q\big)^{-1} \psi^\ell_{\vec{z}}(\vec{n})$.

Having defined left and right eigenfunctions, it remains to demonstrate how to diagonalize $L^{\text{q-Boson}}$ with respect to them, with the ultimate goal of solving $\frac{d}{dt} h(t;\vec{n}) = L^{\text{q-Boson}} h(t;\vec{n})$ for arbitrary initial data. We proceed now by defining a direct and inverse Fourier type transform.

Definition 9.3. Fix the following spaces of functions:

$$\mathcal{W}^k = \Big\{f : \{\vec{n} = (n_1 \geqslant n_2 \geqslant \cdots \geqslant n_k)\} \to \mathbb{C} \text{ of compact support}\Big\},$$
$$\mathcal{C}^k = \mathbb{C}\Big[(z_1 - 1)^{\pm 1}, \dots, (z_k - 1)^{\pm 1}\Big]^{\text{Sym}}.$$

In words, $\mathcal{W}^k$ are all functions of $\vec{n}$ to $\mathbb{C}$ of compact support in $\vec{n}$ and $\mathcal{C}^k$ are all symmetric Laurant polynomials in the variables $(z_1 - 1)$ through $(z_k - 1)$. We may define bilinear pairings on these spaces so that for $f, g \in \mathcal{W}^k$ and $F, G \in \mathcal{C}^k$,

$$\langle f, g\rangle_{\mathcal{W}} = \sum_{n_1\geqslant\cdots\geqslant n_k} f(\vec{n}) g(\vec{n})$$
$$\langle F, G\rangle_{\mathcal{C}} = \oint \cdots \oint d\mu_{(1)^k}(\vec{w}) \prod_{j=1}^{k} \frac{1}{1 - w_j} F(\vec{w}) G(\vec{w}),$$

where the integrals are over circles centered at the origin or radius exceeding one, and the notation $d\mu_\lambda(\vec{w})$ (here $(1)^k$ is the partition with k ones) is recalled from (7.2).

Define the direct transform $\mathcal{F} : \mathcal{W}^k \to \mathcal{C}^k$ and (candidate) inverse transform $\mathcal{J} : \mathcal{C}^k \to \mathcal{W}^k$ as

$$(\mathcal{F}f)(\vec{n}) = \langle f, \psi^r_{\vec{z}} \rangle_{\mathcal{W}}$$
$$(\mathcal{J}G)(\vec{n}) = \langle \psi^\ell(\vec{n}), G \rangle_{\mathcal{W}}.$$

In the second line we have used $\psi^\ell(\vec{n})$ to represent the function which maps $\vec{z}$ to $\psi^\ell_{\vec{z}}(\vec{n})$.

The operator $\mathcal{J}$ can be written in two alternative ways. The first is given by

$$(\mathcal{J}G)(\vec{n}) = \frac{1}{(2\pi i)^k} \oint \cdots \oint \prod_{1 \leqslant A < B \leqslant k} \frac{z_A - z_B}{z_A - q z_B} \prod_{j=1}^k \frac{1}{(1 - z_j)^{n_j + 1}} dz_j$$

with contours such that the z_A contour contains $\{q z_B\}_{B > A}$ and 1. This equivalence follows by the fact that one can deform such contours to all lie on a large circle centered at zero of radius exceeding one. Then, since all contours lie upon the same circle, one can symmetrize the integrand and after an application of the Cauchy determinant formula, we recover the initial expression for $\mathcal{J}$. The second expression comes from unnesting the contours, but onto a single small contour around one. This is accomplished by applying Theorem 7.7, and hence we find that

$$(\mathcal{J}G)(\vec{n}) = \sum_{\lambda \vdash k} \frac{1}{m_1! m_2! \cdots} \oint \cdots \oint d\mu_\lambda(\vec{w}) \prod_{j=1}^{\ell(\lambda)} \frac{1}{(w_j; q)_{\lambda_j}} \psi^\ell_{\vec{w} \circ \lambda}(\vec{n}) G(\vec{w} \circ \lambda),$$

where the $\ell(\lambda)$ integrals occur upon a single contour which contains 1 and has small enough radius so the image under multiplication by q is outside the contour.

THEOREM 9.4 ([**11**]). *On the spaces $\mathcal{W}^k$ and $\mathcal{C}^k$ the operators $\mathcal{F}$ and $\mathcal{J}$ are mutual inverses (i.e. $\mathcal{J}\mathcal{F}$ is the identity on $\mathcal{W}^k$ and $\mathcal{F}\mathcal{J}$ is the identity in $\mathcal{C}^k$). Consequently, the left and right eigenfunctions are biorthogonal so that*

$$\langle \psi^\ell_\bullet(\vec{m}), \psi^r_\bullet(\vec{n}) \rangle_{\mathcal{C}} = \mathbf{1}_{\vec{m} = \vec{n}}$$

and

$$\langle \psi^\ell_{\vec{z}}(\bullet), \psi^r_{\vec{w}}(\bullet) \rangle_{\mathcal{W}} = \frac{1}{k!} \prod_{1 \leqslant A \neq B \leqslant k} \frac{z_A - q z_B}{z_A - z_B} \prod_{j=1}^k \frac{1}{1 - z_j} \det\left[\delta(z_i - w_j)\right]_{i,j=1}^k$$

where this last equality is to be understood in a weak (or integrated) sense.

At the end of this section we will prove part of this theorem, that $\mathcal{J}\mathcal{F}$ is the identity. A generalization of the above theorem (which in fact admits easier proofs) is provided in [**12**]. But first, let us apply this theorem to solve the q-Boson backward equation.

COROLLARY 9.5. *For $h_0 \in \mathcal{W}^k$, the solution to*

$$\frac{d}{dt} h(t, \vec{n}) = \left(L^{q\text{-}Boson} h\right)(t, \vec{n})$$

with initial data h_0 is given by

$$h(t,\vec{n}) = \mathcal{J}\big(e^{t(q-1)(z_1+\cdots+z_k)}\mathcal{F}h_0\big)(t,\vec{n})$$
$$= \frac{1}{(2\pi i)^k}\oint\cdots\oint \prod_{1\leqslant A<B\leqslant k}\frac{z_A-z_B}{z_A-qz_B}\prod_{j=1}^{k} e^{t(q-1)z_j}(1-z_j)^{n_j+1}\big(\mathcal{F}h_0\big)(\vec{z})d\vec{z},$$

with nested contours.

PROOF. From the theorem, $\mathcal{J}\mathcal{F}$ is the identity on $\mathcal{W}^k$. Hence,

$$h(t,\vec{n}) = \big(e^{tL^{\text{q-Boson}}}h_0\big)(\vec{n}) = \big(e^{tL^{\text{q-Boson}}}\mathcal{J}\mathcal{F}h_0\big)(\vec{n})$$

Using the explicit form of the transforms, we find that

$$\big(e^{tL^{\text{q-Boson}}}\mathcal{J}\mathcal{F}h_0\big)(\vec{n}) = e^{tL^{\text{q-Boson}}}\oint\cdots\oint d\mu_{(1)^k}(\vec{w})\prod_{j=1}^{k}\frac{1}{1-w_j}\psi^{\ell}_{\vec{w}}(\vec{n})\big(\mathcal{F}h_0\big)(\vec{z}).$$

We now use the fact that $\psi^{\ell}_{\vec{w}}$ are the left eigenfunctions for $L^{\text{q-Boson}}$ and thus $e^{tL^{\text{q-Boson}}}$ acts diagonally with eigenvalue $e^{t(q-1)(w_1+\cdots w_k)}$. Plugging this in (and going to the nested contours as explained earlier) yields the formula in the corollary. □

There are two apparent limitations of this corollary. The first is that it only applies for $h_0 \in \mathcal{W}^k$. In Section 6 we were concerns with initial data $h_0(\vec{n}) = \prod_{j=1}^{k}\mathbf{1}_{n_j\geqslant 1}$, which does not have compact support. For this initial data it is possible to extend the validity of the corollary with a little work (likely this can be done for a much wider class as well). The second limitation is more serious – the expression $\big(\mathcal{F}h_0\big)(\vec{z})$ involves an infinite summation of initial data against eigenfunctions. However, such an infinite summation is not so useful for asymptotics. We would like to be able to evaluate such summations. Unfortunately, this may be quite hard. However, such a summation is automatic is $h_0 = \mathcal{J}G_0$ for some $G_0 \in \mathcal{C}^k$. In that case, $\mathcal{F}h_0 = \mathcal{F}\mathcal{J}G_0 = G_0$ by the theorem.

For initial data $h_0(\vec{n}) = \prod_{j=1}^{k}\mathbf{1}_{n_j\geqslant 1}$ it can be shown (via residues and extending the fact that $\mathcal{F}\mathcal{J}$ is the identity to some functions outside of $\mathcal{C}^k$) that $h_0(\vec{n}) = \big(\mathcal{J}G_0\big)(\vec{n})$ where $G_0(\vec{z}) = q^{\frac{k(k-1)}{2}}\prod_{j=1}^{k}\frac{z_j-1}{z_j}$. Using this we can recover the solution given in Theorem 6.1 and its corollary. This approach can be applied to some broader classes of initial data (cf. [**13**]), though we will not pursue these here.

Let us close this section by proving that

$$\mathcal{K} := \mathcal{J}\mathcal{F} = \text{Identity}$$

on $\mathcal{W}^k$.

SKETCH OF $\mathcal{J}\mathcal{F}$ = Identity. PT-invariance implies that for $f, g \in \mathcal{W}^k$,

$$\langle \mathcal{K}f, g\rangle_{\mathcal{W}} = \langle f, (PC_q)^{-1}\mathcal{K}(PC_q)g\rangle_{\mathcal{W}}.$$

This can be shown by expanding $\mathcal{K}$ into eigenfunctions and using the relation between left and right eigenfunctions implied by PT-invariance.

In order to prove that $\mathcal{K}$ acts as the identity, it suffices (by linearity) to show that for $f(\vec{n}) = \mathbf{1}_{\vec{n}=\vec{x}}$ for some $\vec{x}$ fixed, $\big(\mathcal{K}f\big)(\vec{y}) = \mathbf{1}_{\vec{y}=\vec{x}}$. Showing that $\big(\mathcal{K}f\big)(\vec{x}) = 1$ involves residue calculus, and we will skip it. We will, however, show that for $\vec{y} \neq \vec{x}$, $\big(\mathcal{K}f\big)(\vec{y}) = 0$.

Set $g(\vec{n}) = \mathbf{1}_{\vec{n}=\vec{y}}$, then

$$\big(\mathcal{K}f\big)(\vec{y}) = \big\langle \mathcal{K}f, g\big\rangle_{\mathcal{W}} = \frac{1}{(2\pi i)^k}\oint\cdots\oint \prod_{1\leqslant A<B\leqslant k}\frac{z_A - z_B}{z_A - qz_B}\prod_{j=1}^{k}(1-z_j)^{-y_j-1}\psi^r_{\vec{z}}(\vec{x}),$$

with integration along nested contours. We wish to prove that this is zero. Consider expanding the z_1 contour to infinity. This can be done without crossing and poles, so evaluating that integral amounts to evaluating the residue of the integrand at infinity. By expanding the right eigenfunction via its definition, we see that the largest exponent of $(1-z_1)$ is $x_1 - y_1 - 1$. Thus, in order that there be a residue at infinity, we should have $x_1 - y_1 - 1 \geqslant -1$, or in other words $x_1 \geqslant y_1$. The implication is that if $x_1 < y_1$, then $\big(\mathcal{K}f\big)(\vec{y}) = 0$.

But, using the PT-invariance relation

$$\big(\mathcal{K}f\big)(\vec{y}) = \big\langle \mathcal{K}f, g\big\rangle_{\mathcal{W}} = \big\langle f, (PC_q)^{-1}\mathcal{K}(PC_q)g\big\rangle_{\mathcal{W}}.$$

This switches the role of x's and y's and the same reasoning as above shows that this is zero if $y_1 < x_1$. The overall implication is that for $\big(\mathcal{K}f\big)(\vec{y})$ to be nonzero, we must have $x_1 = y_1$.

In light of this deduction, assume that $x_1 = y_1$. It is now quite simply to evaluate the residue as z_1 goes to infinity. This yields a similar expression except without the z_1 integration valuable. Thus, in the same manner we establish that in order that $\big(\mathcal{K}f\big)(\vec{y})$ be nonzero, $x_2 = y_2$ and so on until we find that all $x_j = y_j$. □

Appendix A. White noise and stochastic integration

In one sentence, white noise $\xi(t,x), t \geqslant 0, x \in \mathbb{R}$ is the distribution valued Gaussian process with mean zero and covariance

$$\mathbb{E}[\xi(t,x)\xi(s,y)] = \delta(x-y)\delta(s-t).$$

Let us be more precise about what this means. Recall that if g is a distribution and f is a smooth test function, then $\int fg$ is a real number. For example, $\int f(x)\delta(x-y)dx = f(y)$. Therefore, for a smooth function $f(t,x)$ of compact support, the integral

$$\int_{\mathbb{R}_+\times\mathbb{R}} f(t,x)\xi(t,x)dxdt$$

is a random variable. These random variables are jointly Gaussian with mean zero and covariance

$$\begin{aligned}
&\mathbb{E}\left[\int_{\mathbb{R}_+\times\mathbb{R}} f_1(t,x)\xi(t,x)dxdt\int_{\mathbb{R}_+\times\mathbb{R}} f_2(s,y)\xi(s,y)dyds\right]\\
&= \int_{\mathbb{R}_+\times\mathbb{R}} dxdt\int_{\mathbb{R}_+\times\mathbb{R}} \mathbb{E}\left[\xi(s,y)\xi(t,x)\right] f_1(t,x)f_2(s,y)dyds\\
&= \int_{\mathbb{R}_+\times\mathbb{R}} f_1(t,x)f_2(t,x)dxdt.
\end{aligned}$$

There are many ways to construct $\xi(t,x)$. For instance, since the covariance is positive definite and symmetric, general theory of Gaussian processes implies the existence and uniqueness of this process. More concretely, one can choose any orthonormal basis $\{f_n\}$ of $L^2(\mathbb{R}_+ \times \mathbb{R})$ and independent Gaussian random variables $\{Z_n\}$ with mean zero and variance one, and define

$$\xi(t,x) = \sum_{n=1}^{\infty} Z_n f_n(t,x).$$

Within a suitable negative Sobolev space (namely $H_{-1-\delta,\mathrm{loc}}(\mathbb{R}_+ \times \mathbb{R})$ fcr any $\delta > 0$) this construction will yield a unique (in law) element whose covariance can be checked to be as desired. Details on constructing ξ can be found, for instance, in [**23**, **36**].

Assume that $\xi(t,x)$ has been constructed on some probability space $(\Omega, \mathcal{F}, \mathbb{P})$. Now let us construct stochastic integrals with respect to white noise. For non-random functions $f(t,x)$ this is not hard. If $f \in L^2(\mathbb{R}_+ \times \mathbb{R})$ then there are smooth functions f_n with compact support in $\mathbb{R}_+ \times \mathbb{R}$ such that

$$\int_{\mathbb{R}_+ \times \mathbb{R}} |f_n(t,x) - f(t,x)|^2 \, dxdt \to 0.$$

Since

$$\mathbb{E}\left[\left(\int_{\mathbb{R}_+ \times \mathbb{R}} (f_n(t,x) - f_m(t,x))\xi(t,x)dxdt\right)^2\right] = \int_{\mathbb{R}_+ \times \mathbb{R}} |f_n(t,x) - f_m(t,x)|^2 \, dxdt,$$

this means that $\int_{\mathbb{R}_+ \times \mathbb{R}} f_n(t,x)\xi(t,x)dxdt$ is a Cauchy sequence in $L^2(\Omega, \mathcal{F}, \mathbb{P})$. Since $L^2(\Omega, \mathcal{F}, \mathbb{P})$ is complete, then this Cauchy sequence has a limit, which we define to be $\int_{\mathbb{R}_+ \times \mathbb{R}} f(t,x)\xi(t,x)dxdt$.

For random functions, the definition of the integral is a little more complicated. As in the one dimensional case, one has to make a choice of the integral (Itô vs. Stratonovich) . The construction here is essentially the standard Itô integral, but only in the time variable.

Start with smooth functions $\varphi(x)$ on $\mathbb{R}$ with compact support. For $t > 0$ we can define

$$\int_{\mathbb{R}_+ \times \mathbb{R}} \mathbf{1}_{(0,t]}(s)\varphi(x)\xi(s,x)dsdx.$$

This is a Brownian motion in t with variance $\int \varphi^2(x)dx$, since it is Gaussian with mean zero and covariance

$$\begin{aligned}
&\mathbb{E}\left[\int_{\mathbb{R}_+ \times \mathbb{R}} \mathbf{1}_{(0,t_1]}(s)\varphi(x)\xi(x,s)dxds \int_{\mathbb{R}_+ \times \mathbb{R}} \mathbf{1}_{(0,t_2]}(s)\varphi(x)\xi(x,s)dxds\right] \\
&= \int_0^{\min(t_1,t_2)} ds \int_{\mathbb{R}} \varphi^2(x)dx \\
&= \min(t_1,t_2) \int \varphi^2(x)dx.
\end{aligned}$$

(Note: one often hears the statement "white noise is the derivative of Brownian motion").

Let $\mathcal{F}_0 = \varnothing$ and for each $t > 0$ define $\mathcal{F}_t$ to be the σ–field generated by

$$\left\{\int_{\mathbb{R}_+ \times \mathbb{R}} \mathbf{1}_{(0,s]}(u)\varphi(x)\xi(u,x)dxdu : 0 \leqslant s \leqslant t,\right.$$
$$\left.\varphi \text{ a smooth function of compact support in } \mathbb{R}\right\}.$$

It is clear that $\mathcal{F}_t$ is a filtration of $\mathcal{F}$, that is $\mathcal{F}_s \subseteq \mathcal{F}_t$ for $s \leqslant t$.

Now consider slightly more complicated functions. Let $\mathcal{S}$ be the set of functions of the form

$$f(t,x,\omega) = \sum_{i=1}^{n} X_i(\omega)\mathbf{1}_{(a_i,b_i]}(t)\varphi_i(x),$$

where X_i is a bounded $\mathcal{F}_{a_i}$–measurable random variable and φ_i are smooth functions of compact support on $\mathbb{R}$. For functions of this form, define the stochastic integral as

$$\int_{\mathbb{R}_+\times\mathbb{R}} f(t,x)\xi(t,x)dxdt = \sum_{i=1}^{n} X_i \int_{\mathbb{R}_+\times\mathbb{R}} \mathbf{1}_{(a_i,b_i]}(t)\varphi_i(x)\xi(t,x)dxdt.$$

It is easy to check that the integral is linear and an isometry from $L^2(\mathbb{R}_+\times\mathbb{R}\times\Omega,\mathcal{F},\mathbb{P})$ to $L^2(\Omega,\mathcal{F},\mathbb{P})$, that is

$$\int_{\mathbb{R}_+\times\mathbb{R}} \mathbb{E}[f^2(t,x)]dxdt = \mathbb{E}\left[\left(\int_{\mathbb{R}_+\times\mathbb{R}} f(t,x)\xi(t,x)dxdt\right)^2\right].$$

Let $\mathcal{P}$ be the sub–σ–field of $\mathcal{B}(\mathbb{R}_+\times\mathbb{R})\times\mathcal{F}$ generated by $\mathcal{S}$. Let $L^2(\mathbb{R}_+\times\mathbb{R}\times\Omega,\mathcal{F},\mathbb{P})$ be the space of square integrable $\mathbb{P}$–measurable random variables $f(t,x,\omega)$. These will be the integrators. It is important to note that these are non–anticipating in the sense that $f(t,x,\omega)$ only depends on the information $\mathcal{F}_t$ up to time t. This is analogous to the distinction between Itô and Stratonovich integrals in the one–dimensional case. The construction of the stochastic integral will be defined through the isometry and approximation.

LEMMA A.1. *$\mathcal{S}$ is dense in $L^2(\mathbb{R}_+\times\mathbb{R}\times\Omega,\mathcal{F},\mathbb{P})$.*

PROOF. Same as one–dimensional case. □

Thus, if $f \in L^2(\mathbb{R}_+\times\mathbb{R}\times\Omega,\mathcal{F},\mathbb{P})$ there exist $f_n \in \mathcal{S}$ wuch that f_n converges to $f \in L^2(\mathbb{R}_+\times\mathbb{R}\times\Omega,\mathcal{F},\mathbb{P})$. By the isometry,

$$I_n(\omega) := \int_{\mathbb{R}_+\times\mathbb{R}} f_n(t,x,\omega)\xi(t,x)dxdt$$

is a Cauchy sequence in $L^2(\Omega,\mathcal{F},\mathbb{P})$. Hence there is a limit point $I \in L^2(\Omega,\mathcal{F},\mathbb{P})$ which is defined to be the stochastic integral $\int_{\mathbb{R}_+\times\mathbb{R}} f(t,x)\xi(t,x)dxdt$. This is linear in f and the Itô isometry holds.

We can also define multiple stochastic integrals. Let $\Lambda_k = \{0 < t_1 < \cdots < t_k < \infty\}$.

$$\begin{aligned}
&\int_{\Lambda_k}\int_{\mathbb{R}^k} f(\vec{t},\vec{x})\xi^{\otimes k}(\vec{t},\vec{x})d\vec{t}d\vec{x} \\
&\quad := \int_{\Lambda_k}\int_{\mathbb{R}^k} f(t_1,\ldots,t_k,x_1,\ldots,x_k)\xi(t_1,x_1)\cdots\xi(t_k,x_k)dx_1\cdots dx_k dt_1\cdots dt_k.
\end{aligned}$$

This is defined inductively. For example,

$$\begin{aligned}
&\int_{\Lambda_2}\int_{\mathbb{R}^2} f(t_1,t_2,x_1,x_2)\xi(t_1,x_1)\xi(t_2,x_2)dx_1dx_2dt_1dt_2 \\
&\qquad = \int_0^\infty\int_{\mathbb{R}}\left[\int_0^{t_2}\int_{\mathbb{R}} f(t_1,t_2,x_1,x_2)\xi(t_1,x_1)dx_1dt_1\right]\xi(t_2,x_2)dx_2dt_2,
\end{aligned}$$

which is well defined because we just showed that the integrand is progressively measurable. The covariance of these multiple stochastic integrals is

$$\mathbb{E}\left[\int_{\Lambda_k}\int_{\mathbb{R}^k} f(\vec{t},\vec{x})\xi^{\otimes k}(\vec{t},\vec{x})d\vec{t}d\vec{x}\int_{\Lambda_j}\int_{\mathbb{R}^j} g(\vec{t},\vec{x})\xi^{\otimes j}(\vec{t},\vec{x})d\vec{t}d\vec{x}\right] = \langle f,g\rangle_{L^2(\Lambda_k\times\mathbb{R}_k)}\mathbf{1}_{j=k}.$$

It also turns out that they span $L^2(\Omega,\mathcal{F},\mathbb{P})$.

This defines an isometry between $\bigoplus_{k=0}^{\infty} L^2(\Lambda_k\times\mathbb{R}^k,\mathcal{F},\mathbb{P})$ and $L^2(\Omega,\mathcal{F},\mathbb{P})$ by

$$X = \sum_{k=0}^{\infty}\int_{\Lambda_k}\int_{\mathbb{R}^k} f_k(\vec{t},\vec{x})\xi^{\otimes k}(\vec{t},\vec{x})d\vec{t}d\vec{x}.$$

Here f_0 is the constant function $\mathbb{E}X$. This summation is called the Wiener chaos decomposition.

Aside: in a sense, the Wiener chaos decomposition holds more generally (that is, for Gaussian Hilbert spaces, not just for stochastic integrals with respect to white noise). If H is a Gaussian Hilbert space in a probability space $(\Omega,\mathcal{F},\mathbb{P})$ let $\overline{\mathcal{P}}_k(H)$ be the closure in $L^2(\Omega,\mathcal{F},\mathbb{P})$ of

$$\mathcal{P}_k(H) = \{p(\xi_1,\ldots,\xi_m) : p \text{ polynomial of degree } \leqslant k, \xi_1,\ldots,\xi_m \in H, m<\infty\}.$$

Let

$$H^{:k:} = \overline{\mathcal{P}}_k(H)\ominus\overline{\mathcal{P}}_{k-1}(H) = \overline{\mathcal{P}}_k(H)\cap\overline{\mathcal{P}}_{k-1}(H)^{\perp}.$$

Set $H^{:0:}$ to be the constants. The Wiener chaos decomposition is then

$$L^2(\Omega,\mathcal{F}(H),\mathbb{P}) = \bigoplus_{k=0}^{\infty} H^{:k:}.$$

For example, if the probability space is $(\mathbb{R},\mathcal{B},\gamma)$ where $d\gamma = \frac{1}{\sqrt{2\pi}}e^{-x^2/x}dx$ and $H = \{tx : t\in\mathbb{R}\}$ is a one–dimensional Gaussian Hilbert space, then $H^{:k:}$ is also one–dimensional and spanned by the k–th Hermite polynomial h_k. The Wiener chaos decomposition then says that $L^2(d\gamma)$ has $\{h_k\}$ as an orthogonal basis. This is related to the Wiener chaos decomposition above because polynomials of iterated stochastic integrals are themselves iterated stochastic integrals.

Appendix B. Background on Tracy-Widom distribution

The Tracy–Widom distribution [**39**] has cumulative distribution function which can be defined as the Fredholm determinant

$$F_2(s) = \det(1-A)_{L^2(s)},$$

where A is the operator on $L^2(s,\infty)$ with kernel

$$\frac{\mathrm{Ai}(x)\,\mathrm{Ai}'(y)-\mathrm{Ai}'(x)\,\mathrm{Ai}(y)}{x-y}.$$

It can also be defined by

$$F_2(s) = \exp\left(-\int_s^{\infty}(x-s)q(x)^2dx\right),$$

where $q(x)$ solves the Painleve II equation

$$q''(x) = xq(x) + 2q(x)^3$$

with the boundary condition

$$q(x) \sim Ai(x), \quad x \to \infty.$$

This is called the *Hastings-Mcleod* solution.

Appendix C. Asymptotic analysis

The goal is to find asymptotics of expressions of the form

$$\int g(x)e^{nf(x)}dx$$

as $n \to \infty$.

THEOREM C.1. *Let $f, g : [a, b] \to \mathbb{R}$ be smooth. Assume f has a unique global maximum at $c \in (a, b)$ and that $f''(c) < 0$. Define $I(n)$ by*

$$\int_a^b g(x)e^{nf(x)}dx = n^{-1/2}e^{nf(c)}I(n).$$

Then

$$\lim_{n\to\infty} I(n) = \sqrt{-\frac{2\pi}{f''(c)}}g(c).$$

PROOF. (Sketch) Since $f(x)$ has a global maximum at c, then as $n \to \infty$ most of the contributions to the integral come in a small neighbourhood around c. Setting $y = n^{1/2}(x - c)$,

$$\int_a^b g(x)e^{nf(x)}dx = \int_{-n^{1/2}(c-a)}^{n^{1/2}(b-c)} g(c + n^{-1/2}y)e^{nf(c+n^{-1/2}y)}n^{-1/2}dy.$$

Now use the Taylor expansion $f(c + n^{-1/2}y) \approx f(c) + \frac{1}{2}f''(c)n^{-1}y^2$ to get that

$$\lim_{n\to\infty} I(n) = g(c)\int_{-\infty}^{\infty} e^{f''(c)y^2/2}dy = \sqrt{-\frac{2\pi}{f''(c)}}g(c).$$

□

Here is an example to find the asymptotics of $n!$. Use that

$$n! = \Gamma(n+1) = \int_0^\infty t^n e^{-t}dt = n^{n+1}\int_0^\infty x^n e^{-nx}dx$$

by the substitution $t = nx$. This is of the right form, with $f(x) = \log x - x, g(x) = 1$. Solving for the maximum of f on $(0, \infty)$ yields $c = 1$ with $f(c) = -1, f''(c) = -1$. Therefore

$$n! \approx n^n e^{-n}\sqrt{2\pi n},$$

which is Stirling's formula.

We will now use similar ideas to find the behaviour of the tails of the Airy function

$$\mathrm{Ai}(x) = \frac{1}{2\pi}\int_{-\infty}^{\infty} e^{i(t^3/3+xt)}dt.$$

Note that, as written, the integral is only defined conditionally, since the integrand does not converge to 0 at $t \to \pm\infty$. This can be resolved by deforming the contour integral to

$$\mathrm{Ai}(x) = \frac{1}{2\pi} \int_{\infty \cdot e^{5\pi i/6}}^{\infty \cdot e^{\pi i/6}} e^{i(t^3/3 + xt)} dt$$

Note that the integrand is not yet in the form $e^{xf(t)}$. To resolve this, simply take the substitution $t = x^{1/2} z$ to get

$$\mathrm{Ai}(x) = \frac{x^{1/2}}{2\pi} \int_{\infty \cdot e^{5\pi i/6}}^{\infty \cdot e^{\pi i/6}} e^{x^{3/2} f(z)} dz$$

where $f(z) = i(z^3/3 + z)$.

As before, we want to find the saddle points of $f(z)$. Solving for $f'(z) = 0$, we find that the saddle points are at $t = \pm i$. So we know that we want to deform the contour to go through either i or $-i$ (or both). Furthermore, the absolute value of the integrand should decay quickly as it moves away from the saddle point. Since $\left|e^{x^{3/2} f(z)}\right| = e^{Re(x^{3/2} f(z))}$, we should look for contours where $Re(f(z)) < Re(f(i))$ or $Re(f(z) < Re(f(-i))$. The first case is shown in Figure C.1.

FIGURE C.1. Shaded regions indicate where $Re(f(z)) < Re(f(i))$.

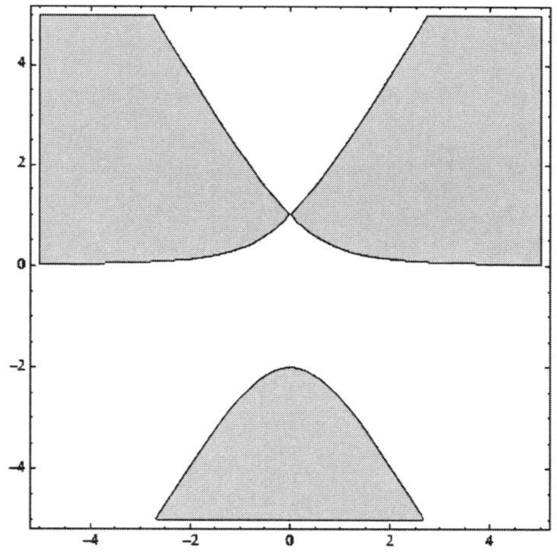

Therefore, the contours can be deformed to begin near $\infty e^{5\pi i/6}$, go through i and then end near $\infty e^{i\pi/6}$. Using the Taylor expansion

$$f(z) \approx -\frac{2}{3} - (z - i)^2 + \cdots$$

and the substitution $u = x^{3/4}(z - i)$ we get

$$\mathrm{Ai}(x) \approx \frac{x^{1/2} e^{-2x^{3/2}/3}}{2\pi} \int e^{-x^{3/2}(z-i)^2} dz \approx \frac{e^{-2x^{3/2}/3}}{2\pi x^{1/4}} \int_{-\infty}^{\infty} e^{-u^2} du,$$

This finally yields

$$\mathrm{Ai}(x) \approx \frac{e^{-\frac{2}{3} x^{3/2}}}{2\sqrt{\pi} x^{1/4}}.$$

EXERCISE C.2. Explain what would go wrong if we tried to use the $-i$ saddle point. Figure C.2 shows a plot of $Re(f(z))$. Hint: it is important to keep in mind the endpoints of the contour.

FIGURE C.2. Shaded regions indicate where $Re(f(z)) < Re(f(-i))$.

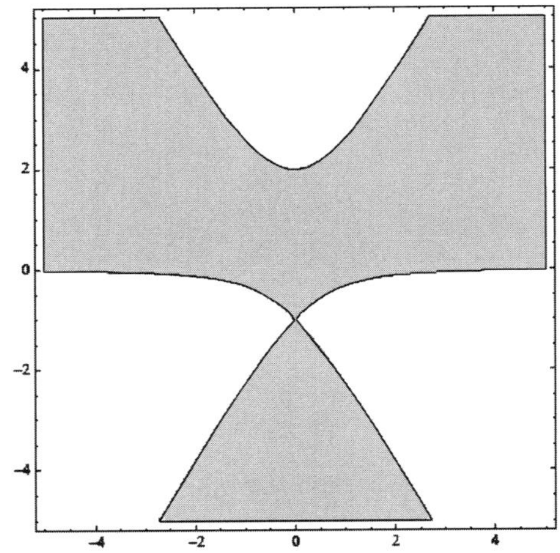

EXERCISE C.3. Use a similar argument to show that as $x \to \infty$

$$\mathrm{Ai}(-x) \approx \frac{1}{\sqrt{\pi} x^{1/4}} \cos\left(\frac{\pi}{4} - \frac{2x^{3/2}}{3}\right).$$

Figure C.3 should be helpful.

FIGURE C.3. Shaded regions indicate where $Re(f(z)) < Re(f(1))$ for the function $f(z) = i(z^3/3 - z)$.

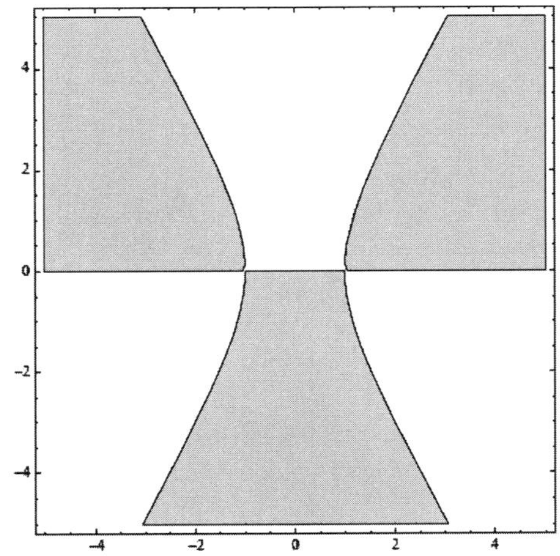

References

[1] Tom Alberts, Konstantin Khanin, and Jeremy Quastel, *The intermediate disorder regime for directed polymers in dimension* 1 + 1, Ann. Probab. **42** (2014), no. 3, 1212–1256, DOI 10.1214/13-AOP858. MR3189070

[2] Lars V. Ahlfors, *Complex analysis*, 3rd ed., McGraw-Hill Book Co., New York, 1978. An introduction to the theory of analytic functions of one complex variable; International Series in Pure and Applied Mathematics. MR510197

[3] Gideon Amir, Ivan Corwin, and Jeremy Quastel, *Probability distribution of the free energy of the continuum directed random polymer in* 1 + 1 *dimensions*, Comm. Pure Appl. Math. **64** (2011), no. 4, 466–537, DOI 10.1002/cpa.20347. MR2796514

[4] George E. Andrews, Richard Askey, and Ranjan Roy, *Special functions*, Encyclopedia of Mathematics and its Applications, vol. 71, Cambridge University Press, Cambridge, 1999. MR1688958

[5] Lorenzo Bertini and Nicoletta Cancrini, *The stochastic heat equation: Feynman-Kac formula and intermittence*, J. Statist. Phys. **78** (1995), no. 5-6, 1377–1401, DOI 10.1007/BF02180136. MR1316109

[6] H. Bethe. Zur Theorie der Metalle. I. Eigenwerte und Eigenfunktionen der linearen Atomkette. (On the theory of metals. I. Eigenvalues and eigenfunctions of the linear atom chain) *Zeitschrift fur Physik*, **71**:205–226, 1931.

[7] Alexei Borodin, Alexey Bufetov, and Ivan Corwin, *Directed random polymers via nested contour integrals*, Ann. Physics **368** (2016), 191–247, DOI 10.1016/j.aop.2016.02.001. MR3485045

[8] Alexei Borodin and Ivan Corwin, *Macdonald processes*, Probab. Theory Related Fields **158** (2014), no. 1-2, 225–400, DOI 10.1007/s00440-013-0482-3. MR3152785

[9] Alexei Borodin and Ivan Corwin, *Moments and Lyapunov exponents for the parabolic Anderson model*, Ann. Appl. Probab. **24** (2014), no. 3, 1172–1198, DOI 10.1214/13-AAP944. MR3199983

[10] Alexei Borodin, Ivan Corwin, and Patrik Ferrari, *Free energy fluctuations for directed polymers in random media in* 1 + 1 *dimension*, Comm. Pure Appl. Math. **67** (2014), no. 7, 1129–1214, DOI 10.1002/cpa.21520. MR3207195

[11] Alexei Borodin, Ivan Corwin, Leonid Petrov, and Tomohiro Sasamoto, *Spectral theory for the q-Boson particle system*, Compos. Math. **151** (2015), no. 1, 1–67, DOI 10.1112/S0010437X14007532. MR3305308

[12] Alexei Borodin, Ivan Corwin, Leonid Petrov, and Tomohiro Sasamoto, *Spectral theory for interacting particle systems solvable by coordinate Bethe ansatz*, Comm. Math. Phys. **339** (2015), no. 3, 1167–1245, DOI 10.1007/s00220-015-2424-7. MR3385995

[13] Alexei Borodin, Ivan Corwin, and Tomohiro Sasamoto, *From duality to determinants for q-TASEP and ASEP*, Ann. Probab. **42** (2014), no. 6, 2314–2382, DOI 10.1214/13-AOP868. MR3265169

[14] Earl A. Coddington and Norman Levinson, *Theory of ordinary differential equations*, McGraw-Hill Book Company, Inc., New York-Toronto-London, 1955. MR0069338

[15] Ivan Corwin, *The Kardar-Parisi-Zhang equation and universality class*, Random Matrices Theory Appl. **1** (2012), no. 1, 1130001, 76, DOI 10.1142/S2010326311300014 MR2930377

[16] Ivan Corwin, *Macdonald processes, quantum integrable systems and the Kardar-Parisi-Zhang universality class*, Proceedings of the International Congress of Mathematicians—Seoul 2014. Vol. III, Kyung Moon Sa, Seoul, 2014, pp. 1007–1034. MR3729062

[17] Ivan Corwin, Jeremy Quastel, and Daniel Remenik, *Renormalization fixed point of the KPZ universality class*, J. Stat. Phys. **160** (2015), no. 4, 815–834, DOI 10.1007/s10955-015-1243-8. MR3373642

[18] D. Forster, David R. Nelson, and Michael J. Stephen, *Large-distance and long-time properties of a randomly stirred fluid*, Phys. Rev. A (3) **16** (1977), no. 2, 732–749, DOI 10.1103/PhysRevA.16.732. MR0459274

[19] Wolfgang Hahn, *Beiträge zur Theorie der Heineschen Reihen. Die* 24 *Integrale der Hypergeometrischen q-Differenzengleichung. Das q-Analogon der Laplace-Transformation* (German), Math. Nachr. **2** (1949), 340–379, DOI 10.1002/mana.19490020604. MR0035344

[20] Martin Hairer, *Solving the KPZ equation*, Ann. of Math. (2) **178** (2013), no. 2, 559–664, DOI 10.4007/annals.2013.178.2.4. MR3071506

[21] T. Halpin Healy, Y.Z. Zhang. Kinetic Roughening, Stochastic Growth, Directed Polymers and all that. *Phys. Rep.*, **254**:215–415, 1995.

[22] D.A. Huse, C. L. Henley. Pinning and roughening of domain walls in Ising systems due to random impurities. *Phys. Rev. Lett.*, **54**:2708–2711, 1985.

[23] Svante Janson, *Gaussian Hilbert spaces*, Cambridge Tracts in Mathematics, vol. 129, Cambridge University Press, Cambridge, 1997. MR1474726

[24] Mehran Kardar, *Replica Bethe ansatz studies of two-dimensional interfaces with quenched random impurities*, Nuclear Phys. B **290** (1987), no. 4, 582–602, DOI 10.1016/0550-3213(87)90203-3. MR922846

[25] M. Kardar, G. Parisi, Y.Z. Zhang. Dynamic scaling of growing interfaces. *Phys. Rev. Lett.*, **56**:889–892, 1986.

[26] Peter D. Lax, *Functional analysis*, Pure and Applied Mathematics (New York), Wiley-Interscience [John Wiley & Sons], New York, 2002. MR1892228

[27] Elliott H. Lieb and Werner Liniger, *Exact analysis of an interacting Bose gas. I. The general solution and the ground state*, Phys. Rev. (2) **130** (1963), 1605–1616. MR0156630

[28] Thomas M. Liggett, *Interacting particle systems*, Classics in Mathematics, Springer-Verlag, Berlin, 2005. Reprint of the 1985 original. MR2108619

[29] Carl Mueller, *On the support of solutions to the heat equation with noise*, Stochastics Stochastics Rep. **37** (1991), no. 4, 225–245, DOI 10.1080/17442509108833738. MR1149348

[30] S. Molchanov, *Lectures on random media*, Lectures on probability theory (Saint-Flour, 1992), Lecture Notes in Math., vol. 1581, Springer, Berlin, 1994, pp. 242–411, DOI 10.1007/BFb0073874. MR1307415

[31] Gregorio R. Moreno Flores, *On the (strict) positivity of solutions of the stochastic heat equation*, Ann. Probab. **42** (2014), no. 4, 1635–1643, DOI 10.1214/14-AOP911. MR3262487

[32] G. Moreno Flores, D. Remenik, J. Quastel. In preparation.

[33] J. Moriarty and N. O'Connell, *On the free energy of a directed polymer in a Brownian environment*, Markov Process. Related Fields **13** (2007), no. 2, 251–266. MR2343849

[34] M. Nica. Intermediate disorder limits for multi-layer semi-discrete directed polymers arXiv:1609.00298.

[35] Neil O'Connell and Marc Yor, *Brownian analogues of Burke's theorem*, Stochastic Process. Appl. **96** (2001), no. 2, 285–304, DOI 10.1016/S0304-4149(01)00119-3. MR1865759

[36] Giovanni Peccati and Murad S. Taqqu, *Wiener chaos: moments, cumulants and diagrams*, Bocconi & Springer Series, vol. 1, Springer, Milan; Bocconi University Press, Milan, 2011. A survey with computer implementation; Supplementary material available online. MR2791919

[37] Jeremy Quastel, *Introduction to KPZ*, Current developments in mathematics, 2011, Int. Press, Somerville, MA, 2012, pp. 125–194. MR3098078

[38] Tomohiro Sasamoto and Miki Wadati, *Exact results for one-dimensional totally asymmetric diffusion models*, J. Phys. A **31** (1998), no. 28, 6057–6071, DOI 10.1088/0305-4470/31/28/019. MR1633078

[39] Craig A. Tracy and Harold Widom, *Level-spacing distributions and the Airy kernel*, Comm. Math. Phys. **159** (1994), no. 1, 151–174. MR1257246

Columbia University, Department of Mathematics, 2990 Broadway, New York, NY 10027, USA

Email address: ivan.corwin@gmail.com

Index

Selected Published Titles in This Series

75 **Michael Damron, Firas Rassoul-Agha, and Timo Seppäläinen, Editors,** Random Growth Models, 2018

73 **Kasso A. Okoudjou, Editor,** Finite Frame Theory, 2016

72 **Van H. Vu, Editor,** Modern Aspects of Random Matrix Theory, 2014

71 **Samson Abramsky and Michael Mislove, Editors,** Mathematical Foundations of Information Flow, 2012

70 **Afra Zomorodian, Editor,** Advances in Applied and Computational Topology, 2012

69 **Karl Sigmund, Editor,** Evolutionary Game Dynamics, 2011

68 **Samuel J. Lomonaco, Jr., Editor,** Quantum Information Science and Its Contributions to Mathematics, 2010

67 **Eitan Tadmor, Jian-Guo Liu, and Athanasios E. Tzavaras, Editors,** Hyperbolic Problems: Theory, Numerics and Applications, 2009

66 **Dorothy Buck and Erica Flapan, Editors,** Applications of Knot Theory, 2009

65 **L. L. Bonilla, A. Carpio, J. M. Vega, and S. Venakides, Editors,** Recent Advances in Nonlinear Partial Differential Equations and Applications, 2007

64 **Reinhard C. Laubenbacher, Editor,** Modeling and Simulation of Biological Networks, 2007

63 **Gestur Ólafsson and Eric Todd Quinto, Editors,** The Radon Transform, Inverse Problems, and Tomography, 2006

62 **Paul Garrett and Daniel Lieman, Editors,** Public-Key Cryptography, 2005

61 **Serkan Hoşten, Jon Lee, and Rekha R. Thomas, Editors,** Trends in Optimization, 2004

60 **Susan G. Williams, Editor,** Symbolic Dynamics and its Applications, 2004

59 **James Sneyd, Editor,** An Introduction to Mathematical Modeling in Physiology, Cell Biology, and Immunology, 2002

58 **Samuel J. Lomonaco, Jr., Editor,** Quantum Computation, 2002

57 **David C. Heath and Glen Swindle, Editors,** Introduction to Mathematical Finance, 1999

56 **Jane Cronin and Robert E. O'Malley, Jr., Editors,** Analyzing Multiscale Phenomena Using Singular Perturbation Methods, 1999

55 **Frederick Hoffman, Editor,** Mathematical Aspects of Artificial Intelligence, 1998

54 **Renato Spigler and Stephanos Venakides, Editors,** Recent Advances in Partial Differential Equations, Venice 1996, 1998

53 **David A. Cox and Bernd Sturmfels, Editors,** Applications of Computational Algebraic Geometry, 1998

52 **V. Mandrekar and P. R.Masani, Editors,** Proceedings of the Norbert Wiener Centenary Congress, 1994, 1997

51 **Louis H. Kauffman, Editor,** The Interface of Knots and Physics, 1996

50 **Robert Calderbank, Editor,** Different Aspects of Coding Theory, 1995

49 **Robert L. Devaney, Editor,** Complex Dynamical Systems: The Mathematics Behind the Mandelbrot and Julia Sets, 1994

48 **Walter Gautschi, Editor,** Mathematics of Computation 1943–1993: A Half-Century of Computational Mathematics, 1995

47 **Ingrid Daubechies, Editor,** Different Perspectives on Wavelets, 1993

46 **Stefan A. Burr, Editor,** The Unreasonable Effectiveness of Number Theory, 1992

45 **De Witt L. Sumners, Editor,** New Scientific Applications of Geometry and Topology, 1992

For a complete list of titles in this series, visit the
AMS Bookstore at **www.ams.org/bookstore/psapmseries/**.

ISBN 978-1-4704-3553-0

9 781470 435530

PSAPM/75